A Multimessenger View of Supermassive Black Holes and the Quasar Main Sequence

A Multimessenger View of Supermassive Black Holes and the Quasar Main Sequence

Guest Editors

Ascension Del Olmo
Paola Marziani

MDPI

Basel • Beijing • Wuhan • Barcelona • Belgrade • Novi Sad • Cluj • Manchester

Guest Editors

Ascension Del Olmo
Institute of Astrophysics of Andalusia
(IAA-CSIC)
Spanish National Research Council
Granada
Spain

Paola Marziani
National Institute for Astrophysics
(INAF)
Astronomical Observatory of Padua
Padua
Italy

Editorial Office
MDPI AG
Grosspeteranlage 5
4052 Basel, Switzerland

This is a reprint of the Special Issue, published open access by the journal *Universe* (ISSN 2218-1997), freely accessible at: https://www.mdpi.com/journal/universe/special_issues/K95B6W5WD1.

For citation purposes, cite each article independently as indicated on the article page online and as indicated below:

Lastname, A.A.; Lastname, B.B. Article Title. *Journal Name* **Year**, *Volume Number*, Page Range.

ISBN 978-3-7258-7230-5 (Hbk)
ISBN 978-3-7258-7231-2 (PDF)
https://doi.org/10.3390/books978-3-7258-7231-2

Contents

About the Editors

Ascension Del Olmo

Ascension Del Olmo is a Senior Scientific Researcher (Investigadora Científica) at the Institute of Astrophysics of Andalusia (IAA-CSIC, Granada, Spain), part of the Spanish National Research Council. Her research mainly focuses on two primary connected areas: nuclear activity in galaxies, ranging from low levels to the extremely high accretion rates observed in quasars, and the influence of gravitational interaction on the evolution of galaxies, from a multi-frequency perspective. Her interest in the highest nuclear accretion rates has centered on the quasar main sequence framework, involving the analysis of the kinematics and physical conditions of the gas-emitting regions around supermassive black holes, and the physical mechanisms linking accretion, outflows, feedback processes and galaxy evolution. Her expertise includes UV, optical and NIR spectroscopy, complemented with radio and X-ray data, for large samples of quasars up to very high redshifts. Her research interests also encompass the relationship between the environment and galaxy evolution, with studies on nuclear, morphological, and kinematic properties as well as the neutral and molecular gas content of galaxies across a wide range of environments, from isolated and galaxy pairs to compact groups and clusters of galaxies. Additionally, she has contributed to technological activities, including the design and development of astronomical instrumentation and scientific software. Beyond her research, she maintains an active role in promoting women in science.

Paola Marziani

Paola Marziani is Prima Ricercatrice (Senior Researcher) with the National Institute for Astrophysics (INAF) at the Padova Astronomical Observatory in Italy and Investigadora Ciéntifica at the IAA-CSIC in Granada, Spain. Her research interests focus on the physics of active galactic nuclei (AGN), with particular emphasis on the structure and kinematics of the line-emitting regions, the quasar main sequence, and the role of accretion and outflows in shaping observed spectral properties. She has extensively studied broad emission lines as diagnostics of the physical conditions, dynamics, and evolution of the gas in the vicinity of supermassive black holes. Her work also addresses the use of quasars as probes of cosmology, including the potential of highly accreting quasars as standardizable candles.

Preface

Over the past three decades, the quasar main sequence has emerged as one of the most powerful organizing principles for understanding the diversity of active galactic nuclei (AGN). What once appeared as bewildering spectral complexity is now recognized as a structured phenomenology, largely governed by fundamental physical parameters such as Eddington ratio, black hole mass, and orientation. The optical plane defined by the width of Hβ and the strength of Fe II emission revealed that quasars do not scatter randomly around an average spectrum: they occupy a well-defined sequence, linking accretion physics to spectral properties in a physically meaningful way (see the introduction in Marziani et al.).

The quasar main sequence has since expanded into a multidimensional framework that connects optical, ultraviolet, infrared, radio, and X-ray diagnostics. It has become a cornerstone for interpreting broad- and narrow-line region dynamics, ionized winds, jet production, and the coupling between nuclear activity and host galaxy evolution (Moiseev & Iksanova). The studies collected in this Special Issue build upon this framework, exploring how accretion-driven processes manifest across wavelengths and how they relate to feedback and cosmic evolution.

In parallel with these conceptual advances, astronomy itself has undergone a profound transformation. For much of the twentieth century, time-domain astronomy was almost exclusively the domain of stellar astrophysics—variable stars, novae, supernovae, and compact binaries. Over the last four decades, however, variability studies have firmly entered the extragalactic domain. Dramatic spectral and photometric changes have been observed in active galaxies (Panda et al.): changing-look quasars (Xu et al., Benitez et al.), extreme outflows, rapid transitions to and from radiatively efficient and inefficient domains (Marziani et al.), and transient absorption events have revealed that AGN are far more dynamic than previously assumed.

Today, these phenomena are investigated in a truly multifrequency approach. Optical spectroscopy tracks the evolution of emission-line regions; ultraviolet observations probe high-ionization winds; infrared data reveal dust reprocessing and obscured phases; radio observations trace relativistic jets and extended feedback; X-ray monitoring constrains the inner accretion flow and coronal physics (Komossa et al.). The contributions assembled in this Reprint reflect this diversity of approaches, each offering a rich and complementary perspective on supermassive black holes and their environments.

Looking ahead, the advent of the Vera C. Rubin Observatory will revolutionize the time-domain exploration of active galaxies. With its unprecedented depth, cadence, and sky coverage, it will detect variability in millions of quasars, enabling systematic studies of changing-look phenomena, transient absorption events, and short-lived periodicities across a wide range of luminosities and redshifts. Such a statistical bonanza would require the support of machine learning and, in general, the support of artificial intelligence (Kovacevic et al.) that will transform episodic discoveries into population studies, allowing variability to be interpreted within the physical framework of the quasar main sequence (Panda et al.). At the same time, spectro-polarimetry offers a uniquely powerful diagnostic of AGN structure, capable of disentangling geometry, scattering regions, and orientation effects that remain hidden in total-flux spectra. Polarimetric measurements can probe equatorial scattering, polar winds, and jet-related components, providing direct constraints on the three-dimensional structure of the central engine (Shablovinskaya et al.)

Together, these works underscore a central theme: the physics of supermassive black holes cannot be fully understood within a single observational window. The future of the field is inherently multifrequency—and increasingly multi-messenger. Indeed, the most dramatic transitions in galactic nuclei are expected to generate gravitational wave signals. The coalescence of black holes, particularly in systems involving a supermassive primary and a less massive companion, will open a new observational frontier. Space-based gravitational wave detectors will unveil populations of black holes that are currently inaccessible to electromagnetic surveys, providing a census of merging systems across cosmic history.

Multi-messenger observations will also shed light on long-standing puzzles such as pseudo-periodicity—or, more accurately, short-lived periodicity—often encountered in spectral and photometric variations of quasars. Whether these signals arise from binary supermassive black holes, disk instabilities, jet precession, or transient accretion phenomena remains an open question. The combined power of electromagnetic and gravitational wave observations promises to disentangle these scenarios.

The contributions gathered in this Special Issue demonstrate the vitality of this research area and point toward a future in which coordinated, multifrequency, and multi-messenger observations will redefine our understanding of supermassive black holes. The journey from spectral correlations and variability to gravitational wave signals reflects the extraordinary evolution of the physical understanding within the field. The coming decade will likely transform it once again.

List of Contributions:

Benítez, E.; Negrete, C.; Ibarra-Medel, H.; Cruz-González, I.; Rodríguez-Espinosa, J. Multi-Epoch Optical Spectroscopy Variability of the Changing-Look AGN Mrk 883. *Universe* **2024**, *10*, 21. https://doi.org/10.3390/universe10010021.

Komossa, S.; Yao, S.; Grupe, D.; Kraus, A. On the Nature of the Radio Calibrator and Gamma-Ray Emitting NLS1 Galaxy 3C 286 and Its Multiwavelength Variability. *Universe* **2024**, *10*, 289. https://doi.org/10.3390/universe10070289.

Kovačević, A.; Ilić, D.; Popović, L.; Andrić Mitrović, N.; Nikolić, M.; Pavlović, M.; Čvorović-Hajdinjak, I.; Knežević, M.; Savić, D. Deep Learning of Quasar Lightcurves in the LSST Era. *Universe* **2023**, *9*, 287. https://doi.org/10.3390/universe9060287.

Marziani, P.; Garnica Luna, K.; Floris, A.; del Olmo, A.; Deconto-Machado, A.; Buendia-Rios, T.; Negrete, C.; Dultzin, D. Super-Eddington Accretion in Quasars. *Universe* **2025**, *11*, 69. https://doi.org/10.3390/universe11020069.

Moiseev, A.; Ikhsanova, A. Gas and Stars in the Teacup Quasar Looking with the 6-m Telescope. *Universe* **2023**, *9*, 66. https://doi.org/10.3390/universe9020066.

Panda, S.; Marziani, P.; Czerny, B.; Rodríguez-Ardila, A.; Pozo Nuñez, F. Spectral Variability Studies in Active Galactic Nuclei: Exploring Continuum and Emission Line Regions in the Age of LSST and JWST. *Universe* **2023**, *9*, 492. https://doi.org/10.3390/universe9120492.

Shablovinskaya, E.; Popović, L.; Uklein, R.; Malygin, E.; Ilić, D.; Ciroi, S.; Oparin, D.; Crepaldi, L.; Slavcheva-Mihova, L.; Mihov, B.; Nikolov, Y. Polarimetric Reverberation Mapping in Medium-Band Filters. *Universe* **2023**, *9*, 52. https://doi.org/10.3390/universe9010052.

Xu, D.; Komossa, S.; Grupe, D.; Wang, J.; Xin, L.; Han, X.; Wei, J.; Bai, J.; Bon, E.; Cangemi, F.; Cordier, B.; Dennefeld, M.; Gallo, L.; Kollatschny, W.; Kong, D.; Ochmann, M.; Qiu, Y.; Schartel, N. Changing-Look Narrow-Line Seyfert 1 Galaxies, their Detection with SVOM, and the Case of NGC 1566. *Universe* **2024**, *10*, 61. https://doi.org/10.3390/universe10020061.

Ascension Del Olmo and Paola Marziani
Guest Editors

universe

MDPI

Review

Super-Eddington Accretion in Quasars

Paola Marziani [1,*], Karla Garnica Luna [2,*], Alberto Floris [1,3,4,*], Ascensión del Olmo [5], Alice Deconto-Machado [5,6], Tania M. Buendia-Rios [2], C. Alenka Negrete [2] and Deborah Dultzin [2]

1 Astronomical Observatory of Padova, National Institute for Astrophysics (INAF), IT-35122 Padova, Italy
2 Instituto de Astronomía, Universidad Nacional Autónoma de México, AP 70-264, Mexico City 04510, Mexico; tbuendia@astro.unam.mx (T.M.B.-R.); alenka@astro.unam.mx (C.A.N.); deborah@astro.unam.mx (D.D.)
3 Department of Physics, University of Crete, Voutes University Campus, 70013 Heraklion, Greece
4 Institute of Astrophysics, Foundation for Research and Technology Hellas, N. Plastira 100, Vassilika Vouton, 70013 Heraklion, Greece
5 Instituto de Astrofísica de Andalucía (IAA), Consejo Superior de Investigaciones Científicas (CSIC), Glorieta de Astronomía s/n, ES18008 Granada, Spain; chony@iaa.es (A.d.O.); alice.deconto@inaf.it (A.D.-M.)
6 Istituto di Astrofisica Spaziale e Fisica Cosmica, National Institute for Astrophysics (INAF), Via Alfonso Corti 12, I-20133 Milano, Italy
* Correspondence: paola.marziani@inaf.it (P.M.); kgarnica@astro.unam.mx (K.G.L.); afloris@physics.uoc.gr (A.F.)

Abstract

This review provides an observational perspective on the fundamental properties of super-Eddington accretion onto supermassive black holes in quasars. It begins by outlining the selection criteria, particularly focusing on optical and UV broad-line intensity ratios, used to identify a population of unobscured super-Eddington candidates. Several defining features place these candidates at the extreme end of the Population A in main sequence of quasars: among them are the highest observed singly-ionized iron emission, extreme outflow velocities in UV resonance lines, and unusually high metal abundances. These key properties reflect the coexistence of a virialized sub-system within the broad-line region alongside powerful outflows, with the observed gas enrichment likely driven by nuclear or circumnuclear star formation. The most compelling evidence for the occurrence of super-Eddington accretion onto supermassive black holes comes from recent observations of massive black holes at early cosmic epochs. These black holes require rapid growth rates that are only achievable through radiatively inefficient super-Eddington accretion. Furthermore, extreme Eddington ratios, close to or slightly exceeding unity, are consistent with the saturation of radiative output per unit mass predicted by accretion disk theory for super-Eddington accretion rates. The extreme properties of super-Eddington candidates suggest that these quasars could make them stable and well-defined cosmological distance indicators, leveraging the correlation between broad-line width and luminosity expected in virialized systems. Finally, several analogies with accretion processes around stellar-mass black holes, particularly in the high/soft state, are explored to provide additional insight into the mechanisms driving super-Eddington accretion.

Keywords: galaxies; active quasars; black hole physics; accretion; galaxies: emission lines; galaxies: evolution; line formation; ISM; abundances

1. Introduction

The astonishing luminosity of quasars (which may reach $\sim 10^{48}$ erg s^{-1}, 10,000 the luminosity of a giant galaxy) suggests a highly efficient energy release mechanism, much

 https://doi.org/10.3390/universe11020069

beyond the efficiency yielded by nuclear reactions in stars. Such high efficiency can only be achieved through infall into the deep gravitational well of a compact object. Any lingering doubt regarding the fundamental tenet of what is now known as the standard model of active galactic nuclei (AGN) (e.g., [1–5] and references therein), namely accretion onto a supermassive black hole, has definitively passed into history [6], while certain aspects of unification schemes based on orientation are still subject to debate [7–13].

As quasar samples from recent surveys grow in size, now reaching the scale of 10^5–10^6 objects [14], new and diverse phenomenologies are beginning to emerge. Super-Eddington accretion involves a scenario that borders the extraordinary: the infalling material onto the black hole is so large that it surpasses the theoretical Eddington limit, the maximum luminosity a black hole can sustain balancing gravitational pull and radiation pressure. The evidence supporting the existence of super-Eddington accretion on stellar-mass black holes has been robust, with examples dating back several decades, particularly in X-ray binaries such as SS433 and GRS 1915+105 [15,16]. In contrast, evidence for quasar activity exceeding the Eddington limit has been more limited and sporadic. For a long time, the concept of super-Eddington accretion in quasars remained primarily theoretical, with no specific ability to identify super-Eddington accretors (e.g., [17]).

Identifying black holes accreting at super-Eddington (SE) rates is important for several reasons. First, while accretion disks are expected to form in basically any realistic scenario [18,19], as matter usually has too much angular momentum to accrete directly, super-Eddington accretion requires consideration of physical processes that are needed to understand the transport of gas within the accretion disk towards the black hole [20,21], which go much beyond the local viscosity parameter considered in the model of steady accretion disks at sub-Eddington rates [3,22]. Second, accurately evaluating the true accretion and mass outflow rates of SE black holes is essential for determining their growth rates, which has significant implications for understanding quasar evolution across all cosmic epochs, especially during the cosmic dawn. Third, SE processes could be among the prime movers of galactic evolution, if accretion is occurring on SMBH powering very luminous quasars [23–25]. Finally, they hold promising potential as cosmological distance indicators because SE accretion leads to some extreme, stable properties [26–29].

This review aims to delineate the observational properties of SE accretion in quasars[1] and to assess some elementary implications for both astrophysical and cosmological contexts. We will first provide a definition of SE regime (Section 2), then select criteria for their identification from optical and UV spectroscopy (Section 3). Our ability to identify at least an (admittedly biased) sample of SE candidates rests on the Eigenvector 1 main sequence (MS [26,30,31]). Several multifrequency properties, from the radio to the X-ray domain (Section 4), are also potential markers of SE accretion. We sketch a structural and dynamical model of the emitting regions involving a virialized and wind system that has been around since the late 1980s, but which still needs to be developed in many of its physical and dynamical aspects (Section 5.1). A major source of uncertainty concerns the accretion disk winds, which appear to be a natural consequence of SE accretion modes (Section 5.2 [32–34]). While evidence of outflows is widespread in AGN, some parameters (for example outflow speed and metallicity) reach extreme values in SE candidates. At the same time, a virialized sub-system offers the potential to use the line widths of prominent emission lines in SE quasars for estimating cosmological distances, thereby contributing to constraints on the Universe expansion history ([26,35], Section 6). Finally, we emphasize the intriguing parallels between stellar-mass and supermassive black holes (Section 7), as well as several unresolved issues that remain open for future investigations (Section 8).

2. Defining Super-Eddington Accretion

Super-Eddington accretion refers to a situation where the mass accretion rate onto a compact object—such as a black hole, neutron star, or white dwarf—is so high that the associated luminosity exceeds the Eddington limit, at which radiation pressure would normally counteract the inward pull of gravity. It can be expressed as the Eddington luminosity

$$L_{\rm Edd} = 4\pi GMc/\kappa, \quad (1)$$

where G is the gravitational constant, c is the speed of light, and κ is the opacity of the accreted material. κ typically represents a measure of how transparent a material is to radiation, and is the amount of absorption and scattering that occurs as light passes through a material. In ionized gases where the number of free electrons plays a role in determining the pressure and other gas properties, the opacity is related to the mean molecular mass per free electron ($\mu_{\rm e}$) which represents the average mass of particles in a gas divided by the number of free electrons. The $\mu_{\rm e}$ can be expressed as $\mu_{\rm e} = 1/\sum_i X_i Z_i/A_i \approx 1.2$, where X_i is the mass fraction of element i, Z_i is the number of protons (and thus, free electrons) in element i, and A_i is the atomic mass of element i (c.f. [5]). In a fully ionized hydrogen gas, where electron scattering dominates the opacity, the specific opacity $\kappa_{\rm es}$ is given by $\kappa_{\rm es} = \sigma_{\rm T}/(m_{\rm p}\mu_{\rm e})$. In this context, κ is inversely proportional to $\mu_{\rm e}$. If $\mu_{\rm e}$ is higher (meaning there are fewer free electrons per unit mass), the opacity due to electron scattering will be lower. Conversely, a lower $\mu_{\rm e}$ (more free electrons per unit mass) results in higher opacity. The Eddington luminosity can then be written as $L_{\rm Edd} = 4\pi GMcm_{\rm p}\mu_{\rm e}/\sigma_{\rm T} \approx 1.26 \cdot 10^{46}(M/10^8 M_\odot)\mu_{\rm e} \approx 1.5 \cdot 10^{46}(M/10^8 M_\odot)$. Here, $\mu_{\rm e}$ is assumed ≈ 1.2 and is obtained by a gas mixture with about 64.7% hydrogen, 33.3% helium, and a small amount of metals (about 2%). Note that for a fully ionized gas $\mu_{\rm e} \approx 1.156$. If a fraction of gas is neutral, the opacity can be much larger and the $L_{\rm Edd}$ correspondingly lower.

The dimensionless accretion rate $\dot{m}$ is a way to express the accretion rate in terms of a unit-less quantity, making it easier to compare across different systems without dependence on specific units. This rate is typically defined by normalizing the actual accretion rate to a characteristic rate related to the Eddington limit:

$$\dot{m} = \dot{M}/\dot{M}_{\rm Edd}, \quad (2)$$

where $\dot{M}$ is the mass accretion rate in units of $M_\odot$ yr^{-1} or g s^{-1} and $\dot{M}_{\rm Edd}$ is the Eddington accretion rate. The $\dot{M}_{\rm Edd}$ is the rate at which the luminosity of the accreting object reaches the Eddington luminosity $L_{\rm Edd}$. It can be defined as follows: $\dot{M}_{\rm Edd} = L_{\rm Edd}/\eta c^2$, where $L_{\rm Edd}$ is the Eddington luminosity, η is the efficiency of converting accreted mass into radiation, and c is the speed of light. Note that there is no consistency in the η used in the literature because η is a function of the accretion rate itself. We assume $\eta = 1$ in the definition of $\dot{M}_{\rm Edd} = L_{\rm Edd}/c^2$ [31], so that the Eddington ratio can be written as follows:

$$L/L_{\rm Edd} = \eta\dot{m}, \quad (3)$$

where L is the bolometric luminosity. The Eddington ratio is a basic observable, as it represents the ratio between the bolometric luminosity and the black hole mass $M_{\rm BH}$ ($L/L_{\rm Edd} \propto L/M_{\rm BH}$), i.e., between radiation and gravitation forces. Its importance is comparable to the M/L ratio of galaxies [36–38]. An estimate of the sub-Eddington $\dot{m}$ can be achieved following the geometrically thin, optically thick accretion disk models, once the luminosity at 5100 Å and the black hole mass are known [22,31,39,40], but this is intrinsically more difficult in the case of a SE disk [41,42].

When the accretion rate surpasses the Eddington limit, the excess radiation pressure should theoretically blow away the infalling material, preventing further accretion. In practice, black holes may accrete at SE rates due to two main factors: first, accretion can occur in a non-spherical manner, such as through a disk, which allows material to funnel onto the object while avoiding direct opposition by radiation pressure [19,43,44]; second, photons can be trapped and advected inward with the gas, reducing the outward radiation pressure that counteracts accretion. Advection-dominated accretion flows (ADAFs) arise in regions where the majority of the energy released by accreting gas is retained and transported inward by the flow, rather than being efficiently radiated away. A "slim disk" is expected to occur at moderately SE accretion rates [45], and have higher radiation pressure leading to a vertical puffing up, making the disk geometrically thick and optically thick. Even when the disk becomes geometrically thick due to increased radiation pressure, the gas remains radiatively inefficient, despite being hot and dense. The local viscosity prescription used in the thin-disk configuration [22] no longer applies, and global considerations, such as angular momentum transport via global torques or outflows, become more important [46,47]. Optically thick, stationary slim disks are non-local solutions [47], obtained by numerical integration of the two-dimensional Navier–Stokes equations with a transsonic point [48,49].

At the highest $\dot{m}$, the flow remains advection dominated, optically thick, and supported by radiation pressure, similar to slim disks, but with even greater radiative inefficiency [50]. The thick disk is expected to take on the form of thick torus with two narrow funnels along the rotation axis (a "Polish doughnut"), which collimates radiation into beams with highly SE luminosities. The thickness of a slim disk is considerable compared to the geometrically thin standard disks [51], with a height to radius H/R ratio in the range of 0.1 to 0.3, but is not as extreme as that of a Polish doughnut ($H/R \gtrsim 1$).

A related prediction of geometrically thick, optically thick accretion disk models is anisotropy in the emission (e.g., [52–55]), and the existence of two regions: one shielded from the hottest continuum [56], and one conical region exposed to the radiation of the funnel's walls (which in the slim disks at moderate accretion may be also of large aperture, see Figure 3 of Wang et al. [56]).

A second observational property predicted by SE disk models, if the accretion flow becomes advection dominated, is the radiative efficiency decrease with $\dot{m}$, for $\dot{m} \gg 1$ [45,48,49,57]. A convenient parameterization for non-rotating black holes $a = 0$ is as follows ([58], c.f. [59]):

$$L/L_{\rm Edd} \approx 2 \cdot [1 + \ln(\dot{m}/50)]. \quad (4)$$

The implication is that the Eddington ratio $L/L_{\rm Edd}$ tends to "saturate" towards limiting values of order unity; correspondingly, the radiative efficiency η rapidly decreases roughly with $1/\sqrt{\dot{m}}$. Slightly larger values can be obtained if the spin parameter is $a > 0$.

Hydrodynamical simulations confirm this basic framework. They illustrate that with increasing accretion rates a flat disk thickens, forming regions where gas velocities exceed local escape velocities [42,60,61]. Initially, the accretion flow remains flat, but with higher accretion rates, the disk thickens and develops regions where gas velocities surpass the local escape velocity. This leads to the generation of powerful winds, which are a hallmark of SE accretion. These winds are intimately connected to the accretion process and play a significant role in the overall dynamics of the system [34].

Three major properties are, therefore, expected to be associated with SE accretion: (1) the development of an inner thick structure (although see [62]), optically thick, with implication for the illumination of the line emitting gas; (2) the extreme value of $L/L_{\rm Edd}$, that is, a radiative output that is growing little with the accretion rate; and (3) presence of high-ionization winds.

3. Identifying Super-Eddington Candidates

3.1. The Quasar Main Sequence: A Synopsis

The main sequence (MS) is a classification scheme that organizes quasars according to their spectral properties, much like the main sequence of stars [30,63–65]. The MS is most commonly represented in the optical plane (Figure 1, see [66] for an alternative), defined by two parameters:

- R_{FeII}: the relative strength of FeII to Hβ emission, measured as the ratio of the FeII complex (around 4570 Å) to the broad Hβ line: R_{FeII} = FeII (4570 Å)/Hβ (Figure 2). The FeII complex, integrated over the range 4434–4684 Å (hereafter FeIIλ4570 for brevity) includes many multiplets, but the strongest contributions around 4570 Å come from multiplets 37 and 38 [67], which consist of lines from transition between terms ^4D and ^4P to ^4F (the F group of Kovačević et al. [68]). Low R_{FeII} values indicate weak global FeII emission, as the type-1 AGN FeII emission is changing intensity with respect to Hβ (Figure 2), but is otherwise self-similar over a broad range of R_{FeII} and AGN luminosity ([65], and references therein). High R_{FeII} values are associated with strong FeII emitters and are linked to high Eddington ratio [63,64,69–73].
- FWHM of Hβ: the FWHM (in km s^{-1}) of the Hβ line is a measure of the velocity of the broad-line region (BLR) gas [74,75], and is in the range of 1000–20,000 km s^{-1} [30]. The FWHM is one of the most widely used "virial broadening estimators" used to infer black hole masses (e.g., [76,77]).

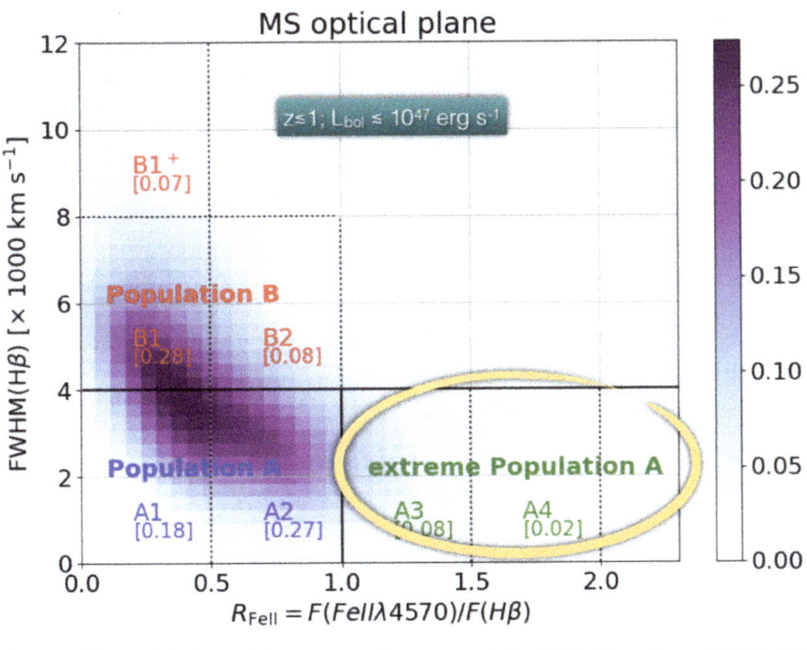

Figure 1. The optical plane of the quasar main sequence, full width half maximum (FWHM) Hβ vs. R_{FeII} (see text for its definition). Spectral types are as identified by Sulentic et al. [78], in intervals of R_{FeII} and FWHM Hβ. The numbers in square brackets are the prevalence of each spectral type in a large optically selected sample of low–z ($\lesssim$1) type 1 AGN. The cartouche emphasizes the area of extreme Pop. A sources with $R_{FeII} \geq 1$.

Quasars can be divided into two broad populations based on their position in the optical plane, with a sub-population of special interest.

- Population A: narrower Hβ lines (FWHM < 4000 km s^{-1}), a range of FeII emission (from weak to high $R_{\rm FeII}$), and often lower CIVλ1549 and [OIII]$\lambda\lambda$4959,5007 equivalent width [69,78,79]. Population A quasars are typically associated with high dimensionless accretion rates and higher Eddington ratios $L/L_{\rm Edd} \approx 0.2$–1.0, compared to the lower values of $L/L_{\rm Edd} \approx 0.01$–0.2 in Population B [63,71,73]. They are overwhelmingly radio-quiet. Radio-loud Narrow Line Seyfert 1 [80–82] are a tiny minority of optically selected Population A samples [83,84]. Meanwhile, broad-absorption line (BAL) QSOs belonging to Population A show more extreme high-velocity BAL troughs and a large BALnicity index [85].
- Extreme Population A ($R_{\rm FeII} \geq 1$): quasars located towards the high $R_{\rm FeII}$ end of the MS (extreme Population A, or xA for brevity) have been found to be characterized by extreme observational parameters among Pop. A sources and highest $\dot{m}$ above the Eddington limit (but still with bolometric luminosity $L \sim L_{\rm Edd}$) [26,31,86].
- Population B: these quasars have broad Hβ lines (FWHM > 4000 km s^{-1}), weak FeII emission (low $R_{\rm FeII}$), and stronger [OIII]$\lambda\lambda$4959,5007 emission [69,78,87]. Population B quasars are located toward the low $R_{\rm FeII}$ and broader Hβ end, and associated with lower accretion rates and higher black hole masses, typically ranging from $10^{8.5}$ to $10^{9.5} M_{\odot}$ [71,88,89]. A defining feature is a strong redward asymmetry in the low-ionization line (LIL) profiles that is attributable to a black hole mass effect [71,88,90,91]. Population B quasars are more likely to be radio-loud, with a probability of about 25%, compared to the 3–4% seen in Population A [83]. Population B is also associated with less extreme BALs and lower density ($\log n_{\rm H} \approx 9.5$–10 [cm^{-3}]) in the BLR [92].

The MS primarily highlights correlations between emission line properties and various physical properties of quasars, such as Eddington ratio, black hole mass, and orientation [63,64,69–73]. Being the projection of a multidimensional space, the optical plane of the quasar MS is affected by several degeneracies. The limit FWHM(Hβ) $\approx$ 4000 km/s, set at low z and in moderate luminosity samples ($L \lesssim 10^{46}$ erg s^{-1}), is by itself luminosity and black hole mass dependent, if it is interpreted as a well-defined limit in $L/L_{\rm Edd}$ and if LILs are predominantly broadened by virial motions [93]. The distribution of data point of Figure 1 applies to low-z ($\lesssim$1) samples. If a broad range of masses and luminosities is considered, the sequence is smeared into a wedge [64], although the main trend associated with $R_{\rm FeII}$ are preserved.

The orientation of the quasar disk axis with respect to the observer's line of sight can also affect the observed width of the emission lines [63,64]. Inclined quasars ($\theta \sim 45$–60) might exhibit broader lines due to the rotational motion of the BLR gas confined in a flattened geometry [65,94–96]. Orientation and black hole mass affect the line width in the same way: an increase of mass and an increase of the angle between the line of sight and the accretion disk axis both lead to an increase in Hβ FWHM [65,97]. The ratio $R_{\rm FeII}$ ultimately depends on the $L/L_{\rm Edd}$, and possibly on orientation [63,73], although the dependence between $L/L_{\rm Edd}$ and $R_{\rm FeII}$ is indirect, with $L/L_{\rm Edd}$ affecting ionization parameter, metallicity, density, and column density of the line emitting gas in a way that is not fully clear as yet ([73,98–100], and references therein).

The spectral appearance of quasars along the MS is retained [101] up to the Cosmic Noon, a period around redshift $z \sim 1$–3 (roughly from 5 to 2 billion years after the Big Bang), when star formation and black hole accretion reached their peak activity [102]. A major survey carried out with the ISAAC instrument at the Very Large Telescope (VLT), covering 53 Hamburg-ESO (HE) quasars (hereafter HEMS), revealed very strong FeII emitters at $1 \lesssim z \lesssim 2$ [90,103,104]. Recent observations using ground-based infrared (IR) spectrometers and James Webb Space Telescope (JWST) have reinforced the stability and distinctiveness of the spectral features defining the MS [105–108]: they confirm that the MS-defining properties are consistent across various luminosities and cosmic epochs [90,109–112]. A higher prevalence of Population A sources at intermediate redshift reflects an evolution in accretion rate of quasars coming down from the Cosmic Noon to the present-day Universe [113–116]. The main difference apparently resides in the broader emission lines of luminous quasars: broader because of the FWHM systematic increase in black hole mass [93], ascribed to a Malmquist-type bias in flux-limited samples [117]. Highly accreting quasars like I Zw 1 and H0557-385 may become undetectable at high redshift if their flux falls below a survey detection threshold. However, their high-luminosity counterparts—which exhibit similar $R_{\rm FeII}$ values—would still be detectable, albeit with broader emission lines due to their larger $M_{\rm BH}$.

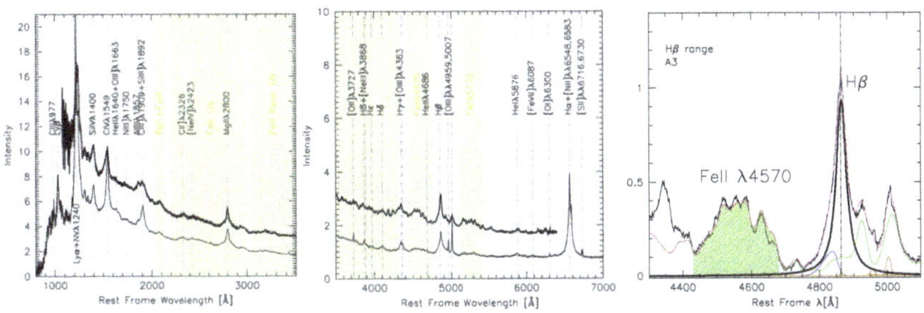

Figure 2. (Left and middle panel) : xA quasar composite ultra-violet (UV) and optical spectrum (thick lines) compared to the median composite spectrum of Vanden Berk et al. [118] for the Sloan Digital Sky Survey (SDSS). Spectral regions with prominent FeII are shaded green. **(Right)** continuum subtracted composite spectrum for spectral type A3 belonging to the xA sub-population [78], with emphasis on the FeIIλ4570 blend measured for the derivation of $R_{\rm FeII}$, again shaded green. A model involving a Lorentzian, unshifted Hβ component and a fainter component skewed and shifted to blue reproduces the Hβ profile.

3.2. *Identification from the Optical Rest Frame*

The optical singly-ionized iron emission is remarkably self-similar among type-1 quasars [87], since significant FeII emission requires very similar conditions: in the photoionization context, a partially ionized zone (PIZ) within the emitting gas with electron temperature $T_{\rm e} \sim 7000$ K [92,119–121], where $\rm Fe^{+}$ is the dominant ionization stage of iron (Figure 3). The relative intensity with respect to Hβ is instead varying by a large factor along the MS, ranging from being undetectable to becoming so prominent that it overwhelms the emission lines within the Hβ spectral range [69,87,122]. The selection criterion [26,31,86]

$$R_{\rm FeII} = {\rm FeII}\ (4570\ {\rm \AA})/{\rm H}\beta \geq 1 \tag{5}$$

isolates quasars with strongest FeII emission and other extreme properties at the tip of the quasar MS. xA sources exhibit distinct, and easily-recognizable spectra compared to typical quasars (Figure 2), due to the dominance of singly ionized iron emission across a wide range from the FUV to the NIR, and specifically around Hβ [31,71].

The parameter $R_{\rm FeII}$ might appear as obscure and irrelevant, but strong FeII emission is actually the main cooling agent in the BLR [123]. FeII emission of xA sources can overwhelm other emission lines and create a pseudo-continuum, exceeding the emission of Lyα and the strongest metal lines. Especially in the UV around the MgII doublet, the contamination of the underlying accretion disk continuum creates difficulties even in assessing their relative contribution (e.g., [100,124,125]). The optical FeII emission is not only a key SE indicator, but also remains remarkably stable over time in most cases (and at least within the temporal extension of monitoring campaigns [126,127]), which is uncommon in other quasar spectra, especially in Pop. B sources where at low accretion rate the changing look phenomenon occurs frequently [128].

The criterion $R_{\text{FeII}} \gtrsim 1$ corresponds to the highest values of the Eddington ratio, close to 1 or slightly above 1 depending on different bolometric corrections and black hole mass estimates [26,31,63,72]. Figure 4 shows the distributions of the L/L_{Edd} as a function of R_{FeII}, in bins of δR_{FeII}= 0.24, for a large sample of almost 700 low-z sources [89]. The violin plots are well-suited to show the overall trend as well as the ample dispersion of the data in each bin. There is a steady increase in the median L/L_{Edd} up to $R_{\text{FeII}} \approx 0.84$; afterwards, the increase is shallower, confirming a threshold around $R_{\text{FeII}} \approx 1$, beyond which, even if R_{FeII} becomes significantly larger than 1, the L/L_{Edd} is only weakly increasing. This behavior reminds the one of L/L_{Edd} as a function of $\dot{m}$. The dispersion is in large part due to uncertainties in the L/L_{Edd} estimates, so that the prominence of FeII could potentially serve as a more reliable proxy for the AGN accretion state.

Other low ionization species are strongest in the spectra of xA quasars, in addition to singly-ionized iron: CaII [98,99,129–131], as well as SiIIλ1263 and λ1816 [132]. In addition, the CIVλ1549 equivalent width is minimized [133–136], with $W \lesssim 10$ Å qualifying most sources as weak lined quasars [137–139]. Why does high Eddington ratio come with an extremely low ionization spectrum? This question was raised almost a quarter century ago. An answer given at that time was that emission line diagnostics suggest very high density and low ionization for the BLR [140]. High column density is also needed for strong FeII emission [141,142], to maintain the PIZ, where Fe^{+} is the dominant ionization stage of iron (Figure 3). It is as if the BLR of xA sources is somehow reduced to a dense, compact remnant, not unlike the Weigelt blobs of η Carinæ [143,144]. Parenthetically, we note that large column density stabilizes the virial motion of the emitting gas against radiation pressure because the radiative acceleration is inversely proportional to the gas column density [145,146]. The right panel of Figure 3 shows three conditions for an orbiting cloud originally in elliptical orbit, for the same L/L_{Edd}: at high N_{c}, there is no noticeable effect (thick trace); for low N_{c} the cloud becomes unbound; and at intermediate N_{c} ($\sim 10^{23}$ cm^{-2}), the cloud remains bound but the orbit undergoes a precessional motion. Large N_{c} gas is expected to be present at the accretion disk surface.

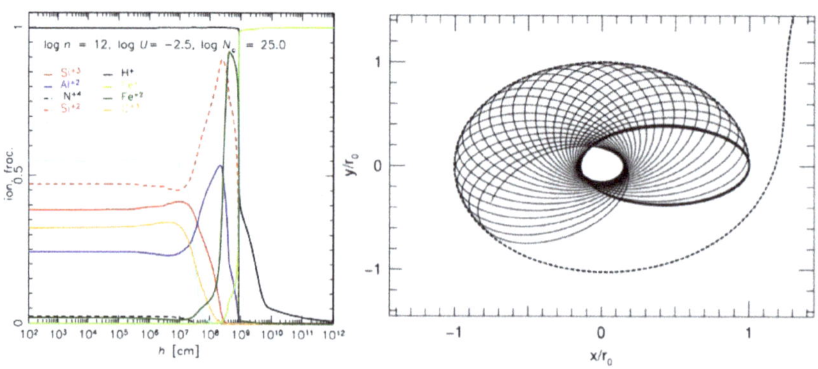

Figure 3. (**Left**) ionization structure as a function of the geometrical depth h in a slab of emitting gas. The continuum is illuminating the slab from the left side ($h = 0$) giving rise to a fully-ionized zone, and a PIZ where hydrogen is partially ionized (black line) and almost all iron is singly-ionized (pale green). (**Right**) Different orbits for clouds of different column density, with the same M_{BH} and L/L_{Edd}. The highest column density clouds remain quasi-virialized with a precessional orbit due to the deviation from the inverse square law for their distance from the central continuum source. The lowest density clouds become unbound (dashed line). Adapted from Refs. [140,146].

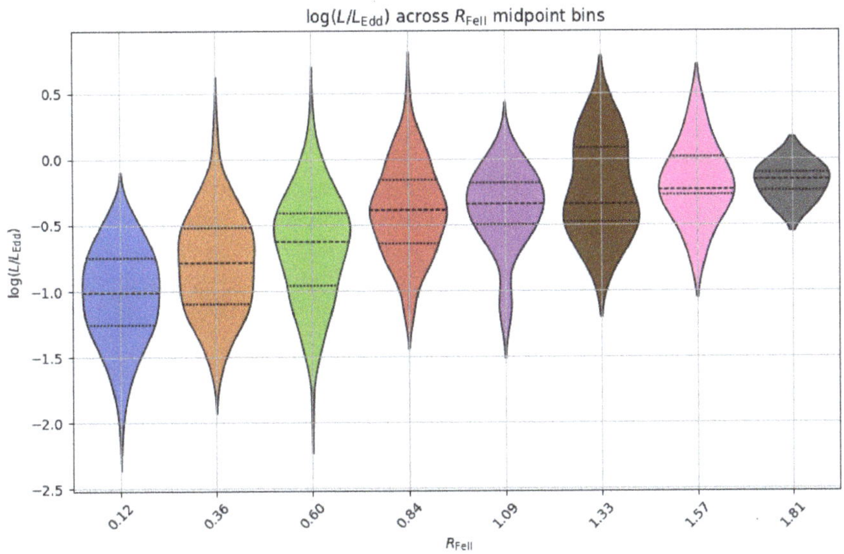

Figure 4. Violin plot of the Eddington ratio $L/L_{\rm Edd}$ as a function of the FeII prominence parameter $R_{\rm FeII}$. Horizontal lines within each range identify median and first and third quartile of the distribution.

3.3. Identification from UV Rest Frame

At redshift above ≈ 1, observations of Hβ become cumbersome because the line is red-shifted into the near IR. Dedicated instrumentation is needed, from ground or space. Ground-based observations are affected by telluric absorption and sky emission. Thus, it would be expedient to have an UV selection criterion at hand. The condition $R_{\rm FeII} \gtrsim 1$ is associated with strong AlIII emission and weak CIII] [147–149]. The weakness of CIII] was one of the major aspects suggesting high density [92,150]. In terms of the Si III]/C III] ratio, extreme Population A quasars show high values ($\gtrsim$1), while this ratio is more modest in the rest of Population A and is never as large in Population B [78,151]. The following conditions

$$\mathrm{SiIII]}\lambda1892/\mathrm{Al\ III}\lambda1860 > 0.5 \,\mathrm{and}\, \mathrm{SiIII]}\lambda1892/\mathrm{CIII]}\lambda1909 < 1 \quad (6)$$

define a criterion that is meant to be equivalent to $R_{\rm FeII} > 1$. The correspondence between $R_{\rm FeII} > 1$ and the UV conditions hold in ≈80% of cases where both spectral ranges are available (Figure 5; [151–153]). Out of a sample of ≈30 xA optically-selected sources, two are located outside of the grey rectangle defining the area in agreement with the UV selection criteria; an additional four have CIII]λ1909 too strong but AlIIIλ1860/SiIII]λ1892 slightly lower than the threshold value 0.5 but consistent within the uncertainties. Data points in discordant location involve mainly borderline discrepancies, especially sources meeting the UV conditions and having $R_{\rm FeII}$ close to 1 but somewhat less than 1 or viceversa, $R_{\rm FeII} > 1$, and CIII]λ1909 stronger than expected, possibly due to an undersubtraction of the FeIIIλ1914, enhanced via Lyα fluorescence [92].

The optical and UV criteria can be supplemented by other indicators. xAs exhibit the highest X-ray variability along the sequence (e.g., [154–158]), while optical variability is more frequent and occurs with greater amplitude in Population B quasars. Other UV intensity ratios can be used to reinforce the classification, typically SiIV + OIV]λ1400/CIVλ1549. In the most extreme quasars, this ratio tends toward unity.

There are several caveats, mainly because both the UV and optical selection criteria are dependent on relative elemental abundances, and such dependence is not established. $R_{\rm FeII}$ is highest in conditions of high density and column density, with low or moderate ionization. Higher iron content makes it easier to achieve $R_{\rm FeII}$ above 1, between $\approx$1.5–2, or even higher [159]. The $R_{\rm FeII}$ ratio is obviously dependent on metallicity, although it is not yet universally accepted that highly super-solar abundances are needed to achieve values $R_{\rm FeII} \gtrsim 1$. Recent works suggest that $R_{\rm FeII} \gtrsim 1$ can be reached with solar metallicity [160,161], albeit stretching other physical parameters to their limits. In addition, UV intermediate ionization line ratios are strongly dependent on density, as they are the ratios of transitions with different critical density [162], right in the domain of the densities expected for the BLR ($n_{\rm H} \sim 10^9$–10^{11} cm^{-3}). Last, silicon is an α element but aluminum is not, and there is a considerable uncertainty about its nucleosynthesis [163,164].

In conclusion, the optical and UV criteria appear to be fairly effective in identifying SE candidates. A significant source of uncertainty in the interpretation of the selection criteria lies in the determination of $L/L_{\rm Edd}$, primarily due to uncertainties in mass estimates (which are influenced by orientation effects) and, even more critically, the bolometric correction. Therefore, we prefer to rely on empirical criteria based on the MS, acknowledging that they likely select only a limited sub-sample of SE candidates—excluding type-1 AGN that may be affected by reddening, embedded in dusty starbursts [165], or type-2. Obscured sources might well be highly accreting [166]. For instance, the prototype type-2 AGN NGC 1068 shows strong FeII emission in polarized light [167,168], and might have been classified as xA if the FeII had been observed in natural light.

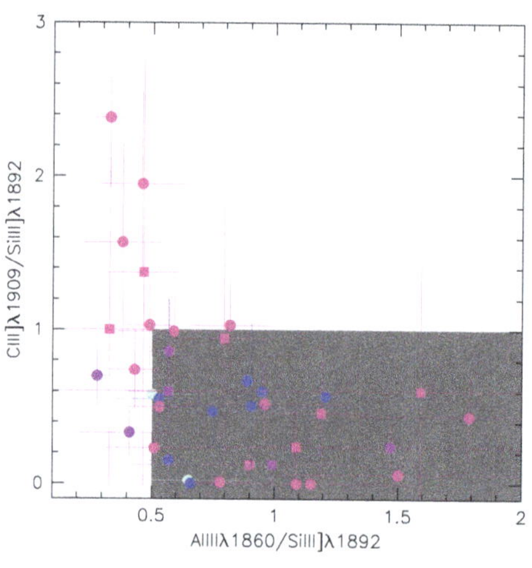

Figure 5. Plane of intensity ratio CIII]λ1909/SiIII]λ1892 vs. AlIIIλ1860/SiIII]λ1892. The data are coded according to their $R_{\rm FeII}$ values. Cyan: A1 ($R_{\rm FeII}$< 0.5), Blue: A2 ($0.5 \leq R_{\rm FeII} < 0.5$); Magenta: A3 ($1 \leq$ rfe < 1.5); Purple: A4 ($1.5 \leq R_{\rm FeII} \lesssim 2$). The shaded area shows the range of ratios used to identify the SE candidates. Circles represent the sources of Marziani et al. [153]; squares are data from Floris et al. [159] and Buendia Rios et al., submitted.

A related issue is whether the selection identifies truly SE quasars. The optical and UV criteria focus on extreme properties observed of xA quasars: enhanced FeII emission, highest velocity outflows, highest chemical abundances, etc. The data do not yet rule out the possibility that an extreme sub-Eddington balance could be at play. Apart from the relation between $\dot{m}$, $L/L_{\rm Edd}$, and $R_{\rm FeII}$ [26,31], the evidence for SE accretion in quasars remains largely circumstantial and based on the large black hole masses at very high redshifts. Therefore, xA quasars, which represent a well-defined observational type, might be more appropriately considered SE candidates, even if, for brevity, we refer to them simply as SE.

4. Properties of Super-Eddington Candidates in the Main Sequence Context

4.1. Multi-Frequency Correlates

Sulentic et al. [169] and Fraix-Burnet et al. [170] list several multifrequency parameters that are systematically different between Population A and B. xA sources can be thought as simply reaching the most extreme values in most observational parameters of Population A. One of the main distinctions is the CIVλ1549 blueshift measured at half peak intensity ($\Delta v_{1/2}$ CIVλ1549) with respect to the quasar rest frame, which is typically largest in xA (in absolute value, $\gtrsim$1000 km s^{-1}) compared to the smaller and null values in the rest of Population A and Population B (-250 to $+70$ km s^{-1} [149,171–173]). xA quasars often show extremely weak CIVλ1549 and [OIII]$\lambda\lambda$4959,5007 emissions compared to Population B, where these lines are more prominent and with a narrower core close to rest frame [173,174]. The [OIII]$\lambda\lambda$4959,5007 lines are found blueshifted by several hundred km s^{-1} among xA (the so-called "blue outliers" [71,175,176]), with shifts that are apparently correlated with those of CIVλ1549 [108,175,177,178]. Additionally, extreme Population A quasars tend to have softer X-ray spectra with large Röntgensatellit (ROSAT) photon indices ($\Gamma_{\rm soft} \gtrsim 2$; $f(\epsilon) \propto \epsilon^{-\Gamma}$) [179], while Population B shows somewhat rarer instances of enhanced soft X-ray emission. A $\Gamma_{\rm soft} \lesssim 2$ indicates a flattish SED as observed in RL and Population B sources; a steeper slope $\Gamma_{\rm soft} \gtrsim 2$ is common among Population A sources, with the most extreme values, $\Gamma_{\rm soft} \approx 3-4$, among xA AGN. Such large $\Gamma_{\rm soft}$ values diagnose a soft X-ray excess [30,179–181] due to the presence of a component superimposed on an ideal power law connecting the near UV emission and the X ray domain, the $\alpha_{\rm ox}$ parameter [182]. The most widely accepted explanation is emission from a Compton-thick corona connected with the innermost accretion disk ([183–185], and references therein)] and possibly even with the bare accretion disk (i.e., without a Comptonizing corona [186]).

Together with $R_{\rm FeII}$ and FWHM Hβ, CIVλ1549 shift and $\Gamma_{\rm soft}$ define the so-called 4DE1 [173,187,188], one of the landmark achievements of the past 30 years of research on AGN [187,189]. The 4DE1 scheme captured the main aspects of the type-1 AGN multifrequency phenomenology. In addition to FWHM Hβ and $R_{\rm FeII}$ increasing with the Eddington ratio [63,71–73], providing the source location in the MS optical plane, the 4DE1 includes two other parameters that are observationally "orthogonal", involving different frequency domains, and that are related to the main parameters of the accretion process in quasars (mass, Eddington ratio, viewing angle, and accretion mode).

4.2. Super-Eddington SED

Figure 6 shows three SEDs obtained for sources satisyfing the criterion $R_{\rm FeII} \geq 1$ (see also Panda and Marziani [190] for a contextualization of the SED along the main sequence). A prominent big blue bump and a soft X-ray excess extending up to few keV has been a distinguishing feature of the SED of sources radiating at high Eddington ratio [183,186,190–194]. Slim disks manage to maintain surface luminosity at the local Eddington limit through radial advection, which alters the expected luminosity distribution. Under conditions of high accretion rates, the emissivity decreases more slowly ($\epsilon(r) \propto r^{-2}$) than as seen in standard sub-Eddington disks ($\propto r^{-3}$). In practice, the predicted SED should peak in the FUV right beyond the Lyman limit, with a flat shape in correspondence of the maximum emission ([41], Figure 6). An observational test is difficult [195], as the FUV is unobservable because of Galactic absorption.

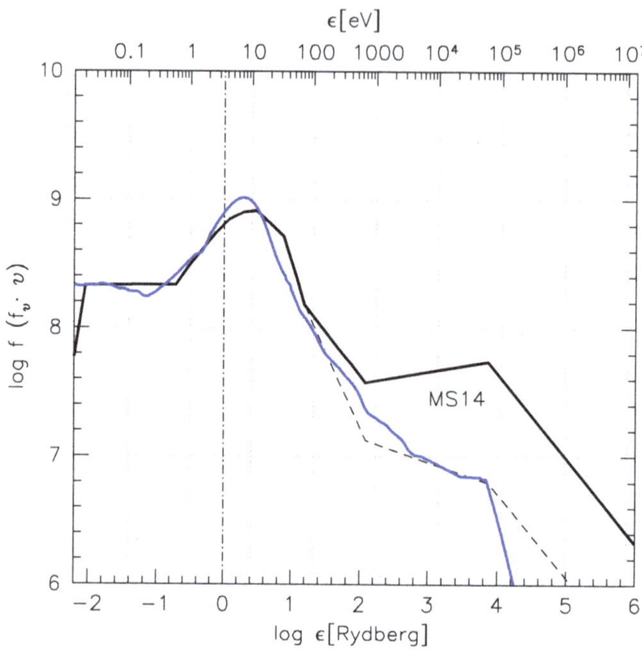

Figure 6. Spectral energy distributions (SEDs) computed by Marziani and Sulentic ([26], MS14) and Garnica et al. (submitted) are presented. In all cases, the sources used to construct the composite SEDs were selected based on the criterion $R_{\rm FeII} \geq 1$. The filled black line represents a composite SED with typical values across various spectral ranges, while the dashed line illustrates the extreme values [26]. The new SED (blue), which is a median of more than 150 sources, shows good agreement with the extreme composite.

On the high energy side, the behavior of the xA SED is even less clear. Up to ≈ 20 keV, the slope of the X-ray continuum was found to be steep, with photon indexes larger than 2 and reaching the most extreme values for the highest $L/L_{\rm Edd}$ sources [26,35,196,197]. Intriguingly, the photon index appears to correlated with the outflow dynamical properties [198,199]. On the one hand, a sizable population of sources are weaker than expected from the slope $\alpha_{\rm ox}$ trend with UV luminosity [197]. On the other hand, beyond 20 keV, several xA sources show a flattening in their nuSTAR spectra. The origin of this flattening is not obvious. A cold Compton-reflection component is observed in Population B sources, and is consistent with scattering toward a geometrically thin, "cold" disk [200–202], where the Fe Kα line is observed at the rest-frame energies expected for low ionization degrees of iron. The accretion geometry is expected to be basically different for super- and sub-Eddington AGN [161,203]; in the SE case, the Fe Kα resonance line is shifted toward higher energies, since the X-ray reflection can only be due to a Compton-thick outflow encircling the funnel of an optically thick, geometrically thick accretion disk, and scattering toward the outer disk hampered by the inner disk geometry. The presence (or absence) of a reflection component might be at the origin of the difference between the SEDs of Figure 6.

On the low energy side, xA sources show a significant far infrared (FIR) excess with respect to the canonical Mathews and Ferland [204] SED, especially prominent for the highest $R_{\rm FeII}$ values. Their host galaxies at low-z are known to undergo vigorous star formation [205], often qualifying them as luminous or ultra-luminous infrared galaxies. They are cold quasars in the interpretation of Kirkpatrick et al. ([206], Section 4.5).

4.3. Chemical Properties and Broad Line Region

Quasar metallicities are consistently reported to be high, typically around 5–10$Z_\odot$ [117,207–210]. Several diagnostic ratios which are easily measurable from the broad line spectrum of quasars show a fairly monotonic dependence on metallicity and other parameters over a wide range of physical conditions: the intensity ratio CIVλ1549/HeIIλ1640 and CIVλ1549/Hβ are sensitive to metallicity as well as to ionization parameter [159,211]; CIVλ1549/CIII]λ1909 is in principle sensitive to the ionization parameter, albeit its use is not appropriate for xA sources because of the CIII]λ1909 intrinsic weakness [26]. The ratio AlIIIλ1860/HeIIλ1640 is monotonically dependent on metallicity; the ratios SiIVλ1397+ OIV]λ1402/CIVλ1549 and SiIVλ1397+ OIV]λ1402/HeIIλ1640 are known metallicity indicators [117,210,212–214]. A value of the SiIVλ1397+OIV]λ1402/CIVλ1549 ratio close to ≈ 1 as found for xA quasars suggests extreme metal abundances [214,215]. Ratios involving the semi-forbidden transitions are sensitive to density [207,210,213,214]. It is a safe approach to rely on UV resonance lines to estimate metallicity and physical properties such as ionization parameters and density (as done in several recent works [140,159,211,215]), for which atomic data are known with high precision, and radiative transfer may be better addressed by the escape probability formalism [216,217]. In the PIZ (Figure 3), several non-local processes play crucial roles [119,121,218–220].

The blends relevant for metallicity estimates in the optical and UV spectrum of the xA quasar HE 1347-2457 are shown in Figure 7. Figure 8 shows the results in the 3D parameter space ionization parameter U, Hydrogen density n_H, and metallicity Z for HE 1347-2457. Measured broad line intensity ratios were compared with the prediction of arrays of `CLOUDY` simulations to cover the entire parameter volume between $7 \leq \log n_H \leq 14$ [cm^{-3}]. $-4.5 \leq \log U \leq 1$, $-2 \leq \log Z \leq 3$, with a step of 0.25 or 0.3 dex, in search for the best model according to a minimum χ^2 criterion. The derived metallicity value is very high, $Z \sim 100 Z_\odot$. The narrow projection on the planes Z-U and Z-density suggest that the Z uncertainty is small, and that the metallicity is well constrained, unlike the other physical parameters, which, however, have a minimum in correspondence of the expected low ionization solution, $\log U \sim -3$, $\log n_H \sim$ 13–14 [cm^{-3}]. The high value of metallicity is supported by the high SiIVλ1397+OIV]λ1402/CIVλ1549 ratio, as well as by the intense AlIIIλ1860, and very prominent FeII (Figure 7 [211,215]). The high metallicity and large outflow velocity that can be inferred from Figures 7 and 8 are consistent with a scenario in which the wind acceleration is driven by the scattering of AGN continuum radiation, through absorption and isotropic re-emission in resonant transitions of the most abundant metals [221,222].

The spectra of xA quasars reveal that the gas in their BLR is consistently enriched in metals at low redshift as well. Metallicity levels reach up to 50 times the solar value in the most extreme xA PHL 1092, and are always at least 10 times solar [159]. Again, quasars such as I Zw 1 and PHL 1092 exhibit significant blueshifts in their high-ionization lines (HILs), particularly evident and easy to measure in the CIVλ1549 line. The diagnostic ratios used to derive metallicities and the UV broad line weakness [136,138] provide a consistent picture of the physical conditions within the BLR, indicating a uniform structure among these SE candidates over several orders of magnitude in luminosity [30,101,134,140,223]. This uniformity suggests that the processes driving the enrichment associated with SE accretion are self-similar and produce consistent BLR characteristics across different xA quasars, with low ionization and high density.

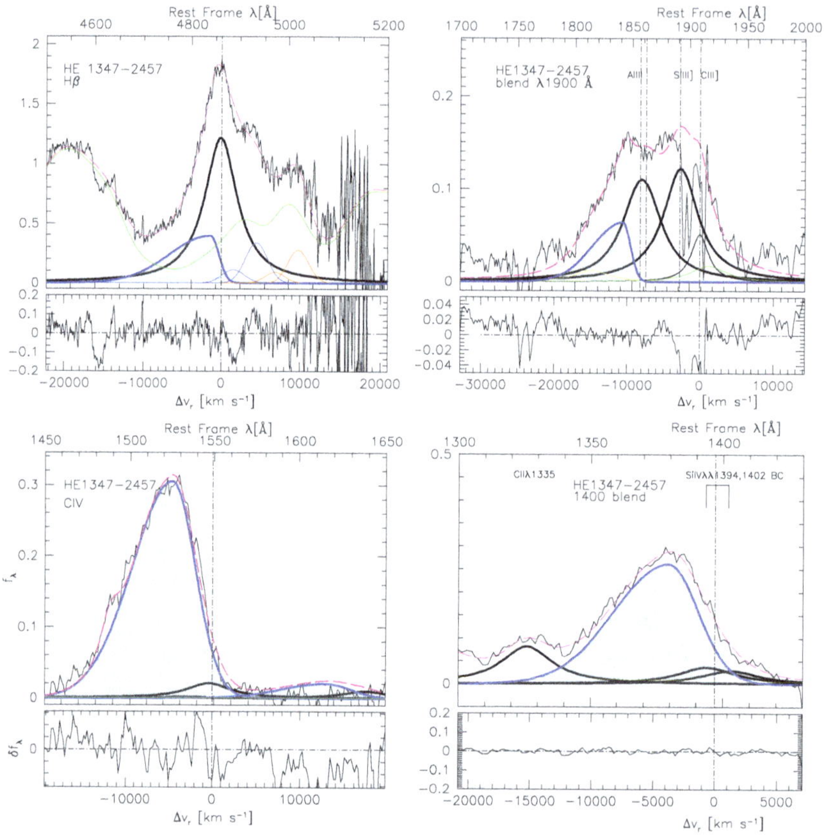

Figure 7. Rest-frame emission line profiles of Hβ, 1,900Å blend, CIVλ1549, and SiIVλ1397+OIV]λ1402, for the HEMS quasar HE1347-2457, after continuum subtraction. The dominance of the outflow emission is emphasized by the cyan shade of the blueshifted components, weakest in the Hβ profile, but overwhelming in the high ionization lines. The quasar rest frame is identified by the dot-dashed vertical line. The virialized components are traced by the thick black lines, at systemic redshift, being extremely weak in the case of the high-ionization CIVλ1549, SiIVλ1397, and HeIIλ1640 lines. The HeIIλ1640 line has been fit self consistently with CIVλ1549. The Hβ line is strongly affected by FeII emission (green line), with $R_{\rm FeII} \gtrsim 1$.

The observed lack of redshift evolution in Z ([224,225], and references therein), the high $Z \gtrsim 10$ inferred for the xA BLR, combined with the steep metallicity gradient from the BLR to the narrow-line region (NLR) and the stellar component [209] suggest that metallicity enrichment occurs in situ [214,226–228]. This inference is further supported by the strong similarities between low-redshift xA quasars [159] and those at the highest redshifts [106,109,229]. Given the relatively short ages at high redshift, there might not be sufficient time for efficient accretion of enriched material produced elsewhere across the host galaxies. Unless BLR Z estimates are wrong by two orders of magnitude, the in situ origin of the enrichment is also consistent with the lower metallicity estimates for the NLR, which reach $Z \sim Z_\odot$ in the most extreme cases, and are usually between sub-solar and slightly super-solar [209,230–232]. A direct comparison with the metallicity in the BLR of xA sources is, however, not appropriate: the NLR estimates refer to different AGN, some of which belong to Population B. Quantifying metallicity in the xA NLR is challenging, as the

O^{2+} lines are usually weak and blueshifted, and other lines are even weaker and difficult to measure. A characterization of the xA NLR metallicity is not yet available.

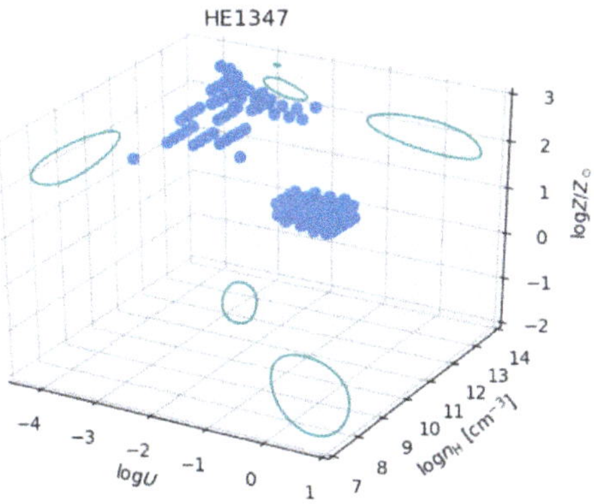

Figure 8. Estimate of physical parameters for the HEMS quasar HE 1347-2457: metallicity Z, density, and ionization parameter. Dots represent photoionization simulations in the parameter space that are in agreement with the minimum χ^2 within 1σ confidence limit. The seagreen lines trace the projection of the 1σ confidence limits in the three parameters after smoothing with a Gaussian kernel.

4.4. Super-Eddington Accretion and Nuclear Star Formation

A starburst–AGN connection has been discussed from the 1980s through the 2000s, whereas several Seyfert galaxies were found in interacting systems with concomitant starbursts, many of them classified as luminous or ultra-luminous IRAS galaxies (e.g., [24,233–240]). The class of cold quasars (blue quasars detected in ultra-luminous IR galaxies because of their emission of cold dust at 250 μm [206]) indicates that there can be a significant range of cosmic epochs where both star formation and AGN activity coexist. In addition, quasars that are hosted in luminous IRAS galaxies are often extreme FeII emitters [241–243]. This result led to propose an evolutionary unification scenario involving supermassive black holes, and starbursts with outflows to explain key observational properties of what we now call SE candidates [233,244,245].

A more recent approach is still focused on explaining several aspects of nuclear activity as mainly associated with massive star formation, and to evaluate the conditions in the inter-stellar medium (ISM) surrounding the black hole and their implications for the accretion process. Nuclear star formation (i.e., *within or around the molecular torus*) [246–249] is expected to be present in most AGN with moderate or high accretion rates (a torus might be absent in low-luminosity AGN [250,251]). The outer self-gravitating part of the disk and the torus provide a reservoir of atomic and molecular gas for the active nucleus fuelling, and star formation in the self-gravitating part of the accretion disk appears a condition naturally associated with high or SE accretion rate [226,227,252–256].

Stars formed within, or embedded by the accretion disk are expected to follow evolutionary tracks different from the ones of stars evolving in a low-density environment. Accretion modified stars, embedded within the dense environment of the AGN accretion disk, can undergo substantial mass gain or experience powerful tidal forces, reshaping their structures, compositions, and lifespans [252,257]. Stars in close orbits around supermassive black holes may accrete disk material, leading to surface enrichment in heavy elements and altered thermal properties [258,259]. Such accretion-modified stars could become more massive and luminous, and may even exhibit enhanced rotation due to the angular momentum transfer from the disk [260]. Stars might even become "immortal", meaning they accrete gas at a rate that balances their mass loss through stellar winds [261–263]. The disk/star interactions not only influence the stars themselves but may also contribute to AGN variability and feedback processes, as accretion-modified stars can fuel further activity within the nucleus through stellar winds or episodic mass loss back into the accretion flow. Over time, the stars in AGN disks can inject elements like nitrogen (the main product of the CNO cycle [264]) into the surrounding medium, and yield peculiar enrichment patterns [265]. These stars can also contribute to phenomena like supernovæ (SNæ) and gamma-ray bursts (GRBs) within the AGN disk.

The connection between nuclear star formation and AGN properties is still sketchy, in part because data cannot resolve nuclear star formation, but also because a full stellar evolution theory for accretion-modified stars in a dense environment is still to be developed [254]. At present, the evidence of star formation as a source of infrared and radio emission in xA sources is associated with the host galaxies: several xA sources tend to follow the FIR—radio correlation of star forming galaxies [84,266–268]. The modest radio-to-optical ratio places xA sources in the radio-intermediate domain [83]. Given the resolution limits, especially severe in the FIR, it has been impossible to probe the extent of nuclear star formation. The origin of the relatively weak radio emission of radio-quiet and radio-intermediate quasars remains unclear [269,270]. The inner X-ray corona and the wind could also produce significant nuclear radio emission [271–273].

It is the heavy enrichment of the BLR emitting gas that makes it necessary to postulate a close association between SE activity and nuclear star formation. Gas enrichment suggested by significant FeII emission in high (and even the highest) redshift quasars [90,106,107,109,274] implies that the enrichment occurs very proximate to or within the accretion disk [228]. Core-collapse supernovæ from massive progenitors have a significant yield in iron [275], along with Al and Si. This makes it possible for highly accreting quasars in Population A not only to exhibit powerful metal-rich winds [247–249,276], but also to enhance the low- and intermediate-ionization lines that are especially prominent in xA sources.

4.5. Evolution Along the Quasar Main Sequence

The scheme of Figure 9 (see also Figure 6 of Marziani et al. [277]) illustrates a possible evolutionary path, emphasizing the transition from highly accreting, young quasars to massive, low Eddington ratio systems. Population A quasars are characterized by lower black hole (BH) masses ($M_{\rm BH}$), higher $L/L_{\rm Edd}$, strong outflows, and low equivalent width [OIII]$\lambda\lambda$4959,5007 and CIVλ1549 HILs. These are typically young or rejuvenated, highly accreting systems, some of which host powerful winds. At the lower-right corner of the sequence, the extreme Population A ($R_{\rm FeII}$ strong NLSy1s at low luminosity) includes younger, highly accreting systems with strong winds, low equivalent width emission lines, and significant blueshifts in both CIVλ1549 and [OIII]$\lambda\lambda$4959,5007 lines. VLTI/MATISSE and GRAVITY observations of H0557-385 and I Zw 1 revealed that the intense radiation field might by at the origin of a flattening in the molecular torus [278]. The K, L, and M band results indicate that the dust structures are compact, and exhibit no significant polar elongation. Notably, in H0557-385, the dust sizes between 3.5 and 8 μm remain constant at approximately 3 to 10 times the sublimation radii, suggesting a direct view of the wind launching region. The absence of polar elongation implies that any winds are likely launched in an equatorial direction or are influenced by the AGN continuum radiation pressure. If equatorial outflows dominate, this could reduce the amount of gas available for accretion, potentially lowering the accretion rate over time, or leading to a stationary configuration contributing to the low ionization part of the BLR [279].

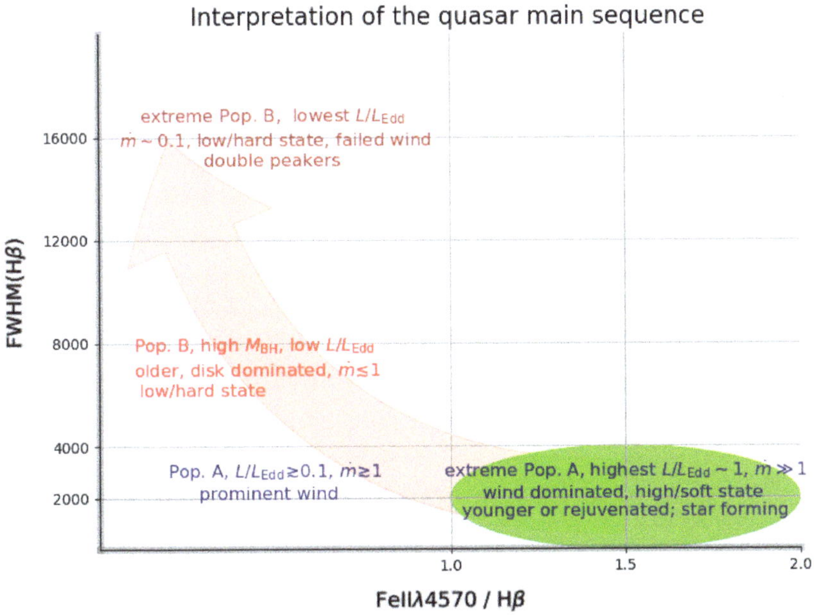

Figure 9. Evolutionary interpretation of the optical plane FWHM Hβ vs. R_{FeII} of the quasar main sequence.

On the other hand, Population B quasars have higher M_{BH}, lower L/L_{Edd}, weaker outflows, and high W [OIII]$\lambda\lambda$4959,5007 lines, representing more evolved, older systems with less active accretion [170]. Powerful radio-loud (jetted [280,281]) sources are located in the upper-left quadrant, associated with older, less active quasars with weaker winds and higher [OIII]$\lambda\lambda$4959,5007 emission. At the upper left end of the MS, we encounter sources with $\dot{m}$ so low that they are at the verge of the inefficient radiative domain. In several cases, Hβ emission is showing a relativistic disk profile, meaning that the emission is associated with a bare, small-size (perhaps truncated) accretion disk [282,283]. These systems are starving black holes [97].

As the quasar evolves and feedback processes (such as winds and other outflows) deplete the surrounding material, the quasar transitions toward lower L/L_{Edd} ratios, with a gradual shift into Population B as the black hole accretion rate decreases and the quasar begins to receive less accretion material. In this phase, outflows weaken, and the quasar [OIII]$\lambda\lambda$4959,5007 emission becomes more prominent [223]. Overall, the sequence encapsulates the idea that quasars evolve from highly accreting (possibly even at SE rate), young, wind-dominated systems in Population A to older, massive, low-Eddington ratio systems in Population B, with feedback and accretion rate playing crucial roles in this transition. In other words, the MS shape of Figure 1 can be well understood as a temporal sequence with black hole mass as a rough clock of accretion history [170].

Massive stars embedded in the accretion disk may be a scenario possible only for young or rejuvenated sources [284]. More evolved systems might involve embedded black holes with extreme mass ratio or neutron stars, yielding gravitational wave chirps that could be detected with LISA up to high redshift [38].

4.6. Accretion Modes

Quasars can be classified into distinct accretion modes based on their position in the mass-luminosity diagram, in addition to their placement along the MS [71,285–287]. In the plane luminosity vs. black hole mass, type 1 AGN occupy a relatively narrow strip constrained between $L/L_{\rm Edd} \approx 0.01$ and ≈ 1. Inefficient radiators are found at or below the lower end of the Eddington ratio range, characterized by low accretion rates and jet-like outflows [288–290]. They are rare in optically selected samples of type 1 AGN, since a cold disk [22] might evaporate into an X-ray hot plasma if $L/L_{\rm Edd} \ll 0.01$ [251]. Population B sources occupy the middle Eddington ratio range, $0.01 \lesssim L/L_{\rm Edd} \lesssim 0.1$–$0.2$, where the accretion disk can be arguably modeled using the classical Shakura–Sunyaev framework [22]. These quasars exhibit moderate accretion rates and relatively stable disks, with limited wind activity (also due to an over-ionization of the inner disk region exposed to the AGN continuum [222,291,292]). The change in structure with increasing dimensionless accretion rate is well-summarized in Figure 1 of Giustini and Proga [293], where all the main accretion modes are sketched. Table 1 provides a slightly modified scheme, in which the onset of an optically thick, geometrically thick region occurs at $L/L_{\rm Edd} \sim 0.1$–0.2, where we see a transition between Pop. A and B properties [71,294]. Above this Eddington ratio, outflows leading to a change in emission line parameters of high ionization lines become frequent, and FeII emission becomes more prominent. It is tempting to think that this critical Eddington ratio is accompanied by a structure change in the inner accretion disk, and a shielded region may develop [56,65], yielding the ideal conditions for the production of the low ionization spectrum emitted in Pop. A and extreme Pop. A. At the same time, radiative forces are higher, and the wind acceleration might become more efficient: low-column density gas is easily ablated away [146]. Toward the high end of the Eddington ratio range, we enter in the domain of xA sources and we find sources which are even more wind-dominated [149,173,177,295,296]. These quasars exhibit significant outflows, as evidenced by the pronounced blue shifts in high ionization lines like CIVλ1549, with radial velocities of the outflow that can reach $\gtrsim$1000 km s^{-1}, up to 5000 km s^{-1} at the highest luminosity (Section 5.2; Figure 7). These are among the outflows that may play a role in regulating the growth of the black hole and the evolution of galaxies [23,297,298].

Table 1. Populations in the luminosity–black hole mass diagrams of type-1 AGN.

Population	Eddington Ratio Range	Prevalence
Inefficient radiators	$L/L_{\rm Edd} \lesssim 10^{-3}$–$10^{-2}$	not included in Figure 1
Population B	10^{-3}–$10^{-2} \lesssim L/L_{\rm Edd} \lesssim 0.1$–$0.3$	≈ 0.46
Population A	0.1–$0.3 \lesssim L/L_{\rm Edd} \lesssim 1$	≈ 0.43
extreme Pop. A	$1 \lesssim L/L_{\rm Edd} \lesssim$ a few	≈ 0.10
highly super-Eddington	$L/L_{\rm Edd} \gg$ a few	not existent?

5. Coexistence of a Virialized System and Powerful Outflows

5.1. A Virialized Low-Ionization System

The HIL and LIL profiles in Population A and xA reflects a dichotomy in the physical and dynamical conditions of the BLR. An impressive, extreme example is shown in Figure 7, which emphasizes the profile differences between CIVλ1549 and Hβ: while Hβ retains a mostly symmetric profile, roughly consistent with the rest frame, most of the CIVλ1549 flux is due to a blueshifted excess. The line component which is shaded blue is ascribed to an outflow, most likely a disk wind driven by radiation pressure [222,291,292,299]. Outflows in AGN are very pervasive [97] and there will be always some low column density gas that will be expelled from the gravitational sphere of influence of the central black hole [145,146,300,301], no matter how low $L/L_{\rm Edd}$ might be. However, it is only above a threshold $L/L_{\rm Edd} \approx 0.1$–0.2 we start seeing prominent outflows in CIVλ1549 and other high ionization lines like SiIVλ1397 and HeIIλ1640 [149,173] such as the one shown in Figure 7. The figure also shows that a symmetric, unshifted component is dominating the emission of Hβ and is still prominent in the intermediate ionization line AlIIIλ1860, although it is overwhelmed in CIVλ1549. The line profiles of Hβ, MgIIλ2800, and AlIIIλ1860 are well modeled by a Lorentzian function, and are consistent with an extended disk profile even if broadened by turbulent motions of the emitting gas [78,302–304], with negligible relativistic effects [305].

This distinction between HILs and LILs is especially pronounced for xA quasars, where the high accretion rates lead to stronger winds for the HILs, while the LILs remain tied to the more virialized region near the disk [27,152]: the CIVλ1549 and Hβ lines appear to behave independently of one another [195,306].

The sketch of Figure 10 illustrates the structure and components of SE AGN focusing on the regions surrounding the supermassive black hole (from 1 to 10^5–10^6) gravitational radii. At the center, the SMBH is enveloped by an ADAF. A mildly relativistic jet may emerge from the black hole along the spin axis, indicating the outflow of material along the polar axis [2,171,307]. The sketch also highlights the presence of a corona above and below the accretion flow, which is responsible for high-energy X-ray emission. The geometry of the corona is actually highly uncertain, with a lamppost geometry usually preferred over a more symmetric configuration [308,309]. A "special" Compton reflection hump representing the interaction of high-energy photons with Compton-thick outflowing material from the very edge of a geometrically-thick accretion disk might be also present: in this case, the emission of Fe Kα is seen displaced by (0.1–0.2) c [161,310]. Ultra-fast outflows (UFOs [311]) are indicated as the inner arrows extending outward from the central engine. The location and the geometry is not fully clear as yet [312–314], but they are believed to be on a scale of 10^2–10^4 gravitational radii, perhaps closer to the SMBH axis than the mildly ionized outflows, which might be associated with the outer ADAF but also with the part of the disk not dominated by radiation pressure (shown as a flatter section in between the ADAF and the torus), shielded (at least in part) by the coronal X-ray radiation. The flattening shown in Figure 10 may not actually occur; however, the thick disk geometry allows for a shielded region (region II of Wang et al. [56]). Among the mildly ionized outflow, there is an interesting relation between the amplitude of the line blueshift and the ionization potential of the parent ionic species: highest blueshifts in CIVλ1549, $\gtrsim$1000 km s^{-1} correspond to moderate blueshifts in AlIIIλ1860 $\approx$ 0.2 the ones of CIVλ1549 [153], and of a few hundred km s^{-1} in MgIIλ2800 [303]. The decreasing velocity could be a consequence of an increase in the launching radius of a radially-stratified wind. Extrapolating the trend to higher energy, one should ascribe the X-ray high speed wind to the innermost region exposed to funnel radiation.

The sketch in Figure 10 highlights an additional layer of complexity in the outer regions of the accretion disk: the potential for star formation within the self-gravitating portion of the disk [226,252,315]. Stars that form within the disk, or are captured from elsewhere, follow evolutionary paths that are influenced by the accretion of disk material. Farther out, a torus of dust and gas surrounds the AGN, with the dust sublimation radius marking the point beyond which dust can survive without being sublimated by the AGN intense radiation. Dust-driven outflows might originate around this transition radius [316–319]. Star-forming regions are indicated where the disk may become self-gravitating. The sketch is meant to suggest that these star-forming regions could affect the accretion process, potentially influencing the overall evolution of the system, and even the development of two main structural components of the active nucleus, the torus and the accretion disk.

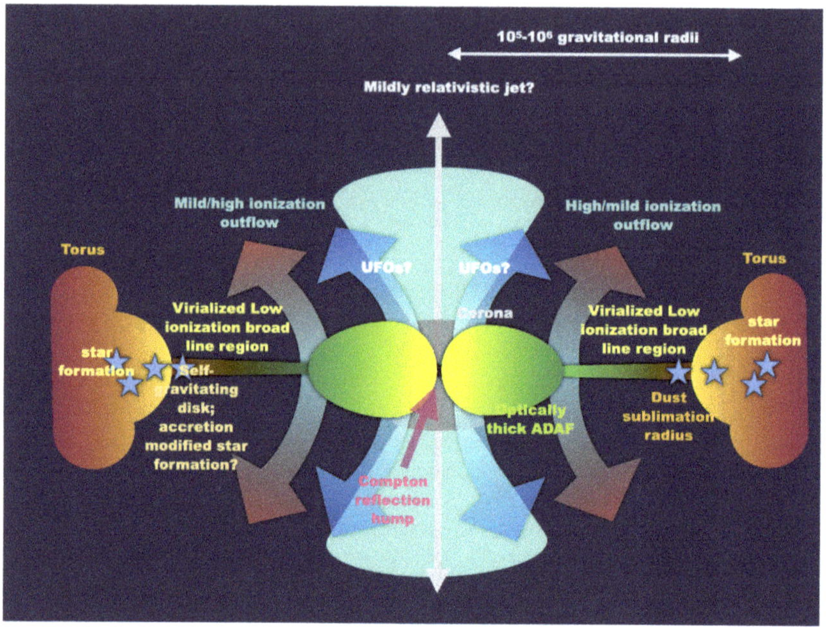

Figure 10. A tentative sketch of the inner working of a super-Eddington quasar. The placement and the configuration of several elements is not drawn to scale, and is largely hypothetical. The scheme is meant to emphasize the existence of a shielded low-ionization region, along with the possibility of concomitant star formation in the outer regions of the accretion disk and in the torus.

5.2. *Powerful Outflows: A Major Factor in Galactic Evolution?*

Recent surveys—such as the Wide-field Infrared Survey Explorer (WISE)/SDSS selected hyper-luminous (WISSH) [320] and the Hamburg-ESO main sequence (HEMS) [90]—have considered the most luminous quasars at the Cosmic Noon. The large blueshifts observed in the CIVλ1549 line are indicative of strong winds generated during their high-accretion phases, producing outflows capable of escaping the BLR and enriching the surrounding regions [108,177,178]. Albeit rare, these sources are of extremely high luminosity, $L \gtrsim 10^{47}$ erg s^{-1}. The estimate of the mechanical parameters of the mildly ionized gas outflows traced by CIVλ1549 and [OIII]$\lambda\lambda$4959,5007 emission permits testing whether AGN may cause a significant feedback effect at least when their outflows are most powerful. The mass outflow rate ($\dot{M}$) and the kinetic power ($\dot{E}_{\rm kin}$) have been estimated adopting a simple biconical geometry where the emitting gas is in radial outflow [177,223,321,322]. In this framework, the relations are formally identical for [OIII] and CIVλ1549, and namely: $M_{\rm ion} \propto L^{\rm out} Z^{-1} n_{\rm H}^{-1}$, where L is the outflow-emitted line luminosity, $n_{\rm H}$ the electron density and Z the metallicity. The mass outflow rate ($\dot{M}^{\rm out}_{\rm ion}$) at a radius r and with an outflow velocity $v_{\rm out}$, might be written as: $\dot{M}^{\rm out}_{\rm ion} \propto L^{\rm out}\, v_{\rm out}\, r^{-1} Z^{-1} n_{\rm H}^{-1}$. The kinetic power, $\dot{E}_{\rm kin}$, is then given by $\dot{E}_{\rm out} \sim \frac{1}{2}\dot{M}_{\rm out} v^2_{\rm out}$, which leads to $\dot{E}_{\rm kin} \propto L^{\rm out} v^3_{\rm out}\, r^{-1} Z^{-1} n_{\rm H}^{-1}$. The values obtained scale with line luminosity and can therefore increase by a factor $\sim 10^3$ going from moderately to very luminous quasars. The increase in radial velocity is more modest, a factor $\sim$5, due to the weak luminosity dependence of the terminal velocity for radiation-driven winds [296,323].

Figure 7 shows an example of xA quasar where the outflow component is clearly dominating the emission in the high ionization lines CIVλ1549, HeIIλ1640, SiIVλ1397, and OIV]λ1402. Figure 11 shows the distribution of mass outflow rate in $M_\odot$ yr^{-1} and of the ratio between kinetic power and bolometric luminosity derived from the CIVλ1549 lines applying the equations written above, for a sample of high-luminosity quasars $L \gtrsim 10^{47}$ erg s^{-1}, mainly from the HEMS survey [90,103,104] and from the recent works by Deconto–Machado et al. [108,111]. There is no obvious systematic difference between the xA sources and the rest of the sample, as confirmed by Kolmogorov–Smirnov and *U*-Mann–Whitney tests. Significant outflows are detected also in Population B, even if the most powerful are restricted to Population A [108,111]. The CIVλ1549 outflows are related to the larger scale outflows traced by [OIII]$\lambda\lambda$4959,5007 [108,111,175,177]. It is tempting to think that the emitting gas might be the one expelled from the active nucleus [324]. At a low redshift, the low [OIII]$\lambda\lambda$4959,5007 equivalent width and its large FWHM are consistent with a compact NLR and hence with a more straightforward physical connection between the CIVλ1549 and the [OIII]$\lambda\lambda$4959,5007 emitting gas: in principle, the [OIII]$\lambda\lambda$4959,5007 could be emitted by gas further downstream in the nuclear outflow [175]. For quasars at high luminosity, the [OIII]$\lambda\lambda$4959,5007 emission is spread over kpc-sized distances [325,326]. The extensive size of the [OIII]$\lambda\lambda$4959,5007 emitting regions in high-luminosity quasars [327,328] may suggest that the emission arises predominantly or even solely from radiative feedback from the AGN on the host galaxy gas.

Even if the Z is measured from diagnostic ratios, the uncertainties are at least $\sim$0.3 dex. The geometry of the outflow introduces a factor of 3, based on the comparison between spherical symmetry and a flat layer. The 1σ uncertainty in the emitting radius is approximately $\pm$30%, considering only the uncertainty of the scaling law parameter, while the uncertainty in the outflow velocity is typically around 30%. Quadratically propagating these uncertainties results in a total factor of $\approx$ 5 at the 1σ confidence level in the kinetic power [108]. The energetics remains extremely high even if conservative lower limits are applied, with $\dot{M}_{out} \sim 100 M_\odot$ yr^{-1}, with typical kinetics powers around $\sim$1% the bolometric luminosity, due to the extreme line luminosities reaching $\sim 10^{45}$ erg s^{-1} for CIVλ1549. Figure 11 shows that the xA quasars have mass outflow rates and kinetic–power ratios to luminosity comparable to non-xA quasars, but that xAs are not extreme (the equivalent width of CIVλ1549 is among the lowest [101,135,138]), even if they are expected to have extreme radiative output per unit black hole mass. The implication is that there is likely a broader population of quasars beyond the SE candidates—essentially encompassing all of Population A—that is capable of exerting significant feedback effects on their host galaxies.

The CIVλ1549 resonance line is representative of the mildly ionized outflows originating from the accretion disk [222,299,301]. This is expected to involve a large fraction of the mass load of the AGN wind, but is not sampling gas in higher ionization stages, specifically the one associated with the UFOs). UFOs are present in jetted AGNs [329], and detected in about half of nearby AGNs, although there are hints of a larger prevalence in highly accreting quasars [330,331]. An UV line-driven disk wind could be at the origin of UFOs in AGN [332]: this appears to be at least the case of the high-Eddington ratio quasar PDS 456 [333]. Acceleration of UFOs likely needs an additional mechanism to yield velocities $\sim$0.1–0.3 c, such as intense magnetic fields [334,335] or thermal pressure gradient [336]. The mass load and the mass outflow rates of the UFO are comparable or superior to the one of the outflows traced by CIVλ1549 [312,337]. UFOs and mildly ionized winds are likely to be associated with coexisting phenomena, i.e., UFOs might not be entirely due to gas downstream or upstream of the BLR outflowing gas. Due to the multiphase nature of quasar outflows, the mass outflow rate and dynamical parameters derived from the CIVλ1549 line may represent only a fraction of the total outflow. If so, it is likely that the combined effect of UFOs and mildly ionized winds could result in a more significant and widespread impact [311,331].

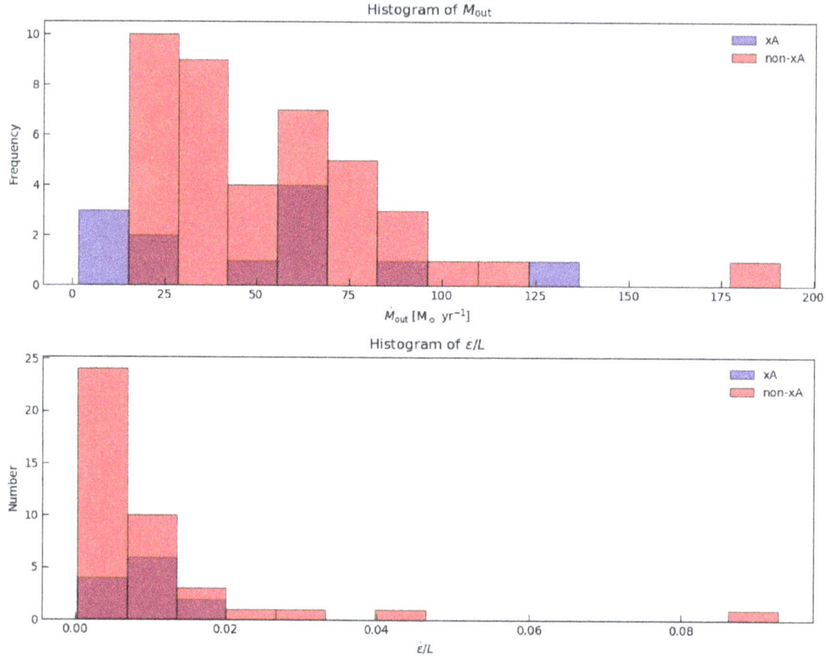

Figure 11. Distribution of mass outflow rate (top) and ratio between the kinetic power $\dot{\epsilon}$ and the bolometric luminosity (bottom) for sources with $L \gtrsim 10^{47}$ erg s^{-1}.

6. Super-Eddington Quasars for Cosmology

6.1. Distance Indicators

The contrast between the behavior of HILs and LILs in Population A and extreme Population A quasars highlights the multiphase nature of quasar outflows and the complex, dichotomic structure of the BLR. The potential to use quasars as reliable cosmic distance indicators is grounded in the well-established understanding that their emitting regions form a virialized low-ionization system [74], governed by the gravitational force of the central supermassive black hole, even in the case of the highest radiative output [140]. Capitalizing on this result, we can consider three basic elements to derive a cosmological luminosity estimators: (1) SE candidates tend to converge toward a limiting value of $L/L_{\rm Edd}$; (2) their optical and UV spectrum is self similar, and easy to recognize, which in turns implies that the radius of the emitting regions should scale $\propto L^{\frac{1}{2}}$, lest significant differences are introduced in the ionization of the line emitting gas; and (3) there is a measure that accurately reflects the virial motion around the central black hole. If we consider the three factors together, it is possible to rewrite $M_{\rm BH}$ in terms of luminosity and virial broadening [26]:

$$L \propto \mathrm{FWHM}^4 \quad (7)$$

The derivation of the Faber-Jackson [338] law followed exactly the same line of reasoning. The caveat is that $L/L_{\rm Edd}$ needs to be constant (in observational terms, have a small scatter around a well-defined value). Paradoxically, this assumption seems to be more easy to achieve for super-Eddington accretors than for galaxies, which are much larger and more complex bodies, each with its own evolutionary path [36].

Virial broadening estimators have been sought for decades [76,77,153,285,339–345]. They allow for the least-inaccurate measurement of the central black hole mass, following the deceptively simple virial mass definition

$$M_{\rm BH} = \mathcal{F}\frac{r_{\rm BLR} \cdot {\rm FWHM}}{G}, \quad (8)$$

where the factor $\mathcal{F}$ incorporates all dependences on geometry, dynamics, and accretion mode. The HI Balmer line Hβ seems to be a good choice (Section 5.1; [76,346]) to construct the Hubble diagram [26,27,29]. At a second stance, the UV AlIIIλ1860 doublet [153] could be a good tool to extend the Hubble diagram to redshifts beyond ≈ 1.4, where Hβ can be observed only with IR spectrometers [28]. Preliminary applications have been carried out employing several samples where the virial broadening estimator has been Hβ FWHM and σ only or Hβ and AlIIIλ1860 jointly [26–29,347]. Despite high dispersion (presently around 0.3 dex in the luminosity estimates), this method provides a promising approach for using quasars as cosmological standard candles, especially for measuring the matter density parameter ($\Omega_{\rm M}$) [27,28]. The main reason resides in the ability of quasars to sample a redshift range around $z \approx 2$: at the corresponding cosmic epochs, the repulsive effect of dark energy was too low, and the main factor in the Universe dynamics was the attractive gravitational force (see Figure 6 of Marziani and Sulentic [26], which shows how differences in cosmological models become quite large, ≈ 0.2 dex in luminosity, around $z \approx 2$).

6.2. *Grown with the Wind, or Saving Λ-Cold Dark Matter*

Some high-z quasars with very massive black hole ($\sim 2 \cdot 10^9 M_\odot$) would have been classified as an unremarkable Pop. B quasar if observed at $z \lesssim 1$. Two examples— along with others at $z \approx 5$ [348]—are PJ308-21 at $z \approx 6.234$ [229] and ULAS J1120+0641, whose black hole mass is $(1.35 \pm 0.04) \times 10^9$ M$_\odot$ based on the MgIIλ2800 line, Eddington ratio ≈ 0.40, and X-ray properties indistinguishable from other bright quasars at $1 < z < 5$ with similar Eddington rates [349]. JWST discovered a population of fairly massive black holes even at the highest z, for example UHZ1 with a mass of $\sim 10^{7.5}$ M$_\odot$ at redshift $z \approx 10.3$ [350]. These newly discovered, extremely massive black holes present a challenge to current growth models, with potential implications for cosmology [351,352], unless a SE accretion phase is considered (e.g., [353], and references therein).

On the one hand, there are no convincing cases with $L/L_{\rm Edd} \gg 1$, in agreement with the solutions computed for SE ADAFs. Even if the balance between infall, radiation trapping, the feedback exerted by radiation forces, magnetic field, and mass loss due to winds [60,61] is not fully understood nor modeled in general relativistic radiation magnetohydrodynamics simulations [59], models do not exclude a SE regime with $\dot{m} \gtrsim 10^3$ and radiative efficiency as low as $\eta \sim 0.01$ [42,354,355] that would make possible a rapid black hole growth. xA quasars with $R_{\rm FeII} \gg 1$ like PG0043+052 and Mark 231 have been found up to $z \approx 6.5$ [101,105,274]. Convincing cases with best virial mass estimates range from 10^6 to 10^9 solar masses: from the case of I Zw 1, PHL 1092, PDS 456, Mark 231 [346,356–358] to the most luminous quasars at the Cosmic Noon (e.g., HE 0359—3959 [90,108,359]).

On the other hand, there are several basic unknowns. The first question is whether accretion is truly SE, or a form of sub-Eddington balance resulting in a radiative output around $L/L_{\rm Edd} \sim 1$. A continuity argument suggests caution, as there is no clear discontinuity between Pop. A and extreme Pop. A, unlike the distinct transition observed between Pop. B and Pop. A. The extreme Pop. A in this respect might be just an extreme once the threshold in Eddington ratio at ≈0.1–0.2 has been passed. The mass outflow rates also raise pressing problems. If the outflow rates estimated in Section 5.2 apply to both sub- and super-Eddington regimes, they may become comparable to the mass accretion rate in the sub-Eddington case, slowing the growth of the black holes. In addition, there is an environmental caveat to consider. ADAF models deal mostly with the innermost few hundreds gravitational radii, and assume that an arbitrary large amount is reaching the disk. On larger scales, the interplay between radiation forces and gravity is not so obvious, especially in the regions beyond the Bondi radius (the radius at which the gravitational pull of the central mass dominates over the thermal pressure of the gas, allowing the gas to move inward and accrete; see, e.g., [360]). It is not clear whether a large amount of matter in excess to hundreds of solar masses per year can be effectively funneled down to the Bondi radius of the black hole [59,361].

We can make a few elementary considerations on the black hole growth with mass loss due to a wind. The rate of change of the black hole mass $M(t)$ is the difference between the accretion rate and the mass loss rate due to winds: $dM/dt = \dot{M}_{\rm acc} - \dot{M}_{\rm wind}$, where $\dot{M}_{\rm acc}$ is the accretion rate of the black hole and $\dot{M}_{\rm wind}$ is mass loss rate due to winds. The accretion rate can be expressed in terms of the Eddington accretion rate $\dot{M}_{\rm Edd}$ and the dimensionless accretion rate $\dot{m}$: $\dot{M}_{\rm acc} = (1-\eta)\dot{m} \cdot \dot{M}_{\rm Edd}$ and the mass loss rate due to winds as a fraction $f_{\rm wind}$ of the accretion rate: $\dot{M}_{\rm wind} = f_{\rm wind} \cdot \dot{M}_{\rm acc}$, so that we can write:

$$dM/dt = \dot{m} \cdot 4\pi G m_{\rm p}/\sigma_T c(1-\eta)(1-f_{\rm wind})M. \quad (9)$$

The solution of this differential equation describes the exponential growth of the black hole mass $M(t)$ considering both accretion and mass loss due to winds is: $M(t) = M_0 \exp\left(t/t_{\rm growth}\right)$, where the characteristic growth timescale $t_{\rm growth}$ is given by $t_{\rm growth} = \sigma_T c/[\dot{m} \cdot 4\pi G m_{\rm p}(1-\eta)(1-f_{\rm wind})] = t_{\rm Edd}/\dot{m}(1-\eta)(1-f_{\rm wind})$.

Figure 12 shows black hole mass estimates as a function of z, assuming standard Λ—cold dark matter cosmology ($H_0 = 70$ km s^{-1} Mpc^{-1}, $\Omega_{\rm M} = 0.3$, $\Omega_\Lambda = 0.7$) ([362], and references therein). The samples include Hβ based measurements for low-z quasars from the SDSS [363], the HEMS survey [90], and several samples of high-z quasars shown by ([340], c.f. [364]). With respect to Marziani and Sulentic [340], the figure adds some recent $M_{\rm BH}$ estimates (brown stars) for the highest redshift quasars, in the range $6 \lesssim z \lesssim 10$. The yellow shaded area represents Eddingon limited growth from seeds between 10^2 M$_\odot$ (Pop. III stars) and 10^6 M$_\odot$ (intermediate mass black holes (IMBHs) produced by direct collapse) formed at $z \approx 20$, assuming $f_{\rm wind} = 0.5$. The new observations fall outside the permitted area; the Eddington limited growth is too slow to account for the newest observations, but was still able to explain the highest $M_{\rm BH}$ up to $z \approx 5$. A modest super-Eddington regime ($\dot{m} \sim 10^2$) is able to explain the highest masses at the highest redshift, since the observed mass outflow rate becomes negligible compared to the accretion rate. This toy scheme is not considering feedback effects which may slow the growth, and an SE phase of $\sim 10^7$ yr is still necessary to explain a $M_{\rm BH} \sim 5 \cdot 10^7M_\odot$ at $z \approx 10$, when the Universe was just half a billion years old, assuming direct collapse with $M_0 \sim 10^5 - 10^6$M$_\odot$.

Systematic searches for IMBHs in the local Universe remain elusive. [365–367]. These objects, with masses ranging from approximately 100 to 100,000 M⊙, are still mostly candidates as definitive dynamical confirmation remains challenging. IMBHs are thought to form through processes such as the merger of smaller black holes or the direct collapse of massive stars in low-metallicity environments [368,369]. Another possible pathway, particularly relevant at high redshifts, involves runaway collisions of massive stars in dense stellar clusters, which could rapidly produce an IMBH of $10^4 - 10^5$ M$_\odot$ in just a few million years [370,371]. At low redshift, a handful of super-Eddington IMBH candidates have been identified, with some exhibiting X-ray brightness or luminosities approaching the Eddington limit [372,373]. However, they appear to be a tiny minority within the Chilingarian et al. [373] sample, and their true nature remains uncertain. It is still unclear whether these low-z candidates serve as valid analogs of the IMBHs expected at the highest redshifts, where a much higher fraction of rapidly accreting black holes is anticipated.

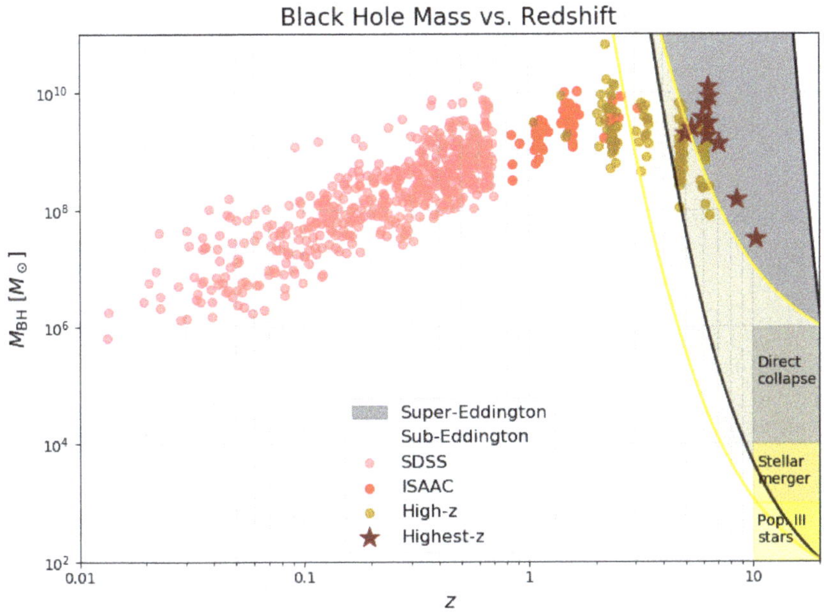

Figure 12. Black hole mass estimates for several samples of quasars as a function of redshift, with a focus on the highest masses at high redshift. The coral data points are from an Hβ SDSS-sample [363], the red ones are the data of the HEMS survey [90], and the dark golden are several high-redshift M_{BH} estimates presented in [340]. The brick-brown stars mostly refer to the most recent observations made possible by JWST [349,350,374–378]. The shaded areas show the range permitted for black hole growth starting from Pop. III star progenitors ($M_{BH} \sim 10^2 M_\odot$) and direct collapse ($M_{BH} \sim 10^5 M_\odot$), for the sub-Eddington (yellow) and super-Eddington regime. See text for details.

7. Analogies Between Stellar Mass and Supermassive Black Holes

7.1. Super-Eddington AGN and Relativistic Jets

The possibility of developing a relativistic jet probably rests on the magnetic field strength: strong magnetic fields lead to a magnetically arrested disk (MAD). The defining feature of a MAD is the presence of a large-scale, ordered magnetic field that is sufficiently strong to exert significant pressure on the inflowing gas. As the gas moves inward toward the black hole, magnetic flux is advected inward and accumulates near the event horizon. When the magnetic pressure near the black hole becomes comparable to or exceeds the ram pressure of the inflowing gas, the accretion flow is arrested [379]. The strong magnetic field near the black hole can efficiently extract rotational energy from the black hole via the Blandford–Znajek process [380]. However, it is not obvious if the thick geometry is preserved [381], nor if the SE regime and a MAD may coexist, or lead to a rapid spin-down of even a maximally rotating supermassive black hole [382,383]. Hydrodynamic simulations of SE accretion disk produce only mildly relativistic jets, $v \approx 0.5c$, close to the disk axis [384–386].

From an observational point of view, do we have evidence of sources that are unambiguously jetted, meeting a strict selection criterion for being radio loud? The criterion requiring a radio-to-optical specific flux ratio in excess of $\log R_K = f_\nu(6\text{cm})/f_\nu(g) \approx 1.8$ [83,84] is a restrictive albeit sufficient condition for the identification of true radio-loud AGN, i.e., with radio emission due to a powerful relativistic jet [280,281]. There is no doubt that several type 1 RL AGN show significant FeII emission. If we apply the radio-to-optical specific flux ratio $\log f_{\nu,1.4\mathrm{GHz}}/f_{\nu,5100} \gtrsim 1.8$ [83,363], along with the SE criterion $R_{\mathrm{FeII}} \gtrsim 1$ accepted at face value, SE radio loud NLSy1s [80,82] may be a very rare phenomenon, if present at all: among the sources listed in the newly published catalog by Paliya et al. [387], they are just $\approx$0.06%. At redshift $z \gtrsim 2$, a recent comparative analysis between radio-quiet and jetted sources did not reveal any jetted quasar with $R_{\mathrm{FeII}} \gtrsim 1$ [108,111]. Nor were they detected in the HEMS survey ([90], and references therein). This might not be surprising as a large part of the accretion power is funnelled into the mechanical power of the jet, which usually exceeds its radiative power by a large factor ($\lesssim 10$ [388]).

7.2. Soft and Hard States of Stellar Mass Black Holes

Galactic black hole candidates frequently experience X-ray nova outbursts, where they transition from a quiescent state to a highly active state (e.g., [389,390]). During these outbursts, a given source can undergo dramatic changes in its accretion rate over the course of a few months. This evolution typically involves the source shifting from low to high accretion rates and then returning to a lower rate as the outburst subsides. The changes in the accretion flow structure and associated emission are well captured in a hardness–intensity diagram, which plots the source X-ray luminosity against its hardness ratio—a measure of the proportion of high-energy (hard) versus low-energy (soft) X-rays [391–393]. In this diagram, high hardness ratios indicate a state where the system is dominated by a hot, extended inner flow or corona and is typically associated with the presence of strong, steady jets [394]. This configuration corresponds to the low/hard state of the black hole binary. Conversely, low hardness ratios signify that the cold, geometrically thin accretion disk extends closer to the black hole's innermost stable circular orbit, leading to a dominance of soft X-ray emission and a reduction or complete suppression of the jet, known as the high/soft state [395]. Ultra-luminous X-ray sources (ULXs) are best cantidates for SE accretors, as they emit X-rays at luminosities exceeding 10^{39} erg/s, significantly higher than the Eddington limit for stellar-mass black holes (e.g., [396,397]). Following the discovery of pulsating ULXs that must contain neutron stars [398], it is now thought that accreting binaries exceeding super-Eddington constitute the majority of the ULX population [399]. A fraction of the ULX (the most luminous ones) might be still more easily explained by intermediate-mass black holes [400], especially if there is a physical limit to the accretion luminosity per unit black hole mass.

It is instructive to focus the attention on a prototypical case, GX 339-4 [401]. GX 339-4 is a well-known black hole X-ray binary that exhibits a wide range of variability across different wavelengths, most notably in the X-ray. GX 339-4 is known for its recurrent outbursts, thought to be caused by increases in the mass accretion rate, possibly due to changes in the accretion disk structure [402]. GX 339-4 undergoes transitions between several distinct spectral states, primarily defined in X-rays: in the low-hard state, GX 339-4 is characterized by a hard X-ray spectrum, dominated by a power-law component with a high-energy cutoff. The emission is likely produced by a hot, optically thin corona. In this state, the system shows strong radio emission from a compact jet [403]. The source transitions to a soft state when the accretion rate increases. It shows a dominant thermal component from the accretion disk, with the X-ray spectrum softening. The radio jet is usually quenched or much weaker in this state [404,405]. During transitions between hard and soft state, GX 339-4 often exhibits quasi-periodic oscillations and rapid variability, suggesting changes in the inner disk configuration and the corona [406,407], in general reproducible by general-relativistic magneto-hydrodynamic simulations [408]. The source often follows a q-shaped hysteresis track in the hardness-intensity diagram, transitioning from the low to high state and back [391,392]. The decline phase does not retrace the same path, as required for a true hysteresis cycle. In the low/hard state, optical and IR emissions are often dominated by synchrotron radiation from the compact jet [409]. GX 339–4 produces steady, compact radio jets, showing a correlation between the radio and X-ray flux, which follow a power-law relation [410,411]. During the transition to the high/soft state, the radio jet is often suppressed, and the radio flux drops significantly.

7.3. Hysteresis Cycles in Binary Stars and AGN

In the context of binary stars, hysteresis cycles occur when the path followed by the system during an increase in accretion rate differs from the path followed during a decrease. This results in different spectral and luminosity states for the same accretion rate, depending on whether the rate is increasing or decreasing. In other words, the transition from one state to another (e.g., from low/hard to high/soft) does not follow the same path as the reverse transition (e.g., from high/soft back to low/hard). This means that the system exhibits a "loop" or "cycle" in the diagram plotting one state variable against another (such as luminosity vs. spectral hardness), rather than a single 1D path [390,412–414].

This cyclical evolution of accretion states and jet formation is characteristic of galactic black hole binaries. In contrast, similar evolutionary patterns in AGN would unfold over much longer timescales (in the order of thousands to millions of years) if the instability associated with a change in viscosity parameter has to propagate over the full disk [415]. Nevertheless, there are several intriguing analogies between SE accretion in stellar-mass black holes and supermassive black holes: (a) a soft state at high accretion rate contrasts with a hard state at low accretion rate; (b) excess radiation pressure in super-Eddington regimes can drive powerful outflows or winds, expelling material from the accretion disk—massive outflows and winds from the disks of SMBH can be observed as blueshifted emission lines, broad absorption lines, or UFOs; (c) the accretion disk around stellar-mass black holes can become geometrically thick and advective, trapping radiation and allowing continued accretion despite high radiation pressures; (d) high variability in X-ray and optical emissions can indicate SE accretion rates—quasars exhibit high variability in their X-ray emission and super soft spectrum, albeit the optical variability of xA sources is apparently of low amplitude [126]; (e) SE super-massive black holes that are relativistically jetted are very rare, if they exist at all.

The analogies between super-Eddington accretion in stellar-mass and supermassive black holes illustrate similar underlying physical processes and phenomena despite the vast differences in scale. The 4DE1 scheme captured the main changes between a soft and hard state in type 1 AGN, as well as the following: (a) $R_{\rm FeII}$ is a tracer of softness, and reaches the highest values in correspondence of SE candidates [26,31]; (b) the soft X-ray photon index $\Gamma_{\rm soft}$ plays a role equivalent to the hardness ratio in X-ray binaries: a flat $\Gamma_{\rm soft} \lesssim 2$ indicated a flattish SED as observed in RL and Population B sources, while a steeper slope $\Gamma_{\rm soft} \gtrsim 2$ is common among Population A sources, with the most extreme values $\Gamma_{\rm soft} \approx 3-4$ in the cases of highest $L/L_{\rm Edd}$. The CIVλ1549 blueshifts represents the prominence of the accretion disk outflow, which is stronger in the high/soft state [173]. The 4DE1 optical relations have been successfully scaled down to stellar-mass sources, predicting the location of two accreting white dwarfs in the $R_{\rm FeII}$–FWHM Hβ plane [416]. This said, the details of the transition between a soft and hard state are as yet very poorly known. It would imply the transition from a Population B sources to a Population A or xA, if we consider SE candidates, giving rise to some changing look phenomenology (Figure 13 and the next section). Apparently, the most extensively monitored sources, such as NGC 5548, remain confined into the B spectral types [417]. In the survey of changing look candidates along the quasar MS carried out by Panda and Śniegowska [418], Population B sources stay Population B (apart from a few, unconvincing cases). Some transient events, however, most likely associated with TDEs [419], have yielded a surge in luminosity leading to SE properties [161,310].

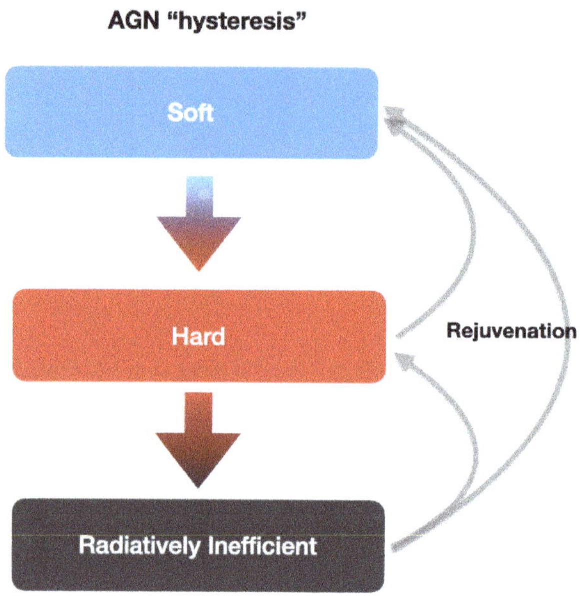

Figure 13. A straight evolutionary path from an initial soft state of high $L/L_{\rm Edd}$and low $M_{\rm BH}$may be followed by a "hard" (disk dominated) state with higher $M_{\rm BH}$and lower $L/L_{\rm Edd}$, down to inefficient accretion if $\dot{m} \ll 1$. The conditions might be reversed in a process that is conceptually similar to the hysteresis cycle in binary stars if the nuclear activity is restarted or enhanced following an increase of accretion rate. The most extreme forms of "rejuvenation" may give rise to a changing look phenomenology.

7.4. *Transient Super-Eddington AGN*

Changing-look AGN are a class of active galactic nuclei (AGN) that exhibit dramatic, rapid changes in their spectral features, often switching between states of high and low activity, and vice versa [128,418]. These transitions can occur over timescales as short as a few weeks to a few months, which are much shorter than the typical dynamical timescales expected for AGN [420]. Transient phenomena such as tidal disruption events (TDE) might induce a SE state [161,203,421,422]. In passing, we note that turn-on transients offer a straightforward test of the main parameter associated with the E1 MS: a transition from high FeII to lower FeII with increasing luminosity might not be possible, if FeII is governed by $L/L_{\rm Edd}$ [422].

A most remarkable case of a changing look AGN is provided by NGC 1566 [423–426]: with increasing luminosity, FeII emission, which was undetectable in the lower luminosity state, appears in the spectrum as if out of nowhere. However, the NGC 1566 high state $R_{\rm FeII}$ is well below 1. The source may qualify as a repeated transition between type-2, type-1 Pop. B, and Pop. A [427], but super-Eddington luminosities are not involved: in the scheme of Table 1, the change of NGC 1566 would correspond to a change between the radiatively inefficient and the Pop. A state. Over time, cases of transition to a stable SE state might be detected. The available data are still insufficient to characterize any hysteresis cycle in AGN definitively. However, AGN are observed to vary more frequently and with larger amplitudes than previously thought [428], and suitable data to detect a transition to a super-Eddington regime (or its turn-off) may become available in due time via the extensive photometric monitoring surveys that have been carried out (from Catalina [429] to the ZTF [430–432], to the PanSTARRs [433], and All Sky Automated Survey for SuperNovae, ASAS-SN [434]) and, from 2025, surveys that will be conducted by the Vera Rubin telescope [435].

8. Conclusions

The main sequence correlations have allowed for the identification of super-Eddington candidates through well-defined optical and UV criteria. These sources, powering quasars and less luminous AGN, are observed over a broad range of black hole masses from the present-day Universe to the earliest cosmic epochs. The SE regime appears to be closely associated with the evolution of a nuclear star forming system, intimately connected to the development of structures in the AGN, in particular the obscuring torus, and the self-gravitation regions of the accretion disk where in situ star formation is possible. This scenario has two main supporting elements: (1) the explanation of the enrichment of the broad line emitting gas over short timescales and (2) the increase of the black hole mass as fast as needed to preserve the current Λ-cold dark matter (CDM) model.

Several open questions remain regarding super-Eddington accretion, concerning the ultimate limits of mass accretion rates, and essentially addressing the following question: how fast can black holes really grow? Does a highly SE regime really exist, and which form may it take [62]? We believe the observational evidence supports a positive answer to the second question, even if it remains largely circumstantial. The powerful winds generated in these systems can significantly influence not only the growth of the black hole but also the evolution of the surrounding galaxy. Accurate estimates of wind parameters and feedback effects over the full range of AGN black hole masses are essential for providing a conclusive answer.

The observed high metallicities suggest efficient metal enrichment processes that need to be understood in the context of nuclear or circum-nuclear star formation. A huge effort in theoretical modeling of accretion-modified stars, nuclear stellar systems, and advection-dominated accretion flows, as well as in the collection of dedicated observational data, are required to elucidate these processes. The connection between low ionization spectra and the super-Eddington phenomenology is another area that requires further exploration, as the virialized low emitting regions might be providing a viable avenue for the use of super-Eddington quasars as cosmological standard candles.

Author Contributions: Writing—original draft preparation, P.M.; writing—review and editing, P.M., A.F. and K.G.L. A.d.O., A.D.-M., C.A.N., T.M.B.-R. and D.D. contributed to the several papers on which this review is based in part. A.d.O., A.D.-M. and T.M.B.-R. contributed with a critical reading and comments. C.A.N. also contributed by signaling helpful references. All authors have read and agreed to the published version of the manuscript.

Funding: A.F. was funded by the European Union ERC-2022-STG-BOOTES-101076343. The views and opinions expressed are, however, those of the author(s) only and do not necessarily reflect those of the European Union or the European Research Council Executive Agency. Neither the European Union nor the granting authority can be held responsible for them. A.d.O. and A.D.M. acknowledge financial support from the Spanish MCIU through project PID2022-140871NB-C21 by "ERDF A way of making Europe", and the Severo Ochoa grant CEX2021-515001131-S funded by MCIN/AEI/10.13039/501100011033.

Data Availability Statement: This review is based entirely on published data or on data that are expected to be published soon.

Conflicts of Interest: The authors declare no conflicts of interest.

Abbreviations

The following abbreviations are used in this manuscript:

4DE1	4D Eigenvector 1
ADAF	Advection Dominated Accretion Flow
AGN	Active Galactic Nucleus/i
BH	Black Hole
BAL	Broad Absorption Line
BLR	Broad Line Region
E1	Eigenvector 1
FIR	Far Infrared
FWHM	Full Width Half Maximum
HE	Hamburg-ESO
HEMS	Hamburg-ESO Main Sequence
HIL	High Ionization Line
ISM	Inter-Stellar Medium
JWST	James Webb Space Telescope
LIL	Low Ionization Line
LISA	Laser Interferometer Space Antenna
MDPI	Multidisciplinary Digital Publishing Institute
MS	Main Sequence
NLR	Narrow Line Region
NLSy1	Narrow Line Seyfert 1
PIZ	Partially Ionized Zone
ROSAT	Röntgensatellit
SDSS	Sloan Digital Sky Survey
SE	Super-Eddington
SMBH	Super-Massive Black Hole
SNR	SuperNova Remnant
UFO	Ultra Fast Outflow
UV	Ultra-Violet
ULX	Ultra-Luminous X-ray
VLT	Very Large Telescope
WISE	Wide-field Infrared Survey Explorer

Notes

1 The term quasar is used here as an umbrella term for type-1 AGN, independent of their luminosity, unless otherwise noted.

References

1. Peterson, B.M. *An Introduction to Active Galactic Nuclei*; Cambridge University Press: Cambridge, UK, 1997.
2. Elvis, M. A Structure for Quasars. *Astrophys. J.* **2000**, *545*, 63–76. [CrossRef]
3. Frank, J.; King, A.; Raine, D.J. *Accretion Power in Astrophysics, Third Edition*; Cambridge University Press: Cambridge, UK, 2002.
4. D'Onofrio, M.; Marziani, P.; Sulentic, J.W. (Eds.) *Fifty Years of Quasars from Early Observations and Ideas to Future Research*; Astrophysics and Space Science Library; Springer: Berlin/Heidelberg, Germany, 2012; Volume 386.
5. Netzer, H. *The Physics and Evolution of Active Galactic Nuclei*; Cambridge University Press: Cambridge, UK, 2013.
6. Event Horizon Telescope Collaboration; Akiyama, K.; Alberdi, A.; Alef, W.; Asada, K.; Azulay, R.; Baczko, A.K.; Ball, D.; Baloković, M.; Barrett, J.; et al. First M87 Event Horizon Telescope Results. IV. Imaging the Central Supermassive Black Hole. *Astrophys. J. Lett.* **2019**, *875*, L4. [CrossRef]
7. Urry, C.M.; Padovani, P. Unified Schemes for Radio-Loud Active Galactic Nuclei. *Publ. Astron. Soc. Pac.* **1995**, *107*, 803. [CrossRef]
8. Dultzin-Hacyan, D.; Krongold, Y.; Fuentes-Guridi, I.; Marziani, P. The Close Environment of Seyfert Galaxies and Its Implication for Unification Models. *Astrophys. J.* **1999**, *513*, L111–L114. [CrossRef]

9. Bianchi, S.; Corral, A.; Panessa, F.; Barcons, X.; Matt, G.; Bassani, L.; Carrera, F.J.; Jiménez-Bailón, E. NGC 3147: A 'true' type 2 Seyfert galaxy without the broad-line region. *Mon. Not. R. Astron. Soc.* **2008**, *385*, 195–199. [CrossRef]
10. Villarroel, B.; Korn, A.J. The different neighbours around Type-1 and Type-2 active galactic nuclei. *Nat. Phys.* **2014**, *10*, 417–420. [CrossRef]
11. Shangguan, J.; Ho, L.C. Testing the Evolutionary Link between Type 1 and Type 2 Quasars with Measurements of the Interstellar Medium. *Astrophys. J.* **2019**, *873*, 90. [CrossRef]
12. Zou, F.; Yang, G.; Brandt, W.N.; Xue, Y. The Host-galaxy Properties of Type 1 versus Type 2 Active Galactic Nuclei. *Astrophys. J.* **2019**, *878*, 11. [CrossRef]
13. López-Navas, E.; Arévalo, P.; Bernal, S.; Graham, M.J.; Hernández-García, L.; Lira, P.; Sánchez-Sáez, P. The Type 1 and Type 2 AGN dichotomy according to their ZTF optical variability. *Mon. Not. R. Astron. Soc.* **2023**, *518*, 1531–1542. [CrossRef]
14. Storey-Fisher, K.; Hogg, D.W.; Rix, H.W.; Eilers, A.C.; Fabbian, G.; Blanton, M.R.; Alonso, D. Quaia, the Gaia-unWISE Quasar Catalog: An All-sky Spectroscopic Quasar Sample. *Astrophys. J.* **2024**, *964*, 69. [CrossRef]
15. Margon, B. The properties of SS 433. *Annu. Rev. Astron. Astrophys.* **1984**, *22*, 507–536. [CrossRef]
16. Greiner, J.; Cuby, J.G.; McCaughrean, M.J. The microquasar GRS 1915+105: Multi-wavelength observations and modeling. *Nature* **2001**, *414*, 522–525. [CrossRef] [PubMed]
17. Teerikorpi, P. The distance scale and Eddington efficiency of luminous quasars. *arXiv* **2005**, arXiv:astro-ph/0510382.
18. Paczyńsky, B.; Wiita, P.J. Thick Accretion Disks and Supercritical Luminosities. *Astron. Astrophys.* **1980**, *88*, 23.
19. Paczynski, B.; Abramowicz, M.A. A model of a thick disk with equatorial accretion. *Astrophys. J.* **1982**, *253*, 897–907. [CrossRef]
20. Balbus, S.A.; Hawley, J.F. A powerful local shear instability in weakly magnetized disks. I—Linear analysis. II—Nonlinear evolution. *Astrophys. J.* **1991**, *376*, 214–233. [CrossRef]
21. Balbus, S.A.; Hawley, J.F. Instability, turbulence, and enhanced transport in accretion disks. *Rev. Mod. Phys.* **1998**, *70*, 1–53. [CrossRef]
22. Shakura, N.I.; Sunyaev, R.A. Black holes in binary systems. Observational appearance. *Astron. Astrophys.* **1973**, *24*, 337–355.
23. Di Matteo, T.; Springel, V.; Hernquist, L. Energy input from quasars regulates the growth and activity of black holes and their host galaxies. *Nature* **2005**, *433*, 604–607. [CrossRef] [PubMed]
24. King, A. The AGN-Starburst Connection, Galactic Superwinds, and M_{BH}-σ. *Astrophys. J. Lett.* **2005**, *635*, L121–L123. [CrossRef]
25. King, A.; Muldrew, S.I. Black hole winds II: Hyper-Eddington winds and feedback. *Mon. Not. R. Astron. Soc.* **2016**, *455*, 1211–1217. [CrossRef]
26. Marziani, P.; Sulentic, J.W. Highly accreting quasars: Sample definition and possible cosmological implications. *Mon. Not. R. Astron. Soc.* **2014**, *442*, 1211–1229. [CrossRef]
27. Dultzin, D.; Marziani, P.; de Diego, J.A.; Negrete, C.A.; Del Olmo, A.; Martínez-Aldama, M.L.; D'Onofrio, M.; Bon, E.; Bon, N.; Stirpe, G.M. Extreme quasars as distance indicators in cosmology. *Front. Astron. Space Sci.* **2020**, *6*, 80. [CrossRef]
28. Czerny, B.; Martínez-Aldama, M.L.; Wojtkowska, G.; Zajaček, M.; Marziani, P.; Dultzin, D.; Naddaf, M.H.; Panda, S.; Prince, R.; Przyluski, R.; et al. Dark Energy Constraintsfrom Quasar Observations. *Acta Phys. Pol. A* **2021**, *139*, 389–393. [CrossRef]
29. Marziani, P.; Dultzin, D.; del Olmo, A.; D'Onofrio, M.; de Diego, J.A.; Stirpe, G.M.; Bon, E.; Bon, N.; Czerny, B.; Perea, J.; et al. The quasar main sequence and its potential for cosmology. In *IAU Symposium S356: Nuclear Activity in Galaxies Across Cosmic Time, Proceedings of the International Astronomical Union, Ababa, Ethiopia, 7–11 October 2019*; Pović, M., Marziani, P., Masegosa, J., Netzer, H., Negu, S.H., Tessema, S.B., Eds.; Cambridge University Press: Cambridge, UK, 2021; pp. 66–71. [CrossRef]
30. Sulentic, J.W.; Marziani, P.; Dultzin-Hacyan, D. Phenomenology of Broad Emission Lines in Active Galactic Nuclei. *Annu. Rev. Astron. Astrophys.* **2000**, *38*, 521–571. [CrossRef]
31. Du, P.; Wang, J.M.; Hu, C.; Ho, L.C.; Li, Y.R.; Bai, J.M. The Fundamental Plane of the Broad-line Region in Active Galactic Nuclei. *Astrophys. J. Lett.* **2016**, *818*, L14. [CrossRef]
32. Paczynski, B. Super-Eddington Winds from Neutron Stars. *Astrophys. J.* **1990**, *363*, 218. [CrossRef]
33. Lupi, A.; Quadri, G.; Volonteri, M.; Colpi, M.; Regan, J.A. Sustained super-Eddington accretion in high-redshift quasars. *Astron. Astrophys.* **2024**, *686*, A256. [CrossRef]
34. Yang, H.; Yuan, F. Wind from the Hot Accretion Flow and Super-Eddington Accretion Flow. *arXiv* **2024**, arXiv:2408.16595. [CrossRef]
35. Wang, J.M.; Du, P.; Valls-Gabaud, D.; Hu, C.; Netzer, H. Super-Eddington Accreting Massive Black Holes as Long-Lived Cosmological Standards. *Phys. Rev. Lett.* **2013**, *110*, 081301. [CrossRef] [PubMed]
36. D'Onofrio, M.; Rampazzo, R.; Zaggia, S. (Eds.) *From the Realm of the Nebulae to Populations of Galaxies*; Astrophysics and Space Science Library; Springer: Cham, Switzerland, 2016; Volume 435. [CrossRef]
37. D'Onofrio, M.; Cariddi, S.; Chiosi, C.; Chiosi, E.; Marziani, P. On the Origin of the Fundamental Plane and Faber-Jackson Relations: Implications for the Star Formation Problem. *Astrophys. J.* **2017**, *838*, 163. [CrossRef]
38. D'Onofrio, M.; Marziani, P. A multimessenger view of galaxies and quasars from now to mid-century. *Front. Astron. Space Sci.* **2018**, *5*, 31. [CrossRef]

39. Davis, S.W.; Laor, A. The Radiative Efficiency of Accretion Flows in Individual Active Galactic Nuclei. *Astrophys. J.* **2011**, *728*, 98. [CrossRef]
40. Panda, S.; Czerny, B.; Adhikari, T.P.; Hryniewicz, K.; Wildy, C.; Kuraszkiewicz, J.; Śniegowska, M. Modeling of the Quasar Main Sequence in the Optical Plane. *Astrophys. J.* **2018**, *866*, 115. [CrossRef]
41. Kubota, A.; Done, C. Modelling the spectral energy distribution of super-Eddington quasars. *Mon. Not. R. Astron. Soc.* **2019**, *489*, 524–533. [CrossRef]
42. Jiang, Y.F.; Stone, J.M.; Davis, S.W. Super-Eddington Accretion Disks around Supermassive Black Holes. *Astrophys. J.* **2019**, *880*, 67. [CrossRef]
43. Inayoshi, K.; Haiman, Z.; Ostriker, J.P. Hyper-Eddington accretion flows on to massive black holes. *Mon. Not. R. Astron. Soc.* **2016**, *459*, 3738–3755. [CrossRef]
44. Toyouchi, D.; Inayoshi, K.; Hosokawa, T.; Kuiper, R. Super-Eddington Mass Growth of Intermediate-Mass Black Holes Embedded in Dusty Circumnuclear Disks. *Astrophys. J.* **2021**, *907*, 74. [CrossRef]
45. Abramowicz, M.A.; Czerny, B.; Lasota, J.P.; Szuszkiewicz, E. Slim accretion disks. *Astrophys. J.* **1988**, *332*, 646–658. [CrossRef]
46. Sirko, E.; Goodman, J. Spectral energy distributions of marginally self-gravitating quasi-stellar object discs. *Mon. Not. R. Astron. Soc.* **2003**, *341*, 501–508. [CrossRef]
47. Thompson, T.A.; Quataert, E.; Murray, N. Radiation Pressure-supported Starburst Disks and Active Galactic Nucleus Fueling. *Astrophys. J.* **2005**, *630*, 167–185. [CrossRef]
48. Sądowski, A. Slim Disks Around Kerr Black Holes Revisited. *Astrophys. J. Supp.* **2009**, *183*, 171–178. [CrossRef]
49. Sądowski, A.; Narayan, R.; McKinney, J.C.; Tchekhovskoy, A. Numerical simulations of super-critical black hole accretion flows in general relativity. *Mon. Not. R. Astron. Soc.* **2014**, *439*, 503–520. [CrossRef]
50. Abramowicz, M.A. Super-Eddington black hole accretion: Polish doughnuts and slim disks. In *Growing Black Holes: Accretion in a Cosmological Context*; Merloni, A., Nayakshin, S., Sunyaev, R.A., Eds.; Springer: Berlin/Heidelberg, Germany, 2005; pp. 257–273. [CrossRef]
51. Abramowicz, M.A.; Straub, O. Accretion discs. *Scholarpedia* **2014**, *9*, 2408. [CrossRef]
52. Madau, P. Thick Accretion Disks around Black Holes and the UV/Soft X-Ray Excess in Quasars. *Astrophys. J.* **1988**, *327*, 116. [CrossRef]
53. Calvani, M.; Marziani, P.; Padovani, P. (Thick) accretion discs—Where are they? In Proceedings of the General Relativity and Gravitational Physics, Boulder, CO, USA, 2–8 July 1989; pp. 102–123.
54. Wang, J.M.; Zhou, Y.Y. Self-Similar Solution of Optically Thick Advection-dominated Flows. *Astrophys. J.* **1999**, *516*, 420–424. [CrossRef]
55. Jin, C.; Done, C.; Ward, M. Super-Eddington QSO RX J0439.6-5311-I. Origin of the soft X-ray excess and structure of the inner accretion flow. *Mon. Not. R. Astron. Soc.* **2017**, *468*, 3663–3681. [CrossRef]
56. Wang, J.M.; Qiu, J.; Du, P.; Ho, L.C. Self-shadowing Effects of Slim Accretion Disks in Active Galactic Nuclei: The Diverse Appearance of the Broad-line Region. *Astrophys. J.* **2014**, *797*, 65. [CrossRef]
57. Mineshige, S.; Kawaguchi, T.; Takeuchi, M.; Hayashida, K. Slim-Disk Model for Soft X-Ray Excess and Variability of Narrow-Line Seyfert 1 Galaxies. *Publ. Astron. Soc. Jpn.* **2000**, *52*, 499–508. [CrossRef]
58. Madau, P.; Haardt, F.; Dotti, M. Super-critical Growth of Massive Black Holes from Stellar-Mass Seeds. *Astrophys. J. Lett.* **2014**, *784*, L38. [CrossRef]
59. Inayoshi, K.; Visbal, E.; Haiman, Z. The Assembly of the First Massive Black Holes. *Annu. Rev. Astron. Astrophys.* **2020**, *58*, 27–97. [CrossRef]
60. Ohsuga, K.; Mori, M.; Nakamoto, T.; Mineshige, S. Supercritical Accretion Flows around Black Holes: Two-dimensional, Radiation Pressure-dominated Disks with Photon Trapping. *Astrophys. J.* **2005**, *628*, 368–381. [CrossRef]
61. Ogawa, T.; Mineshige, S.; Kawashima, T.; Ohsuga, K.; Hashizume, K. Radiation hydrodynamic simulations of a super-Eddington accretor as a model for ultra-luminous sources. *Publ. Astron. Soc. Jpn.* **2017**, *69*, 33. [CrossRef]
62. Pognan, Q.; Trakhtenbrot, B.; Sbarrato, T.; Schawinski, K.; Bertemes, C. Searching for super-Eddington quasars using a photon trapping accretion disc model. *Mon. Not. R. Astron. Soc.* **2020**, *492*, 4058–4079. [CrossRef]
63. Marziani, P.; Sulentic, J.W.; Zwitter, T.; Dultzin-Hacyan, D.; Calvani, M. Searching for the Physical Drivers of the Eigenvector 1 Correlation Space. *Astrophys. J.* **2001**, *558*, 553–560. [CrossRef]
64. Shen, Y.; Ho, L.C. The diversity of quasars unified by accretion and orientation. *Nature* **2014**, *513*, 210–213. [CrossRef]
65. Marziani, P.; Dultzin, D.; Sulentic, J.W.; Del Olmo, A.; Negrete, C.A.; Martínez-Aldama, M.L.; D'Onofrio, M.; Bon, E.; Bon, N.; Stirpe, G.M. A main sequence for quasars. *Front. Astron. Space Sci.* **2018**, *5*, 6. [CrossRef]
66. Wildy, C.; Czerny, B.; Panda, S. Quasar main sequence: A line or a plane. *Astron. Astrophys.* **2019**, *632*, A41. [CrossRef]
67. Moore, C.E. A Multiplet Table of Astrophysical Interest. Revised Edition. Part I—Table of Multiplets. *Contrib. Princet. Univ. Obs.* **1945**, *20*, 1–110.

68. Kovačević, J.; Popović, L.Č.; Dimitrijević, M.S. Analysis of Optical Fe II Emission in a Sample of Active Galactic Nucleus Spectra. *Astrophys. J. Supp.* **2010**, *189*, 15–36. [CrossRef]
69. Boroson, T.A.; Green, R.F. The Emission-Line Properties of Low-Redshift Quasi-stellar Objects. *Astrophys. J. Supp.* **1992**, *80*, 109. [CrossRef]
70. Boroson, T.A. Black Hole Mass and Eddington Ratio as Drivers for the Observable Properties of Radio-loud and Radio-quiet QSOs. *Astrophys. J.* **2002**, *565*, 78–85. [CrossRef]
71. Marziani, P.; Zamanov, R.K.; Sulentic, J.W.; Calvani, M. Searching for the physical drivers of eigenvector 1: Influence of black hole mass and Eddington ratio. *Mon. Not. R. Astron. Soc.* **2003**, *345*, 1133–1144. [CrossRef]
72. Sun, J.; Shen, Y. Dissecting the Quasar Main Sequence: Insight from Host Galaxy Properties. *Astrophys. J. Lett.* **2015**, *804*, L15. [CrossRef]
73. Panda, S.; Marziani, P.; Czerny, B. The Quasar Main Sequence Explained by the Combination of Eddington Ratio, Metallicity, and Orientation. *Astrophys. J.* **2019**, *882*, 79. [CrossRef]
74. Peterson, B.M.; Wandel, A. Keplerian Motion of Broad-Line Region Gas as Evidence for Supermassive Black Holes in Active Galactic Nuclei. *Astrophys. J. Lett.* **1999**, *521*, L95–L98. [CrossRef]
75. Peterson, B.M.; Ferrarese, L.; Gilbert, K.M.; Kaspi, S.; Malkan, M.A.; Maoz, D.; Merritt, D.; Netzer, H.; Onken, C.A.; Pogge, R.W.; et al. Central Masses and Broad-Line Region Sizes of Active Galactic Nuclei. II. A Homogeneous Analysis of a Large Reverberation-Mapping Database. *Astrophys. J.* **2004**, *613*, 682–699. [CrossRef]
76. Vestergaard, M.; Peterson, B.M. Determining Central Black Hole Masses in Distant Active Galaxies and Quasars. II. Improved Optical and UV Scaling Relationships. *Astrophys. J.* **2006**, *641*, 689–709. [CrossRef]
77. Dalla Bontà, E.; Peterson, B.M.; Bentz, M.C.; Brandt, W.N.; Ciroi, S.; De Rosa, G.; Fonseca Alvarez, G.; Grier, C.J.; Hall, P.B.; Hernández Santisteban, J.V.; et al. The Sloan Digital Sky Survey Reverberation Mapping Project: Estimating Masses of Black Holes in Quasars with Single-epoch Spectroscopy. *Astrophys. J.* **2020**, *903*, 112. [CrossRef]
78. Sulentic, J.W.; Marziani, P.; Zamanov, R.; Bachev, R.; Calvani, M.; Dultzin-Hacyan, D. Average Quasar Spectra in the Context of Eigenvector 1. *Astrophys. J.* **2002**, *566*, L71–L75. [CrossRef]
79. Dultzin-Hacyan, D.; Sulentic, J.; Marziani, P.; Calvani, M.; Moles, M. A Correlation Analysis for Emission Lines in 52 AGN. In Proceedings of the IAU Colloq. 159: Emission Lines in Active Galaxies: New Methods and Technique; Peterson, B.M., Cheng, F.Z., Wilson, A.S., Eds.; Astronomical Society of the Pacific Conference Series; Astronomical Society of the Pacific: San Francisco, CA, USA, 1997; Volume 113, p. 262.
80. Komossa, S.; Voges, W.; Xu, D.; Mathur, S.; Adorf, H.M.; Lemson, G.; Duschl, W.J.; Grupe, D. Radio-loud Narrow-Line Type 1 Quasars. *Astron. J.* **2006**, *132*, 531–545. [CrossRef]
81. Abdo, A.A.; Ackermann, M.; Ajello, M.; Baldini, L.; Ballet, J.; Barbiellini, G.; Bastieri, D.; Bechtol, K.; Bellazzini, R.; Berenji, B.; et al. Radio-Loud Narrow-Line Seyfert 1 as a New Class of Gamma-Ray Active Galactic Nuclei. *Astrophys. J. Lett.* **2009**, *707*, L142–L147. [CrossRef]
82. Foschini, L.; Berton, M.; Caccianiga, A.; Ciroi, S.; Cracco, V.; Peterson, B.M.; Angelakis, E.; Braito, V.; Fuhrmann, L.; Gallo, L.; et al. Properties of flat-spectrum radio-loud narrow-line Seyfert 1 galaxies. *Astron. Astrophys.* **2015**, *575*, A13. [CrossRef]
83. Zamfir, S.; Sulentic, J.W.; Marziani, P. New insights on the QSO radio-loud/radio-quiet dichotomy: SDSS spectra in the context of the 4D eigenvector1 parameter space. *Mon. Not. R. Astron. Soc.* **2008**, *387*, 856–870. [CrossRef]
84. Ganci, V.; Marziani, P.; D'Onofrio, M.; del Olmo, A.; Bon, E.; Bon, N.; Negrete, C.A. Radio loudness along the quasar main sequence. *Astron. Astrophys.* **2019**, *630*, A110. [CrossRef]
85. Sulentic, J.W.; Dultzin-Hacyan, D.; Marziani, P.; Bongardo, C.; Braito, V.; Calvani, M.; Zamanov, R. Low Redshift BAL QSOs in the Eigenvector 1 Context. *Rev. Mex. Astron. Y Astrofis.* **2006**, *42*, 23–39.
86. Negrete, C.A.; Dultzin, D.; Marziani, P.; Esparza, D.; Sulentic, J.W.; del Olmo, A.; Martínez-Aldama, M.L.; García López, A.; D'Onofrio, M.; Bon, N.; et al. Highly accreting quasars: The SDSS low-redshift catalog. *Astron. Astrophys.* **2018**, *620*, A118. [CrossRef]
87. Marziani, P.; Sulentic, J.W.; Zamanov, R.; Calvani, M.; Dultzin-Hacyan, D.; Bachev, R.; Zwitter, T. An Optical Spectroscopic Atlas of Low-Redshift Active Galactic Nuclei. *Astrophys. J. Suppl.* **2003**, *145*, 199–211. [CrossRef]
88. Marziani, P. Accretion/Ejection Phenomena and Emission-Line Profile (A)symmetries in Type-1 Active Galactic Nuclei. *Symmetry* **2023**, *15*, 1859. [CrossRef]
89. Mengistue, S.T.; Del Olmo, A.; Marziani, P.; Pović, M.; Martínez-Carballo, M.A.; Perea, J.; Márquez, I. Optical and near-UV spectroscopic properties of low-redshift jetted quasars in the main sequence context. *Mon. Not. R. Astron. Soc.* **2023**, *525*, 4474–4496. [CrossRef]
90. Marziani, P.; Sulentic, J.W.; Stirpe, G.M.; Zamfir, S.; Calvani, M. VLT/ISAAC spectra of the Hβ region in intermediate-redshift quasars. III. Hβ broad-line profile analysis and inferences about BLR structure. *Astron. Astrophys.* **2009**, *495*, 83–112. [CrossRef]
91. Bon, N.; Bon, E.; Marziani, P.; Jovanović, P. Gravitational redshift of emission lines in the AGN spectra. *Astrophys. Space Sci.* **2015**, *360*, 7. [CrossRef]

92. Marziani, P.; Sulentic, J.W.; Negrete, C.A.; Dultzin, D.; Zamfir, S.; Bachev, R. Broad-line region physical conditions along the quasar eigenvector 1 sequence. *Mon. Not. R. Astron. Soc.* **2010**, *409*, 1033–1048. [CrossRef]
93. Marziani, P.; del Olmo, A.; D'Onofrio, M.; Dultzin, D.; Negrete, C.A.; Martínez-Aldama, M.L.; Bon, E.; Bon, N.; Stirpe, G.M. Narrow-line Seyfert 1s: What is wrong in a name? In Proceedings of the Revisiting Narrow-Line Seyfert 1 Galaxies and Their Place in the Universe, Padova Botanical Garden, Padua, Italy, 9–13 April 2018; id.2. SISSA/ISAS, Volume PoS(NLS1-2018) . Available online: https://pos.sissa.it/cgi-bin/reader/conf.cgi?confid=328 (accessed on 1 December 2020) .
94. Wu, X.B.; Han, J.L. Inclinations and Black Hole Masses of Seyfert 1 Galaxies. *Astrophys. J. Lett.* **2001**, *561*, L59–L62. [CrossRef]
95. Bian, W.; Zhao, Y. Masses, accretion rates and inclinations of AGNs. *Astron. Astrophys.* **2002**, *395*, 465–473. [CrossRef]
96. Decarli, R.; Dotti, M.; Treves, A. Geometry and inclination of the broad-line region in blazars. *Mon. Not. R. Astron. Soc.* **2011**, *413*, 39–46. [CrossRef]
97. Marziani, P.; Bon, E.; Bon, N.; D'Onofrio, M.; Punsly, B.; Śniegowska, M.; Czerny, B.; Panda, S.; Martnez Aldama, M.L.; del Olmo, A.; et al. The main sequence of quasars: The taming of the extremes. *Astron. Nachrichten* **2022**, *343*, e210082. [CrossRef]
98. Panda, S.; Małek, K.; Śniegowska, M.; Czerny, B. Strong FeII emission in NLS1s: An unsolved mystery. In *Challenges in Panchromatic Modelling with Next Generation Facilities*; Boquien, M., Lusso, E., Gruppioni, C., Tissera, P., Eds.; IAU Symposium; Cambridge University Press: Cambridge, UK, 2020; Volume 341, pp. 297–298. [CrossRef]
99. Panda, S. The CaFe project: Optical Fe II and near-infrared Ca II triplet emission in active galaxies: Simulated EWs and the co-dependence of cloud size and metal content. *Astron. Astrophys.* **2021**, *650*, A154. [CrossRef]
100. Floris, A.; Pandey, A.; Czerny, B.; Martinez Aldama, M.L.; Panda, S.; Marziani, P.; Prince, R. Dark and bright sides of the Broad Line Region clouds as seen in the FeII emission of SDSS RM 102. *arXiv* **2024**, arXiv:2408.17323. [CrossRef]
101. Chen, Y.; Luo, B.; Brandt, W.N.; Zuo, W.; Dix, C.; Ha, T.; Matthews, B.; Paul, J.D.; Plotkin, R.M.; Shemmer, O. Rest-frame Optical Spectroscopy of Ten $z \sim 2$ Weak Emission-line Quasars. *Astrophys. J.* **2024**, *972*, 191. [CrossRef]
102. Madau, P.; Dickinson, M. Cosmic Star-Formation History. *Annu. Rev. Astron. Astrophys.* **2014**, *52*, 415–486. [CrossRef]
103. Sulentic, J.W.; Stirpe, G.M.; Marziani, P.; Zamanov, R.; Calvani, M.; Braito, V. VLT/ISAAC spectra of the Hβ region in intermediate redshift quasars. *Astron. Astrophys.* **2004**, *423*, 121–132. [CrossRef]
104. Sulentic, J.W.; Repetto, P.; Stirpe, G.M.; Marziani, P.; Dultzin-Hacyan, D.; Calvani, M. VLT/ISAAC spectra of the Hβ region in intermediate-redshift quasars. II. Black hole mass and Eddington ratio. *Astron. Astrophys.* **2006**, *456*, 929–939. [CrossRef]
105. Murayama, T.; Taniguchi, Y.; Evans, A.S.; Sanders, D.B.; Ohyama, Y.; Kawara, K.; Arimoto, N. Near-Infrared Spectroscopy of the High-Redshift Quasar S4 0636+68 at Z = 3.2. *Astron. J.* **1998**, *115*, 2237–2243. [CrossRef]
106. D'Odorico, V.; Bañados, E.; Becker, G.D.; Bischetti, M.; Bosman, S.E.I.; Cupani, G.; Davies, R.; Farina, E.P.; Ferrara, A.; Feruglio, C.; et al. XQR-30: The ultimate XSHOOTER quasar sample at the reionization epoch. *Mon. Not. R. Astron. Soc.* **2023**, *523*, 1399–1420. [CrossRef]
107. Yang, J.; Wang, F.; Fan, X.; Hennawi, J.F.; Barth, A.J.; Bañados, E.; Sun, F.; Liu, W.; Cai, Z.; Jiang, L.; et al. A SPectroscopic Survey of Biased Halos in the Reionization Era (ASPIRE): A First Look at the Rest-frame Optical Spectra of $z > 6.5$ Quasars Using JWST. *Astrophys. J. Lett.* **2023**, *951*, L5. [CrossRef]
108. Deconto–Machado, A.; del Olmo, A.; Marziani, P. Exploring the links between quasar winds and radio emission along the main sequence at high redshift. *Astron. Astrophys.* **2024**, *691*, A15. [CrossRef]
109. Bañados, E.; Venemans, B.P.; Decarli, R.; Farina, E.P.; Mazzucchelli, C.; Walter, F.; Fan, X.; Stern, D.; Schlafly, E.; Chambers, K.C.; et al. The Pan-STARRS1 Distant $z > 5.6$ Quasar Survey: More than 100 Quasars within the First Gyr of the Universe. *Astrophys. J. Supp.* **2016**, *227*, 11. [CrossRef]
110. Bañados, E.; Schindler, J.T.; Venemans, B.P.; Connor, T.; Decarli, R.; Farina, E.P.; Mazzucchelli, C.; Meyer, R.A.; Stern, D.; Walter, F.; et al. The Pan-STARRS1 $z > 5.6$ Quasar Survey. II. Discovery of 55 Quasars at $5.6 < z < 6.5$. *Astrophys. J. Supp.* **2023**, *265*, 29. [CrossRef]
111. Deconto–Machado, A.; del Olmo Orozco, A.; Marziani, P.; Perea, J.; Stirpe, G.M. High-redshift quasars along the main sequence. *Astron. Astrophys.* **2023**, *669*, A83. [CrossRef]
112. Onorato, S.; Hennawi, J.F.; Schindler, J.T.; Yang, J.; Wang, F.; Barth, A.J.; Bañados, E.; Eilers, A.C.; Bosman, S.E.I.; Davies, F.B.; et al. Optical and near-infrared spectroscopy of quasars at $z > 6.5$: Public data release and composite spectrum. *arXiv* **2024**, arXiv:2406.07612. [CrossRef]
113. Cavaliere, A.; Vittorini, V. The Fall of the Quasar Population. *Astrophys. J.* **2000**, *543*, 599–610. [CrossRef]
114. Hopkins, P.F.; Hernquist, L.; Cox, T.J.; Di Matteo, T.; Robertson, B.; Springel, V. A Unified, Merger-driven Model of the Origin of Starbursts, Quasars, the Cosmic X-Ray Background, Supermassive Black Holes, and Galaxy Spheroids. *Astrophys. J. Supp.* **2006**, *163*, 1–49. [CrossRef]
115. Kelly, B.C.; Vestergaard, M.; Fan, X.; Hopkins, P.; Hernquist, L.; Siemiginowska, A. Constraints on Black Hole Growth, Quasar Lifetimes, and Eddington Ratio Distributions from the SDSS Broad-line Quasar Black Hole Mass Function. *Astrophys. J.* **2010**, *719*, 1315–1334. [CrossRef]

116. Hirschmann, M.; Dolag, K.; Saro, A.; Bachmann, L.; Borgani, S.; Burkert, A. Cosmological simulations of black hole growth: AGN luminosities and downsizing. *Mon. Not. R. Astron. Soc.* **2014**, *442*, 2304–2324. [CrossRef]
117. Sulentic, J.W.; Marziani, P.; del Olmo, A.; Dultzin, D.; Perea, J.; Alenka Negrete, C. GTC spectra of $z \approx 2.3$ quasars: Comparison with local luminosity analogs. *Astron. Astrophys.* **2014**, *570*, A96. [CrossRef]
118. Vanden Berk, D.E.; Richards, G.T.; Bauer, A.; Strauss, M.A.; Schneider, D.P.; Heckman, T.M.; York, D.G.; Hall, P.B.; Fan, X.; Knapp, G.R.; et al. Composite Quasar Spectra from the Sloan Digital Sky Survey. *Astron. J.* **2001**, *122*, 549–564. [CrossRef]
119. Kwan, J.; Krolik, J.H. The formation of emission lines in quasars and Seyfert nuclei. *Astrophys. J.* **1981**, *250*, 478–507. [CrossRef]
120. Rees, M.J.; Netzer, H.; Ferland, G.J. Small dense broad-line regions in active nuclei. *Astrophys. J.* **1989**, *347*, 640–655. [CrossRef]
121. Netzer, H. AGN emission lines. In *Active Galactic Nuclei*; Blandford, R.D., Netzer, H., Woltjer, L., Courvoisier, T.J.-L., Mayor, M., Eds.; Springer: Berlin/Heidelberg, Germany, 1990; pp. 57–160.
122. Phillips, M.M. Permitted fe II Emission in Seyfert 1 Galaxies and QSOs I. Observations. *Astrophys. J. Suppl.* **1978**, *38*, 187. [CrossRef]
123. Marinello, A.O.M.; Rodriguez-Ardila, A.; Garcia-Rissmann, A.; Sigut, T.A.A.; Pradhan, A.K. The FeII emission in active galactic nuclei: Excitation mechanisms and location of the emitting region. *Astrophys. J.* **2016**, *820*, 116. [CrossRef]
124. Kovačević-Dojčinović, J.; Popović, L.Č. The Connections Between the UV and Optical Fe ii Emission Lines in Type 1 AGNs. *Astrophys. J. Supp.* **2015**, *221*, 35. [CrossRef]
125. Sexton, R.O.; Matzko, W.; Darden, N.; Canalizo, G.; Gorjian, V. Bayesian AGN Decomposition Analysis for SDSS spectra: A correlation analysis of [O III] λ5007 outflow kinematics with AGN and host galaxy properties. *Mon. Not. R. Astron. Soc.* **2021**, *500*, 2871–2895. [CrossRef]
126. Du, P.; Lu, K.X.; Hu, C.; Qiu, J.; Li, Y.R.; Huang, Y.K.; Wang, F.; Bai, J.M.; Bian, W.H.; Yuan, Y.F.; et al. Supermassive Black Holes with High Accretion Rates in Active Galactic Nuclei. VI. Velocity-resolved Reverberation Mapping of the Hβ Line. *Astrophys. J.* **2016**, *820*, 27. [CrossRef]
127. Du, P.; Zhang, Z.X.; Wang, K.; Huang, Y.K.; Zhang, Y.; Lu, K.X.; Hu, C.; Li, Y.R.; Bai, J.M.; Bian, W.H.; et al. Supermassive Black Holes with High Accretion Rates in Active Galactic Nuclei. IX. 10 New Observations of Reverberation Mapping and Shortened Hβ Lags. *Astrophys. J.* **2018**, *856*, 6. [CrossRef]
128. Komossa, S.; Grupe, D.; Marziani, P.; Popovic, L.C.; Marceta-Mandic, S.; Bon, E.; Ilic, D.; Kovacevic, A.B.; Kraus, A.; Haiman, Z.; et al. The extremes of AGN variability: Outbursts, deep fades, changing looks, exceptional spectral states, and semi-periodicities. *arXiv* **2024**, arXiv:2408.00089. [CrossRef]
129. Dultzin-Hacyan, D.; Taniguchi, Y.; Uranga, L. Where is the Ca II Triplet Emitting Region in AGN? In *Structure and Kinematics of Quasar Broad Line Regions*; Gaskell, C.M., Brandt, W.N., Dietrich, M., Dultzin-Hacyan, D., Eracleous, M., Eds.; Astronomical Society of the Pacific Conference Series; Astronomical Society of the Pacific (ASP): San Francisco, CA, USA, 1999; Volume 175, p. 303.
130. Martínez-Aldama, M.L.; Dultzin, D.; Marziani, P.; Sulentic, J.W.; Bressan, A.; Chen, Y.; Stirpe, G.M. O I and Ca II Observations in Intermediate Redshift Quasars. *Astrophys. J. Suppl.* **2015**, *217*, 3. [CrossRef]
131. Martínez-Aldama, M.L.; Panda, S.; Czerny, B.; Marinello, M.; Marziani, P.; Dultzin, D. The CaFe Project: Optical FeII and Near-Infrared Ca II triplet emission in active galaxies. II. The driver(s) of the Ca II and Fe II and its potential use as a chemical clock. *arXiv* **2021**, arXiv:2101.06999. [CrossRef]
132. Laha, S.; Keenan, F.P.; Ferland, G.J.; Ramsbottom, C.A.; Aggarwal, K.M. Ultraviolet Emission Lines of Si II in Quasars: Investigating the "Si II Disaster". *Astrophys. J.* **2016**, *825*, 28. [CrossRef]
133. Kinney, A.L.; Rivolo, A.R.; Koratkar, A.P. A study of the Baldwin effect in the IUE data set. *Astrophys. J.* **1990**, *357*, 338–345. [CrossRef]
134. Shemmer, O.; Lieber, S. Weak Emission-line Quasars in the Context of a Modified Baldwin Effect. *Astrophys. J.* **2015**, *805*, 124. [CrossRef]
135. Plotkin, R.M.; Shemmer, O.; Trakhtenbrot, B.; Anderson, S.F.; Brandt, W.N.; Fan, X.; Gallo, E.; Lira, P.; Luo, B.; Richards, G.T.; et al. Detection of Rest-frame Optical Lines from X-shooter Spectroscopy of Weak Emission Line Quasars. *Astrophys. J.* **2015**, *805*, 123. [CrossRef]
136. Marziani, P.; Martínez Carballo, M.A.; Sulentic, J.W.; Del Olmo, A.; Stirpe, G.M.; Dultzin, D. The most powerful quasar outflows as revealed by the Civ λ1549 resonance line. *Astrophys. Space Sci.* **2016**, *361*, 29. [CrossRef]
137. Diamond-Stanic, A.M.; Fan, X.; Brandt, W.N.; Shemmer, O.; Strauss, M.A.; Anderson, S.F.; Carilli, C.L.; Gibson, R.R.; Jiang, L.; Kim, J.S.; et al. High-redshift SDSS Quasars with Weak Emission Lines. *Astrophys. J.* **2009**, *699*, 782–799. [CrossRef]
138. Martínez-Aldama, M.L.; del Olmo, A.; Marziani, P.; Sulentic, J.W.; Negrete, C.A.; Dultzin, D.; D'Onofrio, M.; Perea, J. Extreme quasars at high redshift. *Astron. Astrophys.* **2018**, *618*, A179. [CrossRef]
139. Leung, G.C.K.; Finkelstein, S.L.; Pérez-González, P.G.; Morales, A.M.; Taylor, A.J.; Barro, G.; Kocevski, D.D.; Akins, H.B.; Carnall, A.C.; Chávez Ortiz, Ó.A.; et al. Exploring the Nature of Little Red Dots: Constraints on AGN and Stellar Contributions from PRIMER MIRI Imaging. *arXiv* **2024**, arXiv:2411.12005. [CrossRef]

140. Negrete, A.; Dultzin, D.; Marziani, P.; Sulentic, J. BLR Physical Conditions in Extreme Population A Quasars: A Method to Estimate Central Black Hole Mass at High Redshift. *Astrophys. J.* **2012**, *757*, 62. [CrossRef]
141. Collin, S.; Joly, M. The Fe II problem in NLS1s. *New Astron. Rev.* **2000**, *44*, 531–537. [CrossRef]
142. Joly, M.; Véron-Cetty, M.; Véron, P. Fe II emission in AGN. In *Revista Mexicana de Astronomia y Astrofisica*; Conference Series; Instituto de Astronomía: Distrito Federal, México, 2008; Volume 32, pp. 59–61.
143. Johansson, S.; Letokhov, V.S. Anomalous Fe II Spectral Effects and High H I Lyα Temperature in Gas Blobs Near η Carinae. *Astron. Lett.* **2004**, *30*, 58–63. [CrossRef]
144. Johansson, S.; Hartman, H.; Letokhov, V.S. Resonance-enhanced two-photon ionization (RETPI) of Si II and an anomalous, variable intensity of the λ1892 Si III] line in the Weigelt blobs of η Carinae. *Astron. Astrophys.* **2006**, *452*, 253–256. [CrossRef]
145. Marconi, A.; Axon, D.J.; Maiolino, R.; Nagao, T.; Pietrini, P.; Risaliti, G.; Robinson, A.; Torricelli, G. On the Observed Distributions of Black Hole Masses and Eddington Ratios from Radiation Pressure Corrected Virial Indicators. *Astrophys. J. Lett.* **2009**, *698*, L103–L107. [CrossRef]
146. Netzer, H.; Marziani, P. The Effect of Radiation Pressure on Emission-line Profiles and Black Hole Mass Determination in Active Galactic Nuclei. *Astrophys. J.* **2010**, *724*, 318–328. [CrossRef]
147. Hartig, G.F.; Baldwin, J.A. The emission-line regions in broad absorption line quasars. *Astrophys. J.* **1986**, *302*, 64–80. [CrossRef]
148. Bachev, R.; Marziani, P.; Sulentic, J.W.; Zamanov, R.; Calvani, M.; Dultzin-Hacyan, D. Average Ultraviolet Quasar Spectra in the Context of Eigenvector 1: A Baldwin Effect Governed by the Eddington Ratio? *Astrophys. J.* **2004**, *617*, 171–183. [CrossRef]
149. Richards, G.T.; Kruczek, N.E.; Gallagher, S.C.; Hall, P.B.; Hewett, P.C.; Leighly, K.M.; Deo, R.P.; Kratzer, R.M.; Shen, Y. Unification of Luminous Type 1 Quasars through C IV Emission. *Astron. J.* **2011**, *141*, 167. [CrossRef]
150. Aoki, K.; Yoshida, M. The Correlation between SI III[λ1892/CIII] λ1909 and FeII λ 4500/Hβ in Low Redshift QSOs. In *Proceedings of the Quasars and Cosmology*; Ferland, G., Baldwin, J., Eds.; Astronomical Society of the Pacific Conference Series; Astronomical Society of the Pacific: San Francisco, CA, USA, 1999; Volume 162, p. 385.
151. Buendia-Rios, T.M.; Negrete, C.A.; Marziani, P.; Dultzin, D. Statistical analysis of Al III and C III] emission lines as virial black hole mass estimators in quasars. *Astron. Astrophys.* **2023**, *669*, A135. [CrossRef]
152. Marziani, P.; Sulentic, J.W.; Negrete, C.A.; Dultzin, D.; D'Onofrio, M.; Del Olmo, A.; Martinez-Aldama, M.L. Low- and high-z highly accreting quasars in the 4D Eigenvector 1 context. *Astron. Rev.* **2014**, *9*, 6–25.
153. Marziani, P.; Olmo, A.d.; Negrete, C.A.; Dultzin, D.; Piconcelli, E.; Vietri, G.; Martínez-Aldama, M.L.; D'Onofrio, M.; Bon, E.; Bon, N.; et al. The Intermediate-ionization Lines as Virial Broadening Estimators for Population A Quasars. *Astrophys. J. Supp.* **2022**, *261*, 30. [CrossRef]
154. Rani, P.; Stalin, C.S.; Rakshit, S. X-ray flux variability of active galactic nuclei observed using NuSTAR. *Mon. Not. R. Astron. Soc.* **2017**, *466*, 3309–3322. [CrossRef]
155. Rakshit, S.; Stalin, C.S. Optical Variability of Narrow-line and Broad-line Seyfert 1 Galaxies. *Astrophys. J.* **2017**, *842*, 96. [CrossRef]
156. Boller, T.; Liu, T.; Weber, P.; Arcodia, R.; Dauser, T.; Wilms, J.; Nandra, K.; Buchner, J.; Merloni, A.; Freyberg, M.J.; et al. Extreme ultra-soft X-ray variability in an eROSITA observation of the narrow-line Seyfert 1 galaxy 1H 0707-495. *Astron. Astrophys.* **2021**, *647*, A6. [CrossRef]
157. Parker, M.L.; Alston, W.N.; Härer, L.; Igo, Z.; Joyce, A.; Buisson, D.J.K.; Chainakun, P.; Fabian, A.C.; Jiang, J.; Kosec, P.; et al. The nature of the extreme X-ray variability in the NLS1 1H 0707-495. *Mon. Not. R. Astron. Soc.* **2021**, *508*, 1798–1816. [CrossRef]
158. Li, J.; Sun, M.; Wang, T.; He, Z.; Xue, Y. On the origin of the dramatic spectral variability of WPVS 007. *Mon. Not. R. Astron. Soc.* **2019**, *487*, 4592–4602. [CrossRef]
159. Floris, A.; Marziani, P.; Panda, S.; Sniegowska, M.; D'Onofrio, M.; Deconto–Machado, A.; Del Olmo, A.; Czerny, B. Chemical abundances along the quasar main sequence. *arXiv* **2024**, arXiv:2405.04456. [CrossRef]
160. Temple, M.J.; Ferland, G.J.; Rankine, A.L.; Chatzikos, M.; Hewett, P.C. High-ionization emission-line ratios from quasar broad-line regions: metallicity or density? *Mon. Not. R. Astron. Soc.* **2021**, *505*, 3247–3259. [CrossRef]
161. Zhang, Z.; Thomsen, L.L.; Dai, L.; Reynolds, C.S.; García, J.A.; Kara, E.; Connors, R.; Masterson, M.; Yao, Y.; Dauser, T. Modeling X-Ray Multi-Reflection in Super-Eddington Winds. *arXiv* **2024**, arXiv:2407.08596. [CrossRef]
162. Marziani, P.; del Olmo, A.; Perea, J.; D'Onofrio, M.; Panda, S. Broad UV Emission Lines in Type-1 Active Galactic Nuclei: A Note on Spectral Diagnostics and the Excitation Mechanism. *Atoms* **2020**, *8*, 94. [CrossRef]
163. Pignatari, M.; Gallino, R.; Heil, M.; Wiescher, M.; Käppeler, F.; Herwig, F.; Bisterzo, S. The Weak s-Process in Massive Stars and its Dependence on the Neutron Capture Cross Sections. *Astrophys. J.* **2010**, *710*, 1557–1577. [CrossRef]
164. Thielemann, F.K. Explosive Nucleosynthesis: What We Learned and What We Still Do Not Understand. In Proceedings of the Nuclei in the Cosmos XV, Assergi, Italy, 24–29 June 2019; Volume 219, pp. 125–134. [CrossRef]
165. Farrah, D.; Efstathiou, A.; Afonso, J.; Bernard-Salas, J.; Cairns, J.; Clements, D.L.; Croker, K.; Hatziminaoglou, E.; Joyce, M.; Lacy, M.; et al. Stellar and black hole assembly in z < 0.3 infrared-luminous mergers: Intermittent starbursts versus super-Eddington accretion. *Mon. Not. R. Astron. Soc.* **2022**, *513*, 4770–4786. [CrossRef]

166. Miniutti, G.; Saxton, R.D.; Rodríguez-Pascual, P.M.; Read, A.M.; Esquej, P.; Colless, M.; Dobbie, P.; Spolaor, M. A high Eddington-ratio, true Seyfert 2 galaxy candidate: implications for broad-line region models. *Mon. Not. R. Astron. Soc.* **2013**, *433*, 1764–1777. [CrossRef]
167. Antonucci, R.R.J.; Miller, J.S. Spectropolarimetry and the nature of NGC 1068. *Astrophys. J.* **1985**, *297*, 621–632. [CrossRef]
168. Antonucci, R. Unified models for active galactic nuclei and quasars. *Annu. Rev. Astron. Astrophys.* **1993**, *31*, 473–521. [CrossRef]
169. Sulentic, J.; Marziani, P.; Zamfir, S. The Case for Two Quasar Populations. *Balt. Astron.* **2011**, *20*, 427–434. [CrossRef]
170. Fraix-Burnet, D.; Marziani, P.; D'Onofrio, M.; Dultzin, D. The Phylogeny of Quasars and the Ontogeny of Their Central Black Holes. *Front. Astron. Space Sci.* **2017**, *4*, 1. [CrossRef]
171. Marziani, P.; Sulentic, J.W.; Dultzin-Hacyan, D.; Calvani, M.; Moles, M. Comparative Analysis of the High- and Low-Ionization Lines in the Broad-Line Region of Active Galactic Nuclei. *Astrophys. J. Suppl.* **1996**, *104*, 37. [CrossRef]
172. Corbin, M.R.; Boroson, T.A. Combined Ultraviolet and Optical Spectra of 48 Low-Redshift QSOs and the Relation of the Continuum and Emission-Line Properties. *Astrophys. J. Supp.* **1996**, *107*, 69. [CrossRef]
173. Sulentic, J.W.; Bachev, R.; Marziani, P.; Negrete, C.A.; Dultzin, D. C IV λ1549 as an Eigenvector 1 Parameter for Active Galactic Nuclei. *Astrophys. J.* **2007**, *666*, 757–777. [CrossRef]
174. Sulentic, J.W.; Marziani, P. The Intermediate-Line Region in Active Galactic Nuclei: A Region "Præter Necessitatem"? *Astrophys. J.* **1999**, *518*, L9–L12. [CrossRef]
175. Zamanov, R.; Marziani, P.; Sulentic, J.W.; Calvani, M.; Dultzin-Hacyan, D.; Bachev, R. Kinematic Linkage between the Broad- and Narrow-Line-emitting Gas in Active Galactic Nuclei. *Astrophys. J.* **2002**, *576*, L9–L13. [CrossRef]
176. Cracco, V.; Ciroi, S.; Berton, M.; Di Mille, F.; Foschini, L.; La Mura, G.; Rafanelli, P. A spectroscopic analysis of a sample of narrow-line Seyfert 1 galaxies selected from the Sloan Digital Sky Survey. *Mon. Not. R. Astron. Soc.* **2016**, *462*, 1256–1280. [CrossRef]
177. Vietri, G.; Piconcelli, E.; Bischetti, M.; Duras, F.; Martocchia, S.; Bongiorno, A.; Marconi, A.; Zappacosta, L.; Bisogni, S.; Bruni, G.; et al. The WISSH quasars project. IV. Broad line region versus kiloparsec-scale winds. *Astron. Astrophys.* **2018**, *617*, A81. [CrossRef]
178. Vietri, G.; Mainieri, V.; Kakkad, D.; Netzer, H.; Perna, M.; Circosta, C.; Harrison, C.M.; Zappacosta, L.; Husemann, B.; Padovani, P.; et al. SUPER. III. Broad line region properties of AGNs at z $\sim$ 2. *Astron. Astrophys.* **2020**, *644*, A175. [CrossRef]
179. Boller, T.; Freyberg, M.J.; Trümper, J.; Haberl, F.; Voges, W.; Nandra, K. Second ROSAT all-sky survey (2RXS) source catalogue. *Astron. Astrophys.* **2016**, *588*, A103. [CrossRef]
180. Boller, T.; Brandt, W.N.; Fink, H. Soft X-ray properties of narrow-line Seyfert 1 galaxies. *Astron. Astrophys.* **1996**, *305*, 53.
181. Grupe, D. A Complete Sample of Soft X-Ray-selected AGNs. II. Statistical Analysis. *Astron. J.* **2004**, *127*, 1799–1810. [CrossRef]
182. Tananbaum, H.; Avni, Y.; Branduardi, G.; Elvis, M.; Fabbiano, G.; Feigelson, E.; Giacconi, R.; Henry, J.P.; Pye, J.P.; Soltan, A.; et al. X-ray studies of quasars with the Einstein Observatory. *Astrophys. J. Lett.* **1979**, *234*, L9–L13. [CrossRef]
183. Walter, R.; Fink, H.H. The ultraviolet to soft X-ray bump of Seyfert 1 type active galactic nuclei. *Astron. Astrophys.* **1993**, *274*, 105.
184. Wang, J.M.; Netzer, H. Extreme slim accretion disks and narrow line Seyfert 1 galaxies: The nature of the soft X-ray hump. *Astron. Astrophys.* **2003**, *398*, 927–936. [CrossRef]
185. Petrucci, P.O.; Gronkiewicz, D.; Rozanska, A.; Belmont, R.; Bianchi, S.; Czerny, B.; Matt, G.; Malzac, J.; Middei, R.; De Rosa, A.; et al. Radiation spectra of warm and optically thick coronae in AGNs. *Astron. Astrophys.* **2020**, *634*, A85. [CrossRef]
186. Done, C.; Davis, S.W.; Jin, C.; Blaes, O.; Ward, M. Intrinsic disc emission and the soft X-ray excess in active galactic nuclei. *Mon. Not. R. Astron. Soc.* **2012**, *420*, 1848–1860. [CrossRef]
187. Sulentic, J.W.; Marziani, P.; Zwitter, T.; Dultzin-Hacyan, D.; Calvani, M. The Demise of the Classical Broad-Line Region in the Luminous Quasar PG 1416-129. *Astrophys. J.* **2000**, *545*, L15–L18. [CrossRef]
188. Bensch, K.; del Olmo, A.; Sulentic, J.; Perea, J.; Marziani, P. Measures of the Soft X-ray Excess as an Eigenvector 1 Parameter for Active Galactic Nuclei. *J. Astrophys. Astron.* **2015**, *36*, 467–474. [CrossRef]
189. Sulentic, J.; Marziani, P. Quasars in the 4D Eigenvector 1 Context: A stroll down memory lane. *Front. Astron. Space Sci.* **2015**, *2*, 6. [CrossRef]
190. Panda, S.; Marziani, P. High Eddington quasars as discovery tools: Current state and challenges. *Front. Astron. Space Sci.* **2023**, *10*, 1130103. [CrossRef]
191. Shields, G.A. Thermal continuum from acretion disks in quasars. *Nature* **1978**, *272*, 706–708. [CrossRef]
192. Czerny, B.; Elvis, M. Constraints on Quasar Accretion Disks from the Optical/Ultraviolet/Soft X-Ray Big Bump. *Astrophys. J.* **1987**, *321*, 305. [CrossRef]
193. Laor, A.; Fiore, F.; Elvis, M.; Wilkes, B.J.; McDowell, J.C. The Soft X-Ray Properties of a Complete Sample of Optically Selected Quasars. II. Final Results. *Astrophys. J.* **1997**, *477*, 93–113. [CrossRef]
194. Grupe, D.; Komossa, S.; Leighly, K.M.; Page, K.L. The Simultaneous Optical-to-X-Ray Spectral Energy Distribution of Soft X-Ray Selected Active Galactic Nuclei Observed by Swift. *Astrophys. J. Supp.* **2010**, *187*, 64–106. [CrossRef]

195. Capellupo, D.M.; Netzer, H.; Lira, P.; Trakhtenbrot, B.; Mejía-Restrepo, J. Active galactic nuclei at z $\sim$ 1.5 - I. Spectral energy distribution and accretion discs. *Mon. Not. R. Astron. Soc.* **2015**, *446*, 3427–3446. [CrossRef]
196. Ferland, G.J.; Done, C.; Jin, C.; Landt, H.; Ward, M.J. State-of-the-art AGN SEDs for photoionization models: BLR predictions confront the observations. *Mon. Not. R. Astron. Soc.* **2020**, *494*, 5917–5922. [CrossRef]
197. Laurenti, M.; Luminari, A.; Tombesi, F.; Vagnetti, F.; Middei, R.; Piconcelli, E. Location and energetics of the ultra-fast outflow in PG 1448+273. *Astron. Astrophys.* **2021**, *645*, A118. [CrossRef]
198. Grünwald, G.; Boller, T.; Rakshit, S.; Buchner, J.; Dauser, T.; Freyberg, M.; Liu, T.; Salvato, M.; Schichtel, A. The first look at narrow-line Seyfert 1 galaxies with eROSITA. *Astron. Astrophys.* **2023**, *669*, A37. [CrossRef]
199. Boller, T. Unraveling the enigmatic soft X-ray excess: Current understanding and future perspectives. *Astron. Nachrichten* **2023**, *344*, e20230105. [CrossRef]
200. Haardt, F.; Maraschi, L. A two-phase model for the X-ray emission from Seyfert galaxies. *Astrophys. J. Lett.* **1991**, *380*, L51–L54. [CrossRef]
201. Haardt, F.; Maraschi, L. X-Ray Spectra from Two-Phase Accretion Disks. *Astrophys. J.* **1993**, *413*, 507. [CrossRef]
202. Nandra, K.; Pounds, K.A. GINGA observations of the X-ray spectra of Seyfert galaxies. *Mon. Not. R. Astron. Soc.* **1994**, *268*, 405–429. [CrossRef]
203. Kara, E.; Miller, J.M.; Reynolds, C.; Dai, L. Relativistic reverberation in the accretion flow of a tidal disruption event. *Nature* **2016**, *535*, 388–390. [CrossRef] [PubMed]
204. Mathews, W.G.; Ferland, G.J. What heats the hot phase in active nuclei? *Astrophys. J.* **1987**, *323*, 456–467. [CrossRef]
205. Xie, Y.; Ho, L.C.; Zhuang, M.Y.; Shangguan, J. The Infrared Emission and Vigorous Star Formation of Low-redshift Quasars. *Astrophys. J.* **2021**, *910*, 124. [CrossRef]
206. Kirkpatrick, A.; Pope, A.; Roebuck, E.; Yan, L.; Alberts, S.; Herter, T.; Armus, L. Cold Quasars: A New Phase in Quasar Evolution. *Astrophys. J.* **2019**, *879*, 41. [CrossRef]
207. Hamann, F.; Ferland, G. The Chemical Evolution of QSOs and the Implications for Cosmology and Galaxy Formation. *Astrophys. J.* **1993**, *418*, 11. [CrossRef]
208. Hamann, F.; Ferland, G. Elemental Abundances in Quasistellar Objects: Star Formation and Galactic Nuclear Evolution at High Redshifts. *Annu. Rev. Astron. Astrophys.* **1999**, *37*, 487–531. [CrossRef]
209. Xu, F.; Bian, F.; Shen, Y.; Zuo, W.; Fan, X.; Zhu, Z. The evolution of chemical abundance in quasar broad line region. *Mon. Not. R. Astron. Soc.* **2018**, *480*, 345–357. [CrossRef]
210. Shin, J.; Woo, J.H.; Nagao, T.; Kim, S.C. The Chemical Properties of Low-redshift QSOs. *Astrophys. J.* **2013**, *763*, 58. [CrossRef]
211. Śniegowska, M.; Marziani, P.; Czerny, B.; Panda, S.; Martínez-Aldama, M.L.; del Olmo, A.; D'Onofrio, M. High Metal Content of Highly Accreting Quasars. *Astrophys. J.* **2021**, *910*, 115. [CrossRef]
212. Shields, G.A. The abundance of nitrogen in QSOs. *Astrophys. J.* **1976**, *204*, 330–336. [CrossRef]
213. Nagao, T.; Marconi, A.; Maiolino, R. The evolution of the broad-line region among SDSS quasars. *Astron. Astrophys.* **2006**, *447*, 157–172. [CrossRef]
214. Huang, J.; Lin, D.N.C.; Shields, G. Metal enrichment due to embedded stars in AGN discs. *Mon. Not. R. Astron. Soc.* **2023**, *525*, 5702–5718. [CrossRef]
215. Garnica, K.; Negrete, C.A.; Marziani, P.; Dultzin, D.; Śniegowska, M.; Panda, S. High metal content of highly accreting quasars: Analysis of an extended sample. *Astron. Astrophys.* **2022**, *667*, A105. [CrossRef]
216. Hubeny, I.; Blaes, O.; Krolik, J.H.; Agol, E. Non-LTE Models and Theoretical Spectra of Accretion Disks in Active Galactic Nuclei. IV. Effects of Compton Scattering and Metal Opacities. *Astrophys. J.* **2001**, *559*, 680–702. [CrossRef]
217. Dumont, A.M.; Collin, S.; Paletou, F.; Coupé, S.; Godet, O.; Pelat, D. Escape probability methods versus "exact" transfer for modelling the X-ray spectrum of Active Galactic Nuclei and X-ray binaries. *Astron. Astrophys.* **2003**, *407*, 13–30. [CrossRef]
218. Krolik, J.H. *Active Galactic Nuclei: From the Central Black Hole to the Galactic Environment*; Princeton University Press: Princeton, NJ, USA, 1999.
219. Osterbrock, D.E.; Ferland, G.J. *Astrophysics of Gaseous Nebulae and Active Galactic Nuclei*; University Science Books: Mill Valley, CA, USA, 2006.
220. Pradhan, A.K.; Nahar, S.N. *Atomic Astrophysics and Spectroscopy*; Cambridge University Press: Cambridge, UK, 2015.
221. Castor, J.I.; Abbott, D.C.; Klein, R.I. Radiation-driven winds in Of stars. *Astrophys. J.* **1975**, *195*, 157–174. [CrossRef]
222. Murray, N.; Chiang, J.; Grossman, S.A.; Voit, G.M. Accretion Disk Winds from Active Galactic Nuclei. *Astrophys. J.* **1995**, *451*, 498. [CrossRef]
223. Marziani, P.; Sulentic, J.W.; Stirpe, G.M.; Dultzin, D.; Del Olmo, A.; Martínez-Carballo, M.A. Blue outliers among intermediate redshift quasars. *Astrophys. Space Sci.* **2016**, *361*, 3. [CrossRef]
224. Juarez, Y.; Maiolino, R.; Mujica, R.; Pedani, M.; Marinoni, S.; Nagao, T.; Marconi, A.; Oliva, E. The metallicity of the most distant quasars. *Astron. Astrophys.* **2009**, *494*, L25–L28. [CrossRef]

225. Yang, J.; Wang, F.; Fan, X.; Barth, A.J.; Hennawi, J.F.; Nanni, R.; Bian, F.; Davies, F.B.; Farina, E.P.; Schindler, J.T.; et al. Probing Early Supermassive Black Hole Growth and Quasar Evolution with Near-infrared Spectroscopy of 37 Reionization-era Quasars at $6.3 < z < 7.64$. *Astrophys. J.* **2021**, *923*, 262. [CrossRef]
226. Collin, S.; Zahn, J.P. Star formation and evolution in accretion disks around massive black holes. *Astron. Astrophys.* **1999**, *344*, 433–449.
227. Wang, J.M.; Zhai, S.; Li, Y.R.; Songsheng, Y.Y.; Ho, L.C.; Chen, Y.J.; Liu, J.R.; Du, P.; Yuan, Y.F. Star Formation in Self-gravitating Disks in Active Galactic Nuclei. III. Efficient Production of Iron and Infrared Spectral Energy Distributions. *Astrophys. J.* **2023**, *954*, 84. [CrossRef]
228. Fan, X.; Wu, Q. In Situ Star Formation in Accretion Disks and Explanation of Correlation between the Black Hole Mass and Metallicity in Active Galactic Nuclei. *Astrophys. J.* **2023**, *944*, 159. [CrossRef]
229. Loiacono, F.; Decarli, R.; Mignoli, M.; Farina, E.P.; Bañados, E.; Bosman, S.; Eilers, A.C.; Schindler, J.T.; Strauss, M.A.; Vestergaard, M.; et al. A quasar-galaxy merger at $z \sim 6.2$: Black hole mass and quasar properties from the NIRSpec spectrum. *Astron. Astrophys.* **2024**, *685*, A121. [CrossRef]
230. Storchi-Bergmann, T.; Schmitt, H.R.; Calzetti, D.; Kinney, A.L. Chemical Abundance Calibrations for the Narrow-Line Region of Active Galaxies. *Astron. J.* **1998**, *115*, 909–914. [CrossRef]
231. Nagao, T.; Maiolino, R.; Marconi, A. Gas metallicity in the narrow-line regions of high-redshift active galactic nuclei. *Astron. Astrophys.* **2006**, *447*, 863–876. [CrossRef]
232. Armah, M.; Riffel, R.; Dors, O.L.; Oh, K.; Koss, M.J.; Ricci, C.; Trakhtenbrot, B.; Valerdi, M.; Riffel, R.A.; Krabbe, A.C. Oxygen abundances in the narrow line regions of Seyfert galaxies and the metallicity-luminosity relation. *Mon. Not. R. Astron. Soc.* **2023**, *520*, 1687–1703. [CrossRef]
233. Sanders, D.B.; Soifer, B.T.; Elias, J.H.; Madore, B.F.; Matthews, K.; Neugebauer, G.; Scoville, N.Z. Ultraluminous infrared galaxies and the origin of quasars. *Astrophys. J.* **1988**, *325*, 74–91. [CrossRef]
234. Rafanelli, P.; Marziani, P. The complex nature of the interacting system NGC 7592. *Astron. J.* **1992**, *103*, 743–756. [CrossRef]
235. Marziani, P.; Sulentic, J.W.; Calvani, M.; Perez, E.; Moles, M.; Penston, M.V. The peculiar Balmer line profiles of OQ 208. *Astrophys. J.* **1993**, *410*, 56–67. [CrossRef]
236. Heckman, T.M.; González-Delgado, R.; Leitherer, C.; Meurer, G.R.; Krolik, J.; Wilson, A.S.; Koratkar, A.; Kinney, A. A Powerful Nuclear Starburst in the Seyfert Galaxy Markarian 477: Implications for the Starburst-Active Galactic Nucleus Connection. *Astrophys. J.* **1997**, *482*, 114–132. [CrossRef]
237. Dultzin-Hacyan, D.; Krongold, Y.; Marziani, P. The Environment of AGN and AN Evolutionary Scheme. In *Revista Mexicana de Astronomia y Astrofisica;* Conference Series; Reyes-Ruiz, M., Vázquez-Semadeni, E., Eds.; Instituto de Astronomía: Distrito Federal, México, 2003; Volume 18, p. 147.
238. Schweitzer, M.; Lutz, D.; Sturm, E.; Contursi, A.; Tacconi, L.J.; Lehnert, M.D.; Dasyra, K.M.; Genzel, R.; Veilleux, S.; Rupke, D.; et al. Spitzer Quasar and ULIRG Evolution Study (QUEST). I. The Origin of the Far-Infrared Continuum of QSOs. *Astrophys. J.* **2006**, *649*, 79–90. [CrossRef]
239. Wild, V.; Heckman, T.; Charlot, S. Timing the starburst-AGN connection. *Mon. Not. R. Astron. Soc.* **2010**, *405*, 933–947. [CrossRef]
240. Rafanelli, P.; La Mura, G.; Bindoni, D.; Ciroi, S.; Cracco, V.; Di Mille, F.; Vaona, L. The Starburst—AGN Connection: A Critical Review. *Balt. Astron.* **2011**, *20*, 419–426. [CrossRef]
241. Lipari, S.; Terlevich, R.; Macchetto, F. Extreme optical Fe II emission in luminous IRAS active galactic nuclei. *Astrophys. J.* **1993**, *406*, 451–456. [CrossRef]
242. Lipari, S. Galaxies with Extreme Infrared and Fe II Emission. II. IRAS 07598+6508: A Starburst/Young Broad Absorption Line QSO. *Astrophys. J.* **1994**, *436*, 102. [CrossRef]
243. Haas, M.; Klaas, U.; Müller, S.A.H.; Bertoldi, F.; Camenzind, M.; Chini, R.; Krause, O.; Lemke, D.; Meisenheimer, K.; Richards, P.J.; et al. The ISO view of Palomar-Green quasars. *Astron. Astrophys.* **2003**, *402*, 87–111. [CrossRef]
244. Krongold, Y.; Dultzin-Hacyan, D.; Marziani, P. An Evolutionary Sequence for AGN. In *Proceedings of the Active Galactic Nuclei: From Central Engine to Host Galaxy;* Collin, S., Combes, F., Shlosman, I., Eds.; Astronomical Society of the Pacific Conference Series; Astronomical Society of the Pacific: San Francisco, CA, USA, 2003; Volume 290, p. 523.
245. Lípari, S.L.; Terlevich, R.J. Evolutionary unification in composite active galactic nuclei. *Mon. Not. R. Astron. Soc.* **2006**, *368*, 1001–1015. [CrossRef]
246. Wang, J.M.; Yan, C.S.; Li, Y.R.; Chen, Y.M.; Xiang, F.; Hu, C.; Ge, J.Q.; Zhang, S. Evolution of Gaseous Disk Viscosity Driven by Supernova Explosions in Star-Forming Galaxies at High Redshift. *Astrophys. J. Lett.* **2009**, *701*, L7–L11. [CrossRef]
247. Wang, J.; Deng, J.S.; Wei, J.Y. Ongoing star formation in AGN host galaxy discs: A view from core-collapse supernovae. *Mon. Not. R. Astron. Soc.* **2010**, *405*, 2529–2533. [CrossRef]
248. Wang, J.M.; Ge, J.Q.; Hu, C.; Baldwin, J.A.; Li, Y.R.; Ferland, G.J.; Xiang, F.; Yan, C.S.; Zhang, S. Star Formation in Self-gravitating Disks in Active Galactic Nuclei. I. Metallicity Gradients in Broad-line Regions. *Astrophys. J.* **2011**, *739*, 3. [CrossRef]

249. Wang, J.M.; Du, P.; Baldwin, J.A.; Ge, J.Q.; Hu, C.; Ferland, G.J. Star Formation in Self-gravitating Disks in Active Galactic Nuclei. II. Episodic Formation of Broad-line Regions. *Astrophys. J.* **2012**, *746*, 137. [CrossRef]
250. Elitzur, M.; Shlosman, I. The AGN-obscuring Torus: The End of the "Doughnut" Paradigm? *Astrophys. J. Lett.* **2006**, *648*, L101–L104. [CrossRef]
251. Kang, J.L.; Done, C.; Hagen, S.; Temple, M.J.; Silverman, J.D.; Li, J.; Liu, T. Systematic collapse of the accretion disc in AGN confirmed by UV photometry and broad line spectra. *arXiv* **2024**, arXiv:2410.06730. [CrossRef]
252. Artymowicz, P.; Lin, D.N.C.; Wampler, E.J. Star Trapping and Metallicity Enrichment in Quasars and Active Galactic Nuclei. *Astrophys. J.* **1993**, *409*, 592. [CrossRef]
253. Lin, D.N.C. Star/Disk Interaction in the Nuclei of Active Galaxies. In *Proceedings of the IAU Colloq. 159: Emission Lines in Active Galaxies: New Methods and Techniques*; Peterson, B.M., Cheng, F.Z., Wilson, A.S., Eds.; Astronomical Society of the Pacific Conference Series; Astronomical Society of the Pacific: San Francisco, CA, USA, 1997; Volume 113, p. 64.
254. Dittmann, A.J.; Cantiello, M. A Semi-Analytical Model for Stellar Evolution in AGN Disks. *arXiv* **2024**, arXiv:2409.02981. [CrossRef]
255. Fabj, G.; Dittmann, A.J.; Cantiello, M.; Perna, R.; Samsing, J. Mapping the Outcomes of Stellar Evolution in the Disks of Active Galactic Nuclei. *arXiv* **2024**, arXiv:2408.16050. [CrossRef]
256. Liu, J.R.; Wang, Y.L.; Wang, J.M. Accretion-modified Stars in Accretion Disks of Active Galactic Nuclei: Observational Characteristics in Different Regions of the Disks. *Astrophys. J.* **2024**, *969*, 37. [CrossRef]
257. Nayakshin, S.; Cuadra, J.; Springel, V. Simulations of star formation in a gaseous disc around Sgr A*—A failed active galactic nucleus. *Mon. Not. R. Astron. Soc.* **2007**, *379*, 21–33. [CrossRef]
258. Alexander, T.; Livio, M. Tidal Scattering of Stars on Supermassive Black Holes in Galactic Centers. *Astrophys. J. Lett.* **2001**, *560*, L143–L146. [CrossRef]
259. Alexander, T.; Livio, M. Orbital Capture of Stars by a Massive Black Hole via Exchanges with Compact Remnants. *Astrophys. J. Lett.* **2004**, *606*, L21–L24. [CrossRef]
260. Davies, M.B.; Lin, D.N.C. Making massive stars in the Galactic Centre via accretion on to low-mass stars within an accretion disc. *Mon. Not. R. Astron. Soc.* **2020**, *498*, 3452–3456. [CrossRef]
261. Jermyn, A.S.; Dittmann, A.J.; McKernan, B.; Ford, K.E.S.; Cantiello, M. Effects of an Immortal Stellar Population in AGN Disks. *Astrophys. J.* **2022**, *929*, 133. [CrossRef]
262. Dittmann, A.J.; Jermyn, A.S.; Cantiello, M. The Influence of Disk Composition on the Evolution of Stars in the Disks of Active Galactic Nuclei. *Astrophys. J.* **2023**, *946*, 56. [CrossRef]
263. Nasim, S.S.; Fabj, G.; Caban, F.; Secunda, A.; Ford, K.E.S.; McKernan, B.; Bellovary, J.M.; Leigh, N.W.C.; Lyra, W. Aligning Retrograde Nuclear Cluster Orbits with an Active Galactic Nucleus Accretion Disc. *Mon. Not. R. Astron. Soc.* **2023**, *522*, 5393–5401. [CrossRef]
264. Clayton, D.D. *Principles of Stellar Evolution and Nucleosynthesis*; The University of Chicago Press: Chicago, IL, USA, 1968.
265. Napolitano, L.; Castellano, M.; Pentericci, L.; Vignali, C.; Gilli, R.; Fontana, A.; Santini, P.; Treu, T.; Calabrò, A.; Llerena, M.; et al. The dual nature of GHZ9: Coexisting AGN and star formation activity in a remote X-ray source at z = 10.145. *arXiv* **2024**, arXiv:2410.18763. [CrossRef]
266. Zhou, H.Y.; Wang, T.G.; Zhou, Y.Y.; Li, C.; Dong, X.B. Discovery of a Radio-loud Superstrong Fe II-emitting Quasar. *Astrophys. J.* **2002**, *581*, 96–102. [CrossRef]
267. Sani, E.; Lutz, D.; Risaliti, G.; Netzer, H.; Gallo, L.C.; Trakhtenbrot, B.; Sturm, E.; Boller, T. Enhanced star formation in narrow-line Seyfert 1 active galactic nuclei revealed by Spitzer. *Mon. Not. R. Astron. Soc.* **2010**, *403*, 1246–1260. [CrossRef]
268. Caccianiga, A.; Antón, S.; Ballo, L.; Foschini, L.; Maccacaro, T.; Della Ceca, R.; Severgnini, P.; Marchã, M.J.; Mateos, S.; Sani, E. WISE colours and star formation in the host galaxies of radio-loud narrow-line Seyfert 1. *Mon. Not. R. Astron. Soc.* **2015**, *451*, 1795–1805. [CrossRef]
269. Laor, A.; Behar, E. On the origin of radio emission in radio-quiet quasars. *Mon. Not. R. Astron. Soc.* **2008**, *390*, 847–862. [CrossRef]
270. Laor, A.; Baldi, R.D.; Behar, E. What drives the radio slopes in radio-quiet quasars? *Mon. Not. R. Astron. Soc.* **2019**, *482*, 5513–5523. [CrossRef]
271. Panessa, F.; Baldi, R.D.; Laor, A.; Padovani, P.; Behar, E.; McHardy, I. The origin of radio emission from radio-quiet active galactic nuclei. *Nat. Astron.* **2019**, *3*, 387–396. [CrossRef]
272. Yang, X.; Yao, S.; Yang, J.; Ho, L.C.; An, T.; Wang, R.; Baan, W.A.; Gu, M.; Liu, X.; Yang, X.; et al. Radio Activity of Supermassive Black Holes with Extremely High Accretion Rates. *Astrophys. J.* **2020**, *904*, 200. [CrossRef]
273. Chen, S.; Laor, A.; Behar, E.; Baldi, R.D.; Gelfand, J.D. The radio emission in radio-quiet quasars: The VLBA perspective. *Mon. Not. R. Astron. Soc.* **2023**, *525*, 164–182. [CrossRef]
274. Elston, R.; Thompson, K.L.; Hill, G.J. Detection of strong iron emission from quasars at redshift z > 3. *Nature* **1994**, *367*, 250–251. [CrossRef]

275. Chieffi, A.; Limongi, M. Pre-supernova Evolution of Rotating Solar Metallicity Stars in the Mass Range 13–120 $M_\odot$ and their Explosive Yields. *Astrophys. J.* **2013**, *764*, 21. [CrossRef]
276. Wang, T.; Zhou, H.; Yuan, W.; Lu, H.L.; Dong, X.; Shan, H. Metal-Enriched Outflows in the Ultraluminous Infrared Quasar Q1321+058. *Astrophys. J.* **2009**, *702*, 851–861. [CrossRef]
277. Marziani, P.; Floris, A.; Deconto–Machado, A.; Panda, S.; Sniegowska, M.; Garnica, K.; Dultzin, D.; D'Onofrio, M.; Del Olmo, A.; Bon, E.; et al. From Sub-Solar to Super-Solar Chemical Abundances along the Quasar Main Sequence. *Physics* **2024**, *6*, 216–236. [CrossRef]
278. Drewes, F.; Leftley, J.H.; Hönig, S.F.; Tristram, K.R.W.; Kishimoto, M. I Zw 1 and H0557-385: The dusty tori of two high Eddington AGNs observed in the MATISSE LM-Bands. *Mon. Not. R. Astron. Soc.* **2025**, *537*, 1369–1384. [CrossRef]
279. Czerny, B.; Hryniewicz, K. The origin of the broad line region in active galactic nuclei. *Astron. Astrophys.* **2011**, *525*, L8. [CrossRef]
280. Berton, M.; Foschini, L.; Ciroi, S.; Cracco, V.; La Mura, G.; Lister, M.L.; Mathur, S.; Peterson, B.M.; Richards, J.L.; Rafanelli, P. Parent population of flat-spectrum radio-loud narrow-line Seyfert 1 galaxies. *Astron. Astrophys.* **2015**, *578*, A28. [CrossRef]
281. Padovani, P. Active Galactic Nuclei at All Wavelengths and from All Angles. *Front. Astron. Space Sci.* **2017**, *4*, 35. [CrossRef]
282. Strateva, I.V.; Strauss, M.A.; Hao, L.; Schlegel, D.J.; Hall, P.B.; Gunn, J.E.; Li, L.; Ivezić, Ž.; Richards, G.T.; Zakamska, N.L.; et al. Double-peaked Low-Ionization Emission Lines in Active Galactic Nuclei. *Astron. J.* **2003**, *126*, 1720–1749. [CrossRef]
283. Mengistue, S.T.; Marziani, P.; del Olmo, A.; Pović, M.; Perea, J.; Deconto Machado, A. Quasar 3C 47: Extreme Population B jetted source with double-peaked profiles. *Astron. Astrophys.* **2024**, *685*, A116. [CrossRef]
284. Wu, Q. Observational Evidence for Young Radio Galaxies Is Triggered by Accretion Disk Instability. *Astrophys. J. Lett.* **2009**, *701*, L95–L99. [CrossRef]
285. Shen, Y. The mass of quasars. *Bull. Astron. Soc. India* **2013**, *41*, 61–115.
286. D'Onofrio, M.; Marziani, P.; Chiosi, C. Past, Present and Future of the Scaling Relations of Galaxies and Active Galactic Nuclei. *Front. Astron. Space Sci.* **2021**, *8*, 157. [CrossRef]
287. Wu, Q.; Shen, Y. A Catalog of Quasar Properties from Sloan Digital Sky Survey Data Release 16. *Astrophys. J. Suppl. Ser.* **2022**, *263*, 42. [CrossRef]
288. Rees, M.J.; Begelman, M.C.; Blandford, R.D.; Phinney, E.S. Ion-supported tori and the origin of radio jets. *Nature* **1982**, *295*, 17–21. [CrossRef]
289. Narayan, R.; Yi, I. Advection-dominated Accretion: A Self-similar Solution. *Astrophys. J. Lett.* **1994**, *428*, L13. [CrossRef]
290. Narayan, R.; Yi, I. Advection-dominated Accretion: Self-Similarity and Bipolar Outflows. *Astrophys. J.* **1995**, *444*, 231. [CrossRef]
291. Proga, D.; Stone, J.M.; Kallman, T.R. Dynamics of Line-Driven Disk Winds in Active Galactic Nuclei. *Astrophys. J.* **2000**, *543*, 686–696. [CrossRef]
292. Proga, D.; Kallman, T.R. Dynamics of Line-Driven Disk Winds in Active Galactic Nuclei. II. Effects of Disk Radiation. *Astrophys. J.* **2004**, *616*, 688–695. [CrossRef]
293. Giustini, M.; Proga, D. A global view of the inner accretion and ejection flow around super massive black holes. Radiation-driven accretion disk winds in a physical context. *Astron. Astrophys.* **2019**, *630*, A94. [CrossRef]
294. Blandford, R.; Meier, D.; Readhead, A. Relativistic Jets from Active Galactic Nuclei. *Annu. Rev. Astron. Astrophys.* **2019**, *57*, 467–509. [CrossRef]
295. Coatman, L.; Hewett, P.C.; Banerji, M.; Richards, G.T. C iv emission-line properties and systematic trends in quasar black hole mass estimates. *Mon. Not. R. Astron. Soc.* **2016**, *461*, 647–665. [CrossRef]
296. Sulentic, J.W.; del Olmo, A.; Marziani, P.; Martínez-Carballo, M.A.; D'Onofrio, M.; Dultzin, D.; Perea, J.; Martínez-Aldama, M.L.; Negrete, C.A.; Stirpe, G.M.; et al. What does CIVλ1549 tell us about the physical driver of the Eigenvector quasar sequence? *Astron. Astrophys.* **2017**, *608*, A122. [CrossRef]
297. Hopkins, P.F.; Elvis, M. Quasar feedback: More bang for your buck. *Mon. Not. R. Astron. Soc.* **2010**, *401*, 7–14. [CrossRef]
298. Fabian, A.C. Observational Evidence of Active Galactic Nuclei Feedback. *Annu. Rev. Astron. Astrophys.* **2012**, *50*, 455–489. [CrossRef]
299. Murray, N.; Chiang, J. Disk Winds and Disk Emission Lines. *Astrophys. J.* **1997**, *474*, 91. [CrossRef]
300. Ferland, G.J.; Hu, C.; Wang, J.; Baldwin, J.A.; Porter, R.L.; van Hoof, P.A.M.; Williams, R.J.R. Implications of Infalling Fe II-Emitting Clouds in Active Galactic Nuclei: Anisotropic Properties. *Astrophys. J. Lett.* **2009**, *707*, L82–L86. [CrossRef]
301. Risaliti, G.; Elvis, M. A non-hydrodynamical model for acceleration of line-driven winds in active galactic nuclei. *Astron. Astrophys.* **2010**, *516*, A89. [CrossRef]
302. Goad, M.R.; Korista, K.T.; Ruff, A.J. The broad emission-line region: The confluence of the outer accretion disc with the inner edge of the dusty torus. *Mon. Not. R. Astron. Soc.* **2012**, *426*, 3086–3111. [CrossRef]
303. Marziani, P.; Sulentic, J.W.; Plauchu-Frayn, I.; del Olmo, A. Low-Ionization Outflows in High Eddington Ratio Quasars. *Astrophys. J.* **2013**, *764*. [CrossRef]
304. Popović, L.Č.; Kovačević-Dojčinović, J.; Marčeta-Mand ić, S. The structure of the Mg II broad line emitting region in Type 1 AGNs. *Mon. Not. R. Astron. Soc.* **2019**, *484*, 3180–3197. [CrossRef]

305. Chen, K.; Halpern, J.P. Structure of line-emitting accretion disks in active galactic nuclei—ARP 102B. *Astrophys. J.* **1989**, *344*, 115–124. [CrossRef]
306. Mejía-Restrepo, J.E.; Trakhtenbrot, B.; Lira, P.; Netzer, H.; Capellupo, D.M. Active galactic nuclei at z˜1.5: II. Black Hole Mass estimation by means of broad emission lines. *Mon. Not. R. Astron. Soc.* **2016**, *460*, 187–211. [CrossRef]
307. Collin-Souffrin, S.; Dyson, J.E.; McDowell, J.C.; Perry, J.J. The environment of active galactic nuclei. I—A two-component broad emission line model. *Mon. Not. R. Astron. Soc.* **1988**, *232*, 539–550. [CrossRef]
308. Cackett, E.M.; Horne, K.; Winkler, H. Testing thermal reprocessing in active galactic nuclei accretion discs. *Mon. Not. R. Astron. Soc.* **2007**, *380*, 669–682. [CrossRef]
309. Uttley, P.; Cackett, E.M.; Fabian, A.C.; Kara, E.; Wilkins, D.R. X-ray reverberation around accreting black holes. *Astron. Astrophys. Rev.* **2014**, *22*, 72. [CrossRef]
310. Kara, E.; Alston, W.N.; Fabian, A.C.; Cackett, E.M.; Uttley, P.; Reynolds, C.S.; Zoghbi, A. A global look at X-ray time lags in Seyfert galaxies. *Mon. Not. R. Astron. Soc.* **2016**, *462*, 511–531. [CrossRef]
311. Tombesi, F.; Sambruna, R.M.; Reeves, J.N.; Braito, V.; Ballo, L.; Gofford, J.; Cappi, M.; Mushotzky, R.F. Discovery of Ultra-fast Outflows in a Sample of Broad-line Radio Galaxies Observed with Suzaku. *Astrophys. J.* **2010**, *719*, 700–715. [CrossRef]
312. Tombesi, F.; Cappi, M.; Reeves, J.N.; Braito, V. Evidence for ultrafast outflows in radio-quiet AGNs-III. Location and energetics. *Mon. Not. R. Astron. Soc.* **2012**, *422*, L1–L5. [CrossRef]
313. Tombesi, F.; Cappi, M.; Reeves, J.N.; Nemmen, R.S.; Braito, V.; Gaspari, M.; Reynolds, C.S. Unification of X-ray winds in Seyfert galaxies: From ultra-fast outflows to warm absorbers. *Mon. Not. R. Astron. Soc.* **2013**, *430*, 1102–1117. [CrossRef]
314. Laha, S.; Reynolds, C.S.; Reeves, J.; Kriss, G.; Guainazzi, M.; Smith, R.; Veilleux, S.; Proga, D. Ionized outflows from active galactic nuclei as the essential elements of feedback. *Nat. Astron.* **2021**, *5*, 13–24. [CrossRef]
315. Lin, D.N.C.; Pringle, J.E. A viscosity prescription for a self-gravitating accretion disc. *Mon. Not. R. Astron. Soc.* **1987**, *225*, 607–613. [CrossRef]
316. Czerny, B.; Li, Y.R.; Hryniewicz, K.; Panda, S.; Wildy, C.; Sniegowska, M.; Wang, J.M.; Sredzinska, J.; Karas, V. Failed Radiatively Accelerated Dusty Outflow Model of the Broad Line Region in Active Galactic Nuclei. I. Analytical Solution. *Astrophys. J.* **2017**, *846*, 154. [CrossRef]
317. Naddaf, M.H.; Martinez-Aldama, M.L.; Marziani, P.; Panda, S.; Sniegowska, M.; Czerny, B. Dust-driven wind as a model of broad absorption line quasars. *Astron. Astrophys.* **2023**, *675*, A43. [CrossRef]
318. Naddaf, M.H.; Czerny, B.; Szczerba, R. The Picture of BLR in 2.5D FRADO: Dynamics and Geometry. *Astrophys. J.* **2021**, *920*, 30. [CrossRef]
319. Choi, H.; Leighly, K.M.; Terndrup, D.M.; Dabbieri, C.; Gallagher, S.C.; Richards, G.T. The Physical Properties of Low-redshift FeLoBAL Quasars. I. Spectral-synthesis Analysis of the Broad Absorption-line (BAL) Outflows Using SimBAL. *Astrophys. J.* **2022**, *937*, 74. [CrossRef]
320. Bischetti, M.; Piconcelli, E.; Vietri, G.; Bongiorno, A.; Fiore, F.; Sani, E.; Marconi, A.; Duras, F.; Zappacosta, L.; Brusa, M.; et al. The WISSH quasars project. I. Powerful ionised outflows in hyper-luminous quasars. *Astron. Astrophys.* **2017**, *598*, A122. [CrossRef]
321. Cano-Díaz, M.; Maiolino, R.; Marconi, A.; Netzer, H.; Shemmer, O.; Cresci, G. Observational evidence of quasar feedback quenching star formation at high redshift. *Astron. Astrophys.* **2012**, *537*, L8. [CrossRef]
322. Marziani, P.; Negrete, C.A.; Dultzin, D.; Martinez-Aldama, M.L.; Del Olmo, A.; Esparza, D.; Sulentic, J.W.; D'Onofrio, M.; Stirpe, G.M.; Bon, E.; et al. Highly accreting quasars: A tool for cosmology? *Proc. Int. Astron. Union* **2017**, *324*, 245–246. [CrossRef]
323. Laor, A.; Brandt, W.N. The Luminosity Dependence of Ultraviolet Absorption in Active Galactic Nuclei. *Astrophys. J.* **2002**, *569*, 641–654. [CrossRef]
324. Villar Martín, M.; López Cobá, C.; Cazzoli, S.; Pérez Montero, E.; Cabrera Lavers, A. AGN feedback can produce metal enrichment on galaxy scales. *Astron. Astrophys.* **2024**, *690*, A397. [CrossRef]
325. Harrison, C.M.; Alexander, D.M.; Mullaney, J.R.; Swinbank, A.M. Energetic galaxy-wide outflows in high-redshift quasars driven by AGN activity. *Mon. Not. R. Astron. Soc.* **2014**, *426*, 1073–1096. . [CrossRef]
326. Harrison, C.M.; Alexander, D.M.; Mullaney, J.R.; Swinbank, A.M. Kiloparsec-scale outflows are prevalent among luminous AGN: Outflows and feedback in the context of the overall AGN population. *Mon. Not. R. Astron. Soc.* **2014**, *441*, 3306–3347. [CrossRef]
327. Kakkad, D.; Mainieri, V.; Vietri, G.; Carniani, S.; Harrison, C.M.; Perna, M.; Scholtz, J.; Circosta, C.; Cresci, G.; Husemann, B.; et al. SUPER. II. Spatially resolved ionised gas kinematics and scaling relations in z ~ 2 AGN host galaxies. *Astron. Astrophys.* **2020**, *642*, A147. [CrossRef]
328. Singha, M.; Husemann, B.; Urrutia, T.; O'Dea, C.P.; Scharwächter, J.; Gaspari, M.; Combes, F.; Nevin, R.; Terrazas, B.A.; Pérez-Torres, M.; et al. The Close AGN Reference Survey (CARS). Locating the [O III] wing component in luminous local Type 1 AGN. *Astron. Astrophys.* **2022**, *659*, A123. [CrossRef]
329. Tombesi, F.; Tazaki, F.; Mushotzky, R.F.; Ueda, Y.; Cappi, M.; Gofford, J.; Reeves, J.N.; Guainazzi, M. Ultrafast outflows in radio-loud active galactic nuclei. *Mon. Not. R. Astron. Soc.* **2014**, *443*, 2154–2182. [CrossRef]

330. Gofford, J.; Reeves, J.N.; McLaughlin, D.E.; Braito, V.; Turner, T.J.; Tombesi, F.; Cappi, M. The Suzaku view of highly ionized outflows in AGN-II. Location, energetics and scalings with bolometric luminosity. *Mon. Not. R. Astron. Soc.* **2015**, *451*, 4169–4182. [CrossRef]
331. Mestici, S.; Tombesi, F.; Gaspari, M.; Piconcelli, E.; Panessa, F. Unified properties of supermassive black hole winds in radio-quiet and radio-loud AGN. *Mon. Not. R. Astron. Soc.* **2024**, *532*, 3036–3055. [CrossRef]
332. Mizumoto, M.; Nomura, M.; Done, C.; Ohsuga, K.; Odaka, H. UV line-driven disc wind as the origin of UltraFast Outflows in AGN. *Mon. Not. R. Astron. Soc.* **2021**, *503*, 1442–1458. [CrossRef]
333. Matzeu, G.A.; Reeves, J.N.; Braito, V.; Nardini, E.; McLaughlin, D.E.; Lobban, A.P.; Tombesi, F.; Costa, M.T. Evidence for a radiatively driven disc-wind in PDS 456? *Mon. Not. R. Astron. Soc.* **2017**, *472*, L15–L19. [CrossRef]
334. Fukumura, K.; Tombesi, F.; Kazanas, D.; Shrader, C.; Behar, E.; Contopoulos, I. Stratified Magnetically Driven Accretion-disk Winds and Their Relations to Jets. *Astrophys. J.* **2014**, *780*, 120. [CrossRef]
335. Fukumura, K.; Kazanas, D.; Shrader, C.; Behar, E.; Tombesi, F.; Contopoulos, I. Variable Nature of Magnetically Driven Ultra-fast Outflows. *Astrophys. J. Lett.* **2018**, *864*, L27. [CrossRef]
336. Begelman, M.C.; McKee, C.F.; Shields, G.A. Compton heated winds and coronae above accretion disks. I. Dynamics. *Astrophys. J.* **1983**, *271*, 70–88. [CrossRef]
337. Gianolli, V.E.; Bianchi, S.; Petrucci, P.O.; Brusa, M.; Chartas, G.; Lanzuisi, G.; Matzeu, G.A.; Parra, M.; Ursini, F.; Behar, E.; et al. Supermassive Black Hole Winds in X-rays: SUBWAYS. III. A population study on ultra-fast outflows. *Astron. Astrophys.* **2024**, *687*, A235. [CrossRef]
338. Tully, R.B.; Fisher, J.R. A new method of determining distances to galaxies. *Astron. Astrophys.* **1977**, *54*, 661–673.
339. Marziani, P.; Dultzin-Hacyan, D.; Sulentic, J.W. Accretion onto Supermassive Black Holes in Quasars: Learning from Optical/UV Observations. In *New Developments in Black Hole Research*; Kreitler, P.V., Ed.; Nova Press: New York, NY, USA, 2006; p. 123.
340. Marziani, P.; Sulentic, J.W. Estimating black hole masses in quasars using broad optical and UV emission lines. *New Astron. Rev.* **2012**, *56*, 49–63. [CrossRef]
341. Runnoe, J.C.; Shang, Z.; Brotherton, M.S. The orientation dependence of quasar spectral energy distributions. *Mon. Not. R. Astron. Soc.* **2013**, *435*, 3251–3261. [CrossRef]
342. Brotherton, M.S.; Runnoe, J.C.; Shang, Z.; DiPompeo, M.A. Bias in C IV-based quasar black hole mass scaling relationships from reverberation mapped samples. *Mon. Not. R. Astron. Soc.* **2015**, *451*, 1290–1298. [CrossRef]
343. Savić, D.; Goosmann, R.; Popović, L.Č.; Marin, F.; Afanasiev, V.L. AGN black hole mass estimates using polarization in broad emission lines. *Astron. Astrophys.* **2018**, *614*, A120. [CrossRef]
344. Afanasiev, V.L.; Popović, L.Č.; Shapovalova, A.I. Spectropolarimetry of Seyfert 1 galaxies with equatorial scattering: black hole masses and broad-line region characteristics. *Mon. Not. R. Astron. Soc.* **2019**, *482*, 4985–4999. [CrossRef]
345. Savić, Đ.; Popović, L.Č.; Shablovinskaya, E.; Afanasiev, V.L. Estimating supermassive black hole masses in active galactic nuclei using polarization of broad Mg II, H α, and H β lines. *Mon. Not. R. Astron. Soc.* **2020**, *497*, 3047–3054. [CrossRef]
346. Marziani, P.; del Olmo, A.; Martínez-Carballo, M.A.; Martínez-Aldama, M.L.; Stirpe, G.M.; Negrete, C.A.; Dultzin, D.; D'Onofrio, M.; Bon, E.; Bon, N. Black hole mass estimates in quasars. A comparative analysis of high- and low-ionization lines. *Astron. Astrophys.* **2019**, *627*, A88. [CrossRef]
347. D'Onofrio, M.; Marziani, P.; Chiosi, C.; Negrete, C.A. The Correlation Luminosity-Velocity Dispersion of Galaxies and Active Galactic Nuclei. *Universe* **2024**, *10*, 254. [CrossRef]
348. Wolf, C.; Lai, S.; Onken, C.A.; Amrutha, N.; Bian, F.; Hon, W.J.; Tisserand, P.; Webster, R.L. The accretion of a solar mass per day by a 17-billion solar mass black hole. *Nat. Astron.* **2024**, *8*, 520–529. [CrossRef]
349. Bosman, S.E.I.; Álvarez-Márquez, J.; Colina, L.; Walter, F.; Alonso-Herrero, A.; Ward, M.J.; Ã-stlin, G.; Greve, T.R.; Wright, G.; Bik, A.; et al. A mature quasar at cosmic dawn revealed by JWST rest-frame infrared spectroscopy. *Nat. Astron.* **2024**, *8*, 1054–1065. [CrossRef]
350. Bogdán, Á.; Goulding, A.D.; Natarajan, P.; Kovács, O.E.; Tremblay, G.R.; Chadayammuri, U.; Volonteri, M.; Kraft, R.P.; Forman, W.R.; Jones, C.; et al. Evidence for heavy-seed origin of early supermassive black holes from a z $\approx$ 10 X-ray quasar. *Nat. Astron.* **2024**, *8*, 126–133. [CrossRef]
351. Melia, F. The cosmic timeline implied by the JWST reionization crisis. *Astron. Astrophys.* **2024**, *689*, A10. [CrossRef]
352. Melia, F. Strong observational support for the Rh=ct timeline in the early universe. *Phys. Dark Universe* **2024**, *46*, 101587. [CrossRef]
353. Huško, F.; Lacey, C.G.; Roper, W.J.; Schaye, J.; Briggs, J.M.; Schaller, M. The effects of super-Eddington accretion and feedback on the growth of early supermassive black holes and galaxies. *arXiv* **2024**, arXiv:2410.09450. [CrossRef]
354. Jiang, Y.F.; Stone, J.M.; Davis, S.W. A Global Three-dimensional Radiation Magneto-hydrodynamic Simulation of Super-Eddington Accretion Disks. *Astrophys. J.* **2014**, *796*, 106. [CrossRef]

355. Sądowski, A.; Narayan, R.; Tchekhovskoy, A.; Abarca, D.; Zhu, Y.; McKinney, J.C. Global simulations of axisymmetric radiative black hole accretion discs in general relativity with a mean-field magnetic dynamo. *Mon. Not. R. Astron. Soc.* **2015**, *447*, 49–71. [CrossRef]
356. Huang, Y.K.; Hu, C.; Zhao, Y.L.; Zhang, Z.X.; Lu, K.X.; Wang, K.; Zhang, Y.; Du, P.; Li, Y.R.; Bai, J.M.; et al. Reverberation Mapping of the Narrow-line Seyfert 1 Galaxy I Zwicky 1: Black Hole Mass. *Astrophys. J.* **2019**, *876*, 102. [CrossRef]
357. Marinello, M.; Rodríguez-Ardila, A.; Marziani, P.; Sigut, A.; Pradhan, A. Panchromatic Properties of the Extreme Fe ii Emitter PHL 1092. *Mon. Not. R. Astron. Soc.* **2020**, *494*, 4187–4202. [CrossRef]
358. Li, Y.R.; Hu, C.; Yao, Z.H.; Chen, Y.J.; Bai, H.R.; Yang, S.; Du, P.; Fang, F.N.; Fu, Y.X.; Liu, J.R.; et al. Spectroastrometry and Reverberation Mapping of Active Galactic Nuclei. I. The Hβ Broad-line Region Structure and Black Hole Masses of Five Quasars. *Astrophys. J.* **2024**, *974*, 86. [CrossRef]
359. Martínez-Aldama, M.L.; Del Olmo, A.; Marziani, P.; Negrete, C.A.; Dultzin, D.; Martínez-Carballo, M.A. HE0359-3959: An Extremely Radiating Quasar. *Front. Astron. Space Sci.* **2017**, *4*, 29. [CrossRef]
360. Mushano, T.; Ogawa, T.; Ohsuga, K.; Yajima, H.; Omukai, K. Impact of Lyα radiation force on super-Eddington accretion onto a massive black hole. *arXiv* **2024**, arXiv:2410.04378. [CrossRef]
361. Milosavljević, M.; Couch, S.M.; Bromm, V. Accretion Onto Intermediate-Mass Black Holes in Dense Protogalactic Clouds. *Astrophys. J. Lett.* **2009**, *696*, L146–L149. [CrossRef]
362. D'Onofrio, M.; Burigana, C. *Questions of Modern Cosmology: Galileo's Legacy*; Springer: Berlin/Heidelberg, Germany, 2009. [CrossRef]
363. Zamfir, S.; Sulentic, J.W.; Marziani, P.; Dultzin, D. Detailed characterization of Hβ emission line profile in low-z SDSS quasars. *Mon. Not. R. Astron. Soc.* **2010**, *403*, 1759. [CrossRef]
364. Trakhtenbrot, B.; Netzer, H. Black hole growth to z = 2 - I. Improved virial methods for measuring M_{BH} and L/L_{Edd}. *Mon. Not. R. Astron. Soc.* **2012**, *427*, 3081–3102. [CrossRef]
365. Mapelli, M. Binary black hole mergers: Formation and populations. *Front. Astron. Space Sci.* **2020**, *7*, 38. [CrossRef]
366. Greene, J.E.; Strader, J.; Ho, L.C. Intermediate-Mass Black Holes. *Annu. Rev. Astron. Astrophys.* **2020**, *58*, 257–312. [CrossRef]
367. Davis, B.L.; Graham, A.W.; Soria, R.; Jin, Z.; Karachentsev, I.D.; Karachentseva, V.E.; D'Onghia, E. Identification of Intermediate-mass Black Hole Candidates among a Sample of Sd Galaxies. *Astrophys. J.* **2024**, *971*, 123. [CrossRef]
368. Ebisuzaki, T.; Makino, J.; Tsuru, T.G.; Funato, Y.; Portegies Zwart, S.; Hut, P.; McMillan, S.; Matsushita, S.; Matsumoto, H.; Kawabe, R. Missing Link Found? The "Runaway" Path to Supermassive Black Holes. *Astrophys. J. Lett.* **2001**, *562*, L19–L22. [CrossRef]
369. Mapelli, M. Massive black hole binaries from runaway collisions: The impact of metallicity. *Mon. Not. R. Astron. Soc.* **2016**, *459*, 3432–3446. [CrossRef]
370. Portegies Zwart, S.F.; McMillan, S.L.W. The Runaway Growth of Intermediate-Mass Black Holes in Dense Star Clusters. *Astrophys. J.* **2002**, *576*, 899–907. [CrossRef]
371. Portegies Zwart, S.F.; Baumgardt, H.; Hut, P.; Makino, J.; McMillan, S.L.W. Formation of massive black holes through runaway collisions in dense young star clusters. *Nature* **2004**, *428*, 724–726. [CrossRef]
372. Reines, A.E.; Greene, J.E.; Geha, M. Dwarf Galaxies with Optical Signatures of Active Massive Black Holes. *Astrophys. J.* **2013**, *775*, 116. [CrossRef]
373. Chilingarian, I.V.; Katkov, I.Y.; Zolotukhin, I.Y.; Grishin, K.A.; Beletsky, Y.; Boutsia, K.; Osip, D.J. A Population of Bona Fide Intermediate-mass Black Holes Identified as Low-luminosity Active Galactic Nuclei. *Astrophys. J.* **2018**, *863*, 1. [CrossRef]
374. Yue, M.; Eilers, A.C.; Simcoe, R.A.; Mackenzie, R.; Matthee, J.; Kashino, D.; Bordoloi, R.; Lilly, S.J.; Naidu, R.P. EIGER. V. Characterizing the Host Galaxies of Luminous Quasars at z $\gtrsim$ 6. *Astrophys. J.* **2024**, *966*, 176. [CrossRef]
375. Stone, M.A.; Lyu, J.; Rieke, G.H.; Alberts, S.; Hainline, K.N. Undermassive Host Galaxies of Five z $\sim$ 6 Luminous Quasars Detected with JWST. *Astrophys. J.* **2024**, *964*, 90. [CrossRef]
376. Kokorev, V.; Fujimoto, S.; Labbe, I.; Greene, J.E.; Bezanson, R.; Dayal, P.; Nelson, E.J.; Atek, H.; Brammer, G.; Caputi, K.I.; et al. UNCOVER: A NIRSpec Identification of a Broad-line AGN at z = 8.50. *Astrophys. J. Lett.* **2023**, *957*, L7. [CrossRef]
377. Mazzucchelli, C.; Bischetti, M.; D'Odorico, V.; Feruglio, C.; Schindler, J.T.; Onoue, M.; Bañados, E.; Becker, G.D.; Bian, F.; Carniani, S.; et al. XQR-30: Black hole masses and accretion rates of 42 z $\gtrsim$ 6 quasars. *Astron. Astrophys.* **2023**, *676*, A71. [CrossRef]
378. Übler, H.; Maiolino, R.; Curtis-Lake, E.; Pérez-González, P.G.; Curti, M.; Perna, M.; Arribas, S.; Charlot, S.; Marshall, M.A.; D'Eugenio, F.; et al. GA-NIFS: A massive black hole in a low-metallicity AGN at z $\sim$ 5.55 revealed by JWST/NIRSpec IFS. *Astron. Astrophys.* **2023**, *677*, A145. [CrossRef]
379. Narayan, R.; Igumenshchev, I.V.; Abramowicz, M.A. Magnetically Arrested Disk: An Energetically Efficient Accretion Flow. *Publ. Astron. Soc. Jpn.* **2003**, *55*, L69–L72. [CrossRef]
380. Blandford, R.D.; Znajek, R.L. Electromagnetic extraction of energy from Kerr black holes. *Mon. Not. R. Astron. Soc.* **1977**, *179*, 433–456. [CrossRef]
381. McKinney, J.C.; Dai, L.; Avara, M.J. Efficiency of super-Eddington magnetically-arrested accretion. *Mon. Not. R. Astron. Soc.* **2015**, *454*, L6–L10. [CrossRef]

382. Ricarte, A.; Narayan, R.; Curd, B. Recipes for Jet Feedback and Spin Evolution of Black Holes with Strongly Magnetized Super-Eddington Accretion Disks. *Astrophys. J. Lett.* **2023**, *954*, L22. [CrossRef]
383. Lowell, B.; Jacquemin-Ide, J.; Tchekhovskoy, A.; Duncan, A. Rapid Black Hole Spin-down by Thick Magnetically Arrested Disks. *Astrophys. J.* **2024**, *960*, 82. [CrossRef]
384. Sądowski, A.; Narayan, R. Powerful radiative jets in supercritical accretion discs around non-spinning black holes. *Mon. Not. R. Astron. Soc.* **2015**, *453*, 3213–3221. [CrossRef]
385. Yang, H.; Yuan, F.; Kwan, T.; Dai, L. The properties of wind and jet from a super-Eddington accretion flow around a supermassive black hole. *Mon. Not. R. Astron. Soc.* **2023**, *523*, 208–220. [CrossRef]
386. Czerny, B.; Sniegowska, M.; Janiuk, A.; You, B. Accretion processes onto black holes: Theoretical problems, observational constraints. *arXiv* **2023**, arXiv:2312.02911. [CrossRef]
387. Paliya, V.S.; Stalin, C.S.; Domínguez, A.; Saikia, D.J. Narrow-line Seyfert 1 galaxies in Sloan Digital Sky Survey: A new optical spectroscopic catalogue. *Mon. Not. R. Astron. Soc.* **2024**, *527*, 7055–7069. [CrossRef]
388. Foschini, L.; Dalla Barba, B.; Tornikoski, M.; Andernach, H.; Marziani, P.; Marscher, A.P.; Jorstad, S.G.; Järvelä, E.; Antón, S.; Dalla Bontà, E. The Power of Relativistic Jets: A Comparative Study. *Universe* **2024**, *10*, 156. [CrossRef]
389. Lewin, W.H.G.; van Paradijs, J.; van den Heuvel, E.P.J. *X-Ray Binaries*; Cambridge University Press: Cambridge, UK, 1997.
390. Remillard, R.A.; McClintock, J.E. X-ray properties of black-hole binaries. *Annu. Rev. Astron. Astrophys.* **2006**, *44*, 49–92. [CrossRef]
391. Fender, R.P.; Belloni, T.M.; Gallo, E. Towards a unified model for black hole X-ray binary jets. *Mon. Not. R. Astron. Soc.* **2004**, *355*, 1105–1118. [CrossRef]
392. Homan, J.; Belloni, T. The Evolution of Black Hole States. *Astrophys. Space Sci.* **2005**, *300*, 107–117. [CrossRef]
393. Belloni, T.M. States and Transitions in Black Hole Binaries. In *The Jet Paradigm*; Belloni, T., Ed.; Lecture Notes in Physics; Springer: Berlin/Heidelberg, Germany, 2010; Volume 794, p. 53. [CrossRef]
394. Koljonen, K.I.I.; Maccarone, T.; McCollough, M.L.; Gurwell, M.; Trushkin, S.A.; Pooley, G.G.; Piano, G.; Tavani, M. The hypersoft state of Cygnus X-3. A key to jet quenching in X-ray binaries? *Astron. Astrophys.* **2018**, *612*, A27. [CrossRef]
395. Done, C.; Gierliński, M.; Kubota, A. Modelling the behaviour of accretion flows in X-ray binaries. Everything you always wanted to know about accretion but were afraid to ask. *Astron. Astroph. Rev.* **2007**, *15*, 1–66. [CrossRef]
396. Gladstone, J.C.; Roberts, T.P.; Done, C. The ultraluminous state. *Mon. Not. R. Astron. Soc.* **2009**, *397*, 1836–1851. [CrossRef]
397. Middleton, M.J.; Miller-Jones, J.C.A.; Markoff, S.; Fender, R.; Henze, M.; Hurley-Walker, N.; Scaife, A.M.M.; Roberts, T.P.; Walton, D.; Carpenter, J.; et al. Bright radio emission from an ultraluminous stellar-mass microquasar in M 31. *Nature* **2013**, *493*, 187–190. [CrossRef] [PubMed]
398. Bachetti, M.; Harrison, F.A.; Walton, D.J.; Grefenstette, B.W.; Chakrabarty, D.; Fürst, F.; Barret, D.; Beloborodov, A.; Boggs, S.E.; Christensen, F.E.; et al. An ultraluminous X-ray source powered by an accreting neutron star. *Nature* **2014**, *514*, 202–204. [CrossRef]
399. Kaaret, P.; Feng, H.; Roberts, T.P. Ultraluminous X-Ray Sources. *Annu. Rev. Astron. Astrophys.* **2017**, *55*, 303–341. [CrossRef]
400. MacKenzie, A.D.A.; Roberts, T.P.; Walton, D.J. The hyperluminous X-ray source population. *Astron. Nachrichten* **2023**, *344*, e20230028. [CrossRef]
401. Makishima, K.; Maejima, Y.; Mitsuda, K.; Bradt, H.V.; Remillard, R.A.; Tuohy, I.R.; Hoshi, R.; Nakagawa, M. Simultaneous X-Ray and Optical Observations of GX 339-4 in an X-Ray High State. *Astrophys. J.* **1986**, *308*, 635. [CrossRef]
402. Dunn, R.J.H.; Fender, R.P.; Körding, E.G.; Belloni, T.; Cabanac, C. A global spectral study of black hole X-ray binaries. *Mon. Not. R. Astron. Soc.* **2010**, *403*, 61–82. [CrossRef]
403. Corbel, S.; Fender, R.P.; Tzioumis, A.K.; Nowak, M.; McIntyre, V.; Durouchoux, P.; Sood, R. Coupling of the X-ray and radio emission in the black hole candidate and compact jet source GX 339-4. *Astron. Astrophys.* **2000**, *359*, 251–268. [CrossRef]
404. Homan, J.; Buxton, M.; Markoff, S.; Bailyn, C.D.; Nespoli, E.; Belloni, T. Multiwavelength Observations of the 2002 Outburst of GX 339-4: Two Patterns of X-Ray-Optical/Near-Infrared Behavior. *Astrophys. J.* **2005**, *624*, 295–306. [CrossRef]
405. Fender, R.P.; Homan, J.; Belloni, T.M. Jets from black hole X-ray binaries: Testing, refining and extending empirical models for the coupling to X-rays. *Mon. Not. R. Astron. Soc.* **2009**, *396*, 1370–1382. [CrossRef]
406. Belloni, T.; Homan, J.; Casella, P.; van der Klis, M.; Nespoli, E.; Lewin, W.H.G.; Miller, J.M.; Méndez, M. The evolution of the timing properties of the black-hole transient GX 339-4 during its 2002/2003 outburst. *Astron. Astrophys.* **2005**, *440*, 207–222. [CrossRef]
407. Motta, S.; Muñoz-Darias, T.; Casella, P.; Belloni, T.; Homan, J. Low-frequency oscillations in black holes: A spectral-timing approach to the case of GX 339-4. *Mon. Not. R. Astron. Soc.* **2011**, *418*, 2292–2307. [CrossRef]
408. Dihingia, I.K.; Mizuno, Y.; Sharma, P. High-soft to Low-hard State Transition in Black Hole X-Ray Binaries with GRMHD Simulations. *Astrophys. J.* **2023**, *958*, 105. [CrossRef]
409. Corbel, S.; Fender, R.P. Near-Infrared Synchrotron Emission from the Compact Jet of GX 339-4. *Astrophys. J. Lett.* **2002**, *573*, L35–L39. [CrossRef]

410. Corbel, S.; Nowak, M.A.; Fender, R.P.; Tzioumis, A.K.; Markoff, S. Radio/X-ray correlation in the low/hard state of GX 339-4. *Astron. Astrophys.* **2003**, *400*, 1007–1012. [CrossRef]
411. Gallo, E.; Fender, R.P.; Pooley, G.G. A universal radio-X-ray correlation in low/hard state black hole binaries. *Mon. Not. R. Astron. Soc.* **2003**, *344*, 60–72. [CrossRef]
412. Esin, A.A.; McClintock, J.E.; Narayan, R. Models of black hole X-ray binaries: Spectral energy distributions of LHSs and VHSs. *Astrophys. J.* **1997**, *489*, 865–889. [CrossRef]
413. Maccarone, T.J. On the evolution of the compact jet and the nature of X-ray states in black hole binaries. *Astron. Astrophys.* **2003**, *409*, 697–706. [CrossRef]
414. Done, C.; Gierliński, M.; Kubota, A. Modelling the behaviour of accretion flows in X-ray binaries: A review. *Astron. Astrophys. Rev.* **2007**, *15*, 1–66. [CrossRef]
415. Siemiginowska, A.; Czerny, B.; Kostyunin, V. Evolution of an Accretion Disk in an Active Galactic Nucleus. *Astrophys. J.* **1996**, *458*, 491. [CrossRef]
416. Zamanov, R.; Marziani, P. Searching for the Physical Drivers of Eigenvector 1: From Quasars to Nanoquasars. *Astrophys. J.* **2002**, *571*, L77–L80. [CrossRef]
417. Bon, N.; Bon, E.; Marziani, P. AGN Broad Line Region variability in the context of Eigenvector 1: case of NGC 5548. *Front. Astron. Space Sci.* **2018**, *5*, 3. [CrossRef]
418. Panda, S.; Śniegowska, M. Changing-look Active Galactic Nuclei. I. Tracking the Transition on the Main Sequence of Quasars. *Astrophys. J. Supp.* **2024**, *272*, 13. [CrossRef]
419. Komossa, S.; Greiner, J. Discovery of a giant and luminous X-ray outburst from the optically inactive galaxy pair RX J1242.6-1119. *Astron. Astrophys.* **1999**, *349*, L45–L48. [CrossRef]
420. Sniegowska, M.; Czerny, B.; Bon, E.; Bon, N. Possible mechanism for multiple changing-look phenomena in active galactic nuclei. *Astron. Astrophys.* **2020**, *641*, A167. [CrossRef]
421. Dai, L.; McKinney, J.C.; Roth, N.; Ramirez-Ruiz, E.; Miller, M.C. A Unified Model for Tidal Disruption Events. *Astrophys. J. Lett.* **2018**, *859*, L20. [CrossRef]
422. Petrushevska, T.; Leloudas, G.; Ilić, D.; Bronikowski, M.; Charalampopoulos, P.; Jaisawal, G.K.; Paraskeva, E.; Pursiainen, M.; Rakić, N.; Schulze, S.; et al. The rise and fall of the iron-strong nuclear transient PS16dtm. *Astron. Astrophys.* **2023**, *669*, A140. [CrossRef]
423. Ilić, D.; Oknyansky, V.; Popović, L.Č.; Tsygankov, S.S.; Belinski, A.A.; Tatarnikov, A.M.; Dodin, A.V.; Shatsky, N.I.; Ikonnikova, N.P.; Rakić, N.; et al. A flare in the optical spotted in the changing-look Seyfert NGC 3516. *Astron. Astrophys.* **2020**, *638*, A13. [CrossRef]
424. Oknyansky, V.L.; Tsygankov, S.S.; Dodin, A.V.; Tatarnikov, A.M.; Belinski, A.A.; Ikonnikova, N.P.; Burlak, M.A.; Fedoteva, A.A.; Shatsky, N.I.; Mishin, E.O.; et al. Changing Look AGN NGC 3516 brightens again after long being in a very low state. *Astron. Telegr.* **2020**, *13691*, 1.
425. Oknyansky, V.L.; Tsygankov, S.S.; Lipunov, V.M.; Gorbovskoy, E.S.; Tyurina, N.V. Discovery of new changing look in NGC 1566. In Proceedings of the Nuclear Activity in Galaxies Across Cosmic Time, Addis Ababa, Ethiopia, 7–11 October 2019; Pović, M., Marziani, P., Masegosa, J., Netzer, H., Negu, S.H., Tessema, S.B., Eds.; IAU Symposium; Cambridge University Press: Cambridge, UK, 2021; Volume 356, pp. 127–131. [CrossRef]
426. Oknyansky, V. Changing looks of the nucleus of the Seyfert galaxy NGC 1566 compared with other changing-look AGNs. *Astron. Nachrichten* **2022**, *343*, e210080. [CrossRef]
427. Xu, D.W.; Komossa, S.; Grupe, D.; Wang, J.; Xin, L.P.; Han, X.H.; Wei, J.Y.; Bai, J.Y.; Bon, E.; Cangemi, F.; et al. Changing-Look Narrow-Line Seyfert 1 Galaxies, their Detection with SVOM, and the Case of NGC 1566. *Universe* **2024**, *10*, 61. [CrossRef]
428. Ulrich, M.H.; Maraschi, L.; Urry, C.M. Variability of Active Galactic Nuclei. *Annu. Rev. Astron. Astrophys.* **1997**, *35*, 445–502. [CrossRef]
429. Drake, A.J.; Djorgovski, S.G.; Mahabal, A.; Beshore, E.; Larson, S.; Graham, M.J.; Williams, R.; Christensen, E.; Catelan, M.; Boattini, A.; et al. First Results from the Catalina Real-Time Transient Survey. *Astrophys. J.* **2009**, *696*, 870–884. [CrossRef]
430. Graham, M.J.; Kulkarni, S.R.; Bellm, E.C.; Adams, S.M.; Barbarino, C.; Blagorodnova, N.; Bodewits, D.; Bolin, B.; Brady, P.R.; Cenko, S.B.; et al. The Zwicky Transient Facility: Science Objectives. *Publ. Astron. Soc. Pac.* **2019**, *131*, 078001. [CrossRef]
431. Masci, F.J.; Laher, R.R.; Rusholme, B.; Shupe, D.L.; Groom, S.; Surace, J.; Jackson, E.; Monkewitz, S.; Beck, R.; Flynn, D.; et al. The Zwicky Transient Facility: Data Processing, Products, and Archive. *Publ. Astron. Soc. Pac.* **2019**, *131*, 018003. [CrossRef]
432. Bellm, E.C.; Kulkarni, S.R.; Graham, M.J.; Dekany, R.; Smith, R.M.; Riddle, R.; Masci, F.J.; Helou, G.; Prince, T.A.; Adams, S.M.; et al. The Zwicky Transient Facility: System Overview, Performance, and First Results. *Publ. Astron. Soc. Pac.* **2019**, *131*, 018002. [CrossRef]
433. Chambers, K.C.; Magnier, E.A.; Metcalfe, N.; Flewelling, H.A.; Huber, M.E.; Waters, C.Z.; Denneau, L.; Draper, P.W.; Farrow, D.; Finkbeiner, D.P.; et al. The Pan-STARRS1 Surveys. *arXiv* **2016**, arXiv:1612.05560. [CrossRef]

434. Kochanek, C.S.; Shappee, B.J.; Stanek, K.Z.; Holoien, T.W.S.; Thompson, T.A.; Prieto, J.L.; Dong, S.; Shields, J.V.; Will, D.; Britt, C.; et al. The All-Sky Automated Survey for Supernovae (ASAS-SN) Light Curve Server v1.0. *Publ. Astron. Soc. Pac.* **2017**, *129*, 104502. [CrossRef]
435. Ivezić, Ž.; Kahn, S.M.; Tyson, J.A.; Abel, B.; Acosta, E.; Allsman, R.; Alonso, D.; AlSayyad, Y.; Anderson, S.F.; Andrew, J.; et al. LSST: From Science Drivers to Reference Design and Anticipated Data Products. *Astrophys. J.* **2019**, *873*, 111. [CrossRef]

universe

MDPI

Article

On the Nature of the Radio Calibrator and Gamma-Ray Emitting NLS1 Galaxy 3C 286 and Its Multiwavelength Variability

S. Komossa [1,*], S. Yao [2,1], D. Grupe [3] and A. Kraus [1]

1 Max Planck Institut für Radioastronomie, Auf dem Hügel 69, 53121 Bonn, Germany; syao@mpifr.de (S.Y.)
2 National Astronomical Observatories, Chinese Academy of Sciences, Beijing 100101, China
3 Department of Physics, Geology, and Engineering Technology, Northern Kentucky University, 1 Nunn Drive, Highland Heights, KY 41099, USA
* Correspondence: skomossa@mpifr.de

Abstract: The quasar 3C 286, a well-known calibrator source in radio astronomy, was found to exhibit exceptional multiwavelength properties. Its rich and complex optical emission-line spectrum revealed its narrow-line Seyfert 1 (NLS1) nature. Given its strong radio emission, this makes 3C 286 one of the radio-loudest NLS1 galaxies known to date. 3C 286 is also one of very few known compact steep-spectrum (CSS) sources detected in the gamma-ray regime. Observations in the X-ray regime, rarely carried out so far, revealed evidence for variability, raising the question whether it is driven by the accretion disk or jet. 3C 286 is also well known for its damped Lyman alpha system from an intervening absorber at $z = 0.692$, triggering a search for the corresponding X-ray absorption along the line-of-sight. Here, we present new observations in the radio, X-ray, optical, and UV bands. The nature of the X-ray variability is addressed. Spectral evidence suggests that it is primarily driven by the accretion disk (not the jet), and the X-ray spectrum is well fit by a powerlaw plus soft excess model. The radio flux density and polarization remain constant at the Effelsberg telescope resolution, reconfirming the use of 3C 286 as radio calibrator. The amount of reddening/absorption along the line-of-sight *intrinsic* to 3C 286 is rigorously assessed. None is found, validating the derivation of a high Eddington ratio ($L/L_{\rm Edd} \sim 1$) and of the very high radio-loudness index of 3C 286. Based on the first deep Chandra image of 3C 286, tentative evidence for hard X-ray emission from the SW radio lobe is reported. A large variety of models for the gamma-ray emission of 3C 286 are briefly discussed.

Keywords: active galactic nuclei; 3C 286; quasars; narrow-line Seyfert 1 galaxies; compact steep spectrum sources; supermassive black holes; accretion disks; jets; gamma-ray emission; X-ray spectra

1. Introduction: 3C 286

The quasar 3C 286 serves as one of the most important calibrator sources in the centimeter radio bands. Its remarkable and puzzling multiwavelength (MWL) characteristics raise fundamental questions about the interplay between the jet and accretion disk, as well as the enigmatic nature of (variable) gamma-ray emission in steep-spectrum radio sources with jets not pointing at the observer.

After a summary of the intriguing MWL characteristics of 3C 286, combining the properties of compact steep-spectrum source (CSS), very radio-loud Narrow-line Seyfert 1 (NLS1) galaxy, gamma-ray detection despite a misaligned jet, and presence of a damped Lyman alpha absorber (DLA) despite an apparent lack of X-ray absorption, rarely seen together in a single system, we present new observations across the electromagnetic spectrum from the radio to high-energy regime and discuss several key open questions: how much variability in each spectral band, how much reddening and X-ray absorption, and which mechanism drives the gamma-ray emission?

We use a cosmology with $H_0 = 70\ \mathrm{km\,s^{-1}\,Mpc^{-1}}$, $\Omega_M = 0.3$ and $\Omega_\Lambda = 0.7$, and the cosmology calculator of Wright [1]. This corresponds to a scale of 7.66 kpc/$''$ at the redshift of 3C 286 (z = 0.849).

1.1. Radio Properties and Radio Calibrator

3C 286 is part of the 3rd Cambridge Catalogue of radio sources from 1959 [2]. After an improved radio position was taken with the Owens Valley Radio Observatory in the early 1960s, it was possible to identify its optical counterpart initially thought to be a star [3].

Later, 3C 286 was one of the first quasars imaged with the technique of Very Long Baseline Interferometry (VLBI). Radio images at sub-arcsecond resolution revealed a core region with a jet and counter jet, dominated by the central component [4–6]. On large scales, two lobes extend 0.7$''$ to the East and 2.6$''$ to the South-West [6–8]. At tens-of-parsec-scale resolution, the central region is resolved into two bright components of comparable flux density (at 15 GHz), interpreted as two hot spots or a core–jet structure [9–13]. An et al. [12] favored the core–jet interpretation, where the NE component (C1) with an inverted spectrum below $\sim$8 GHz represents the core, and the SW component (C2) represents the brightest knot in the jet. Based on that interpretation, our viewing angle towards the inner jet of 3C 286 on tens-of-parsecs scale was estimated at 48°. The inner jet is only mildly relativistic with $v \sim 0.5c$ [12].

The (integrated) radio spectrum of 3C 286 is steep, between 1.4 and 50 GHz ($\alpha = -0.6$; $S_\nu \propto \nu^\alpha$). Given its compactness, 3C 286 was therefore identified as CSS [14].

3C 286 is one of the strongest extragalactic radio sources in polarized emission [15,16] ($\sim$1 Jy at 5 GHz). Its fractional polarization, about 10% at 20 cm, increases with increasing frequency up to 17% at 1.3 mm observed with ALMA [17].

Its very constant (integrated) radio properties established 3C 286 as a radio calibrator in both total flux and linear polarization, [18–20].

1.2. Optical Spectroscopy and NLS1 Classification

The optical spectrum of 3C 286 exhibits bright permitted and forbidden emission lines at redshift $z = 0.849$ ([21–23], our Table 1). The first optical spectra were taken by Maarten Schmidt in 1962, when 3C 286 was still considered a 'radio star' and its extragalactic nature was not yet known [24]. Three years later, the first redshift measurement of 3C 286 was published [25], implying its cosmological distance.

Table 1. Properties of 3C 286. Column entries are as follows: (1) source coordinates in right ascension (RA) and declination (Decl), (2) redshift, (3) SDSS g magnitude, (4) Hβ line width, (5) SMBH mass, (6) Eddington ratio, and (7) radio loudness at 5 GHz.

J2000 Coordinates RA, Decl (1)	z (2)	g Magnitude (3)	FWHM (H$\beta_{\rm broad}$) km/s (4)	SMBH Mass $M_\odot$ (5)	$L/L_{\rm Edd}$ (6)	Radio Loudness $R_{\rm 5\,GHz}$ (7)
13 h 31 m 08.2879 s, +30°30′32.958″	0.849	17.33	2001	$1.3\ 10^8$	1.0	$10^{4.4}$

The NLS1 classification of 3C 286 was first suggested in 2016 based on new SDSS [26] spectroscopy, which included the Hβ–[OIII] regime for the first time at high signal-to-noise ratio [27,28]. A detailed analysis of the optical spectrum was presented by Yao and Komossa [29], firmly establishing the NLS1 classification[1]. NLS1 galaxies are defined as AGN with narrow widths of their Balmer lines from the broad-line region (BLR) with FWHM(H$\beta_{\rm broad}$) $<$ 2000 km/s, [OIII]5007/H$\beta_{\rm totl}$ $<$ 3, and FeII/H$\beta_{\rm totl}$ $>$ 0.5 [30–32]. They show a wealth of exceptional MWL characteristics at one extreme end of AGN correlation space [33–36] (for recent reviews, see NLS1 properties in the MWL bands [37], the X-rays [38], the radio [39], the gamma-rays [40,41], and the MWL properties of the radio-loud sources [42]). Initially mostly studied in the optical [30] and X-ray [43] regimes, the radio properties of NLS1s revealed that only 7% of them are radio-loud [44], with a mix of CSS and blazar-like sources [44,45]. A subset of them are detected in the gamma-ray regime [46], some highly variable, re-confirming the presence of relativistic jets. Perhaps the most remarkable recent finding was the detection of high-amplitude radio flaring of several previously radio-silent NLS1 galaxies [47–49].

3C 286 fulfills all of the optical spectroscopic NLS1 galaxy classification criteria: an FWHM($H\beta_{\rm broad}$) = 2000 km/s, a line ratio [OIII]5007/$H\beta_{\rm totl}$ = 1.1, and a line ratio FeII 4570/$H\beta_{\rm totl}$ = 0.76, where FeII 4570 is calculated by integrating the FeII complex between 4434 and 4684 Å[2]. In the quasar main-sequence diagram [FWHM($H\beta$) vs. FeII/$H\beta$]; [50,51], 3C 286 is located well within the region of population A sources (category A2).

The high-ionization lines in the optical spectrum of 3C 286, [OIII]5007, [NeIII]3870, and [NeV]3426, exhibit strong blueshifted components (blue wings), which are highly broadened up to 2398 km/s in the case of [NeIII] [29], indicating strong outflows and/or jet–cloud interactions. Based on its FWHM($H\beta_{\rm broad}$), 3C 286 hosts an SMBH with mass 1.3×10^8 $M_\odot$ and accretes at an Eddington ratio $L/L_{\rm Edd} = 1$ [29] consistent with the value of $L/L_{\rm Edd} \sim 0.9$ obtained from SED modeling [52].

The spectral energy distribution (SED) of 3C 286 rises steeply in the optical [29,52], interpreted as emission from the accretion disk dominating over the non-thermal emission from the jet at optical–UV wavelengths.

3C 286 is very radio-loud [29]. Given its Kellermann [53] radio-loudness index of $R_{5\,\rm GHz} = f_{5\,\rm GHz}/f_{4400\rm A} = 10^{4.4}$, 3C 286 is among the radio-loudest NLS1 galaxies known to date (for comparison, see Figure 8 of the largest SDSS NLS1 sample of [54]).

1.3. UV Properties and DLA System

The UV spectrum of 3C 286 shows a damped Lyman alpha absorber (DLA) system at redshift $z = 0.692$ [55]. A foreground low-surface brightness galaxy at $\sim$2.5$''$ distance from 3C 286 [56,57] is thought to be responsible for the UV absorption as well as for the HI 21 cm absorption detected at that redshift [58]. Based on the UV spectrum, a column density of the intervening absorber of $N_{\rm H} \sim 2 \times 10^{21}$ cm^{-2} was estimated [55].

1.4. Fermi Gamma-Ray Detection of 3C 286 despite Its Misaligned Jet

3C 286 is associated with the Fermi-LAT gamma-ray source 4FGL J1331.0+3032 [59,60]. Based on the first 11 years of Fermi observations, Zhang et al. [52] reported an average gamma-ray flux in the energy range 100 MeV–300 GeV of $f_\gamma = 2.32 \times 10^{-12}$ erg cm^{-2} s^{-1}. The significance of the gamma-ray signal based on the maximum likelihood Test Statistics (TS) is TS = 54.2, and there is evidence for gamma-ray variability at $\rm TS_{var}$ = 64.3, where $\rm TS_{var}$ is the variability index [52]. The evidence for variability depends on the bin size of the light curve, and lower values of $\rm TS_{var}$, consistent with constant emission, have also been reported in the literature [60,61].

The majority of long-lived extragalactic high-energy emitters detected in the gamma-ray regime are blazars [59,62,63]. 3C 286 is one of only a small fraction of misaligned sources detected with Fermi, raising the question about the gamma-ray emission mechanism. A variety of different models have been discussed in the literature, but no single emission mechanism valid for the whole population has been robustly established yet (see our Section 6.5)[3].

1.5. X-ray Properties

The gamma-ray-detected NLS1 galaxies show a range of spectral shapes in X-rays [64–66]. Some are dominated by a flat jet component, while others show a significant soft excess, commonly interpreted as emission from the accretion disk.

Until recently, 3C 286 was rarely observed in the X-ray band. It was detected with Einstein [67] and in the ROSAT all-sky survey [68]. Snapshots of 1–2 ks with Chandra [29,69] and the Neil Gehrels Swift observatory [29] (Swift hereafter) revealed a flat X-ray spectrum with $\Gamma_x \approx 2$, but more complex spectral models were not constrained and could not be excluded. Yao and Komossa [29] also reported evidence for X-ray variability, raising the question whether the X-rays were dominated by the disk or jet; particularly important because of the use of 3C 286 as a radio calibrator.

Given its remarkable MWL properties, and the evidence for X-ray variability in particular, we obtained new observations with Swift, the Effelsberg radio telescope, and Chandra that we present in the next sections. Table 2 provides a summary of the main classifications of 3C 286.

Table 2. MWL classifications of 3C 286.

Classification/Property	Waveband	Comment
NLS1 galaxy	optical	among radio-loudest known
CSS	radio	spectral index $\alpha = -0.6$
misaligned Fermi source	gamma-rays	
DLA absorber	UV/radio	intervening; $z = 0.692$

2. Swift Observations

In order to search for X-ray flux and spectral variability of 3C 286 on both short and long time scales of days to years, we have observed the quasar seven times with Swift [70] between 2020 August and 2024 April (Table 3).

Table 3. Log of our Swift observations of 3C 286 between 2020 and 2023 with observation ids (OBSIDs) 13644–1 to 13644–8 (except 13644–5, not carried out). The Swift instrument is listed in the first column, the UVOT filters are reported in the second column, the energy band of the XRT and the central wavelengths of the UVOT filters [71] are given in the third column, and the exposure times t are reported in the remaining columns.

	Filter	ΔE or λ_c	t (ks) 2020-08-16	2020-08-21	2023-11-04	2023-11-14	2023-11-29	2023-12-03	2024-04-18
XRT		0.3–10 keV	2.03	1.39	1.93	1.68	1.66	2.18	1.51
UVOT	W2	1928 Å	0.65	0.42	0.63	0.55	0.57	0.71	0.48
	M2	2246 Å	0.49	0.37	0.45	0.38	0.23	0.12	0.33
	W1	2600 Å	0.32	0.21	0.31	0.28	0.35	0.35	0.24
	U	3465 Å	0.16	0.10	0.16	0.14	0.10	0.18	0.12
	B	4392 Å	0.16	0.10	0.16	0.14	0.10	0.18	0.12
	V	5468 Å	0.16	0.10	0.16	0.14	0.10	0.18	0.12

2.1. XRT Data Reduction and Spectral Analysis

The X-ray data were obtained with the Swift X-ray telescope (XRT) [72] operating in photon counting mode [73]. To carry out the timing and spectral analysis, we selected source photons within a circular area with a radius of 20 detector pixels (one pixel corresponds to a scale of 2.36″). Background photons were extracted in a nearby circular region with an extraction radius of 100 detector pixels. X-ray spectra of the source and background in the energy band 0.3–10 keV were generated. The spectral analysis was then carried out with the package XSPEC, version 12.14.0b [74]. The X-ray light curve is shown in Figure 1. Spectral fits were performed on the unbinned data and based on W-statistics [75,76] (W-stat hereafter).

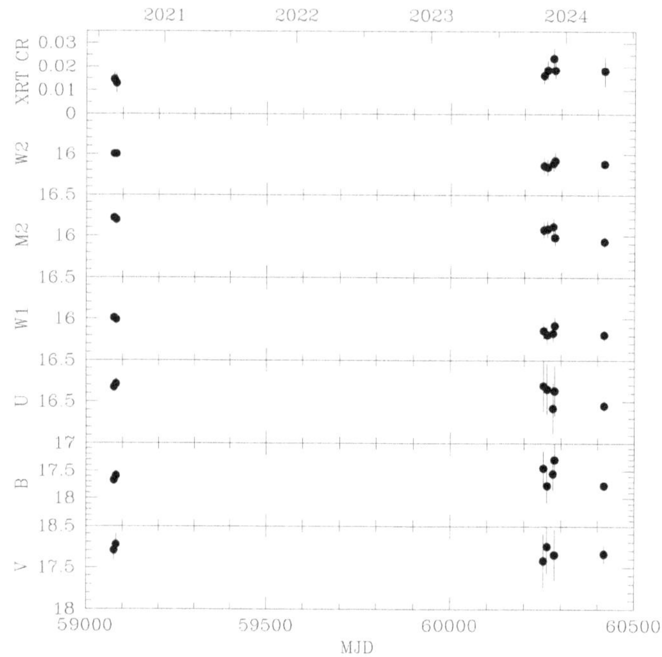

Figure 1. Swift light curve of 3C 286 between 2020 and 2024. UV–optical magnitudes were corrected for Galactic extinction.

In order to measure the average X-ray spectral slopes of 3C 286, we have split the Swift observations of 3C 286 into low-state (2020) and high-state (2023) data and then merged all data sets of each state in order to obtain two spectra. For the auxiliary response files (arfs), we applied the FTOOLS command *addarf* and weighted the contribution of each arf by the exposure time of the individual spectra divided by the total exposure times of the merged spectra. The response file *swxpc0to12s6–20130101v014.rmf* was applied to the spectra.

Both spectra were then first fit with a single powerlaw subject to Galactic absorption of $N_H = 1.11 \times 10^{20}$ cm^{-2} [77]. This gives a powerlaw photon index of $\Gamma_x = 1.83 \pm 0.27$ (W-stat/d.o.f. = 32.9/36) in low-state and $\Gamma_x = 2.11 \pm 0.15$ (W-stat/d.o.f. = 72.4/90) in high-state (Figure 2). If the column density N_H is left as free parameter at $z = 0$, it underpredicts the Galactic value but comes with large uncertainties and was therefore fixed at the Galactic N_H. Given the differences in the two spectra, next, both were fit simultaneously. The spectral index Γ_x was tied to the same value in this approach, and the normalizations were left free to vary. This again implies significant variability, with $K_{pl} = 2.6 \pm 0.5 \times 10^{-4}$ (2020) and $4.1 \pm 0.5 \times 10^{-4}$ (2023), where K_{pl} is the normalization at 1 keV in units of photons keV^{-1} cm^{-2} s^{-1}.

Next, in order to test for more realistic and complex spectral models and evaluate the cause for variability, we have used a powerlaw plus soft excess description. The soft excess was parameterized as blackbody, and additional cold absorption from an intervening absorber at redshift $z = 0.692$ and with $N_H = 1.75 \times 10^{21}$ cm^{-2} was added. The XSPEC model used was *TBabs * zTBabs * (zpowerlaw + zbbdy)*, where *TBabs* and *zTBabs* are the absorption models as described by Wilms et al. [78] in our Galaxy and at a redshift $z = 0.692$, respectively; *zpowerlaw* is the powerlaw model at $z = 0.849$; and *zbbdy* is the blackbody model at $z = 0.849$. The powerlaw model is defined as $A(E) = K_{pl}[E(1+z)]^{-\Gamma_x}$, where K_{pl} is the normalization at 1 keV in units of photons keV^{-1} cm^{-2} s^{-1}. The blackbody model is defined as $A(E) = K_{bb} \frac{8.0525[E(1+z)]^2 dE}{(1+z)(kT)^4[\exp[E(1+z)/kT]-1]}$, where kT is the temperature in keV, and K_{bb} is the normalization in $L_{39}/[D_{10}(1+z)]^2$, where L_{39} is the source luminosity in units of 10^{39} erg s^{-1} and D_{10} is the distance to the source in units of 10 kpc [76]. Since multi-parameter fitting is not warranted for the short Swift observations, we have fixed the intervening absorption, powerlaw index, and the temperature of the blackbody component to the preliminary values obtained from our recent XMM-Newton observation of 3C 286 [79] (Γ_x = 2.09 and kT_{bb} = 0.177 keV) and left the normalizations of both components free to vary. The difference between the two spectra can then be tentatively explained by an increase in the soft (blackbody) component in 2023. Results of the spectral fits are reported in Table 4. The XMM-Newton spectral fit results based on the same model are shown for comparison. With Swift, the following 0.3–10 kev fluxes (corrected for Galactic absorption and for intervening absorption) of the powerlaw plus blackbody model are measured: $f = 5.1 \times 10^{-13}$ erg cm^{-2} s^{-1} (2020) and 8.3×10^{-13} erg cm^{-2} s^{-1} (2023), respectively. [For comparison, the simple single powerlaw fit gives $f = 4.5 \times 10^{-13}$ erg cm^{-2} s^{-1} (2020) and 6.3×10^{-13} erg cm^{-2} s^{-1} (2023), respectively. These values are lower than from the two-component fit because they miss the blackbody component and the correction for the extra absorption.]

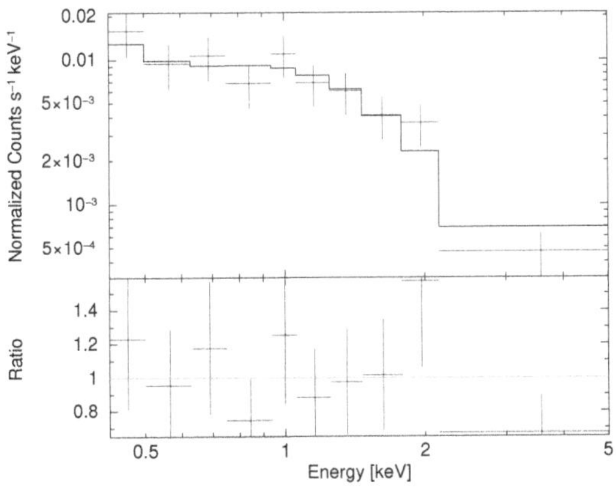

Figure 2. Composite binned Swift X-ray spectrum of 3C 286 from 2023 based on a single powerlaw fit. (Note that the actual spectral fits were performed on the unbinned spectrum.)

Table 4. Results of Swift X-ray spectral fits to 3C 286 for the 2020 low-state and 2023 high-state. XMM-Newton results are shown for comparison. Columns are as follows: (1) column density of the Galactic absorption, (2) column density of the intervening absorber at z = 0.692, (3) X-ray powerlaw photon index Γ_x, (4) powerlaw normalization K_{pl} at 1 keV in units of photons keV^{-1} cm^{-2} s^{-1}, (5) blackbody temperature kT in eV, and (6) normalization K_{bb} of the blackbody component in units of $L_{39}/[D_{10}(1+z)]^2$ (see Section 2.1).

Date	**N_{Gal} 10^{20} cm^{-2} (1)**	**$N_{z=0.692}$ 10^{21} cm^{-2} (2)**	**Γ_x (3)**	**K_{pl} 10^{-4} (4)**	**kT eV (5)**	**K_{bb} 10^{-5} (6)**
2020 low-state	1.11	–	1.83 ± 0.27	2.23 ± 0.63	–	–
	1.11	1.75	2.09	3.03 ± 0.67	177	$0.42^{+0.59}_{-0.42}$
2023 high-state	1.11	–	2.11 ± 0.15	4.31 ± 0.64	–	–
	1.11	1.75	2.09	4.21 ± 0.55	177	1.43 ± 0.60
2023 XMM	1.11	$1.75^{+0.62}_{-0.49}$	2.09 ± 0.10	$3.40^{+0.49}_{-0.46}$	177^{+28}_{-23}	$1.13^{+0.92}_{-0.43}$

A source detection was performed within XIMAGE on the co-added 2020–2023 data sets. All other X-ray sources are much fainter than 3C 286 itself. In Appendix A.1 of the Appendix A, all sources detected above 3σ are listed.

2.2. UVOT Data Reduction and Analysis

The UV–optical telescope (UVOT) [80] was used to observe 3C 286 seven times in three optical and three UV filters. For further analysis, the data sets of each filter were first co-added in each single observation. The source counts were then extracted in a region of circular or elliptical shape centered on 3C 286. A nearby area of 20″ radius was used to extract the background counts. The radius of the source extraction region was the standard 5″ during the 2020 observations. However, the data in 2023 are affected by a slight, uncorrected drift motion of the Swift satellite [81] such that sources appear no longer point-like but slightly elongated.

Therefore, different source extraction methods and tests were performed. First, a larger source extraction region of 15″ radius was used. Overall, this gives values consistent with the 2020 measurements, except for occasional single-filter deviations by up to 0.3 mag, which are not physical. Next, we used elliptical source extraction regions of varying radius and orientation, individually matching the source elongation in each filter and observation. This gives very similar results. Next, we checked if any of the UVOT observations fell onto the known small-scale, low-sensitivity areas (https://www.swift.ac.uk/analysis/uvot/sss.php) [accessed on 15 June 2024] of the detector, but that was not the case. Finally, we also checked photometry of several other brighter and fainter sources in the field of view, and overall their emission is constant. In conclusion, while high-precision photometry is not possible for the 2023 epochs, the optical and UV magnitudes of 3C 286 are found to be consistent with being constant within an uncertainty of ± 0.3 mag in the optical and ± 0.1 mag in the UV bands.

The background-corrected counts were converted into magnitudes in the Vega system using the latest calibration, as described by Poole et al. [71] and Breeveld et al. [82]. The magnitudes in each filter were determined with the UVOT tool *uvotsource* with the parameter *apercorr* set to *curveofgrowth* in order to compensate for the deviations from the calibrated circular 5″ radius source extraction region in 2023. The data were corrected for Galactic reddening based on the reddening curves of Cardelli et al. [83] and using $E_{B-V} = 0.01$. The extinction-corrected magnitudes are reported in Appendix A.1 of the Appendix A.

3. Effelsberg Radio Flux Density and Polarimetry Observations

3C 286 was observed in 2023 on November 3 and 16 in the radio regime, quasi-simultaneous with the Swift X-ray, UV, and optical observations of November 4 and 14. Observations of 3C 286 with the Effelsberg 100 m radio telescope are routinely performed within the MOMO (Multi-wavelength Observations and Modelling of OJ 287) project; e.g., [84,85]. Usually, 3C 286 serves as one of several radio calibrators during these observations, and its constant radio emission is always carefully checked against other calibration observations and against the theoretically expected telescope sensitivity. However, for the analysis of the data in November 2023, 3C 286 was treated as the main target, and additional other radio calibrators (e.g., 3C161 and NGC 7027) were observed and used to measure the flux density and polarization of 3C 286 itself in order to re-confirm the constancy of the radio flux of 3C 286 quasi-simultaneous with the new Swift MWL data.

3.1. Flux Densities

Flux densities were measured between 2.6 and 41 GHz. Data acquisition and reduction was carried out as follows [86]. The observations were carried out in form of cross-scans. Several sub-scans were performed in azimuth and elevation over the source position. The amplitude (antenna temperature) and pointing offsets were determined by fitting a Gaussian profile to the individual sub-scans. After correction for the usually small pointing offset, the amplitudes of all sub-scans were averaged. In the next step, the atmospheric attenuation as well as the telescope's gain-elevation effect (loss of sensitivity due to the gravitational deformation of the main dish when tilted) were corrected. Finally, the corrected antenna temperatures were compared to the expected flux densities using the calibrator sources and telescope sensitivities for the conversion into the Jansky scale.

Typical errors are in the range of 0.5–2% for the lower frequencies and about 5–10% for the high frequencies (depending mostly on the weather conditions).

3.2. Polarimetry

Polarimetry was conducted with the Effelsberg telescope at several frequencies between 2.6 and 41 GHz. For the analysis of the linear polarization, the values of Stokes Q and U were taken from the corresponding channels of the cross-scan data at the position of the maximum of the Stokes I channel. For the correction of the instrumental effects, the Müller matrix method [87] was applied using average Müller matrices from the calibration database, i.e., from former observations. As with Stokes I, additional calibrators were used to check the consistency of the results. Typical measurement errors are $\sim$5% for the polarized flux density P and between 3° and 5° for the polarization angle χ (depending on the frequency and weather conditions).

Results (Figure 3) show that the radio emission of 3C 286 is constant when comparing with past radio measurements of 3C 286 at similar resolution. No variability is detected in the radio regime. The fractional polarization p shows a small dependence on frequency increasing from 11 to 13% (see Appendix A.1 of Appendix A).

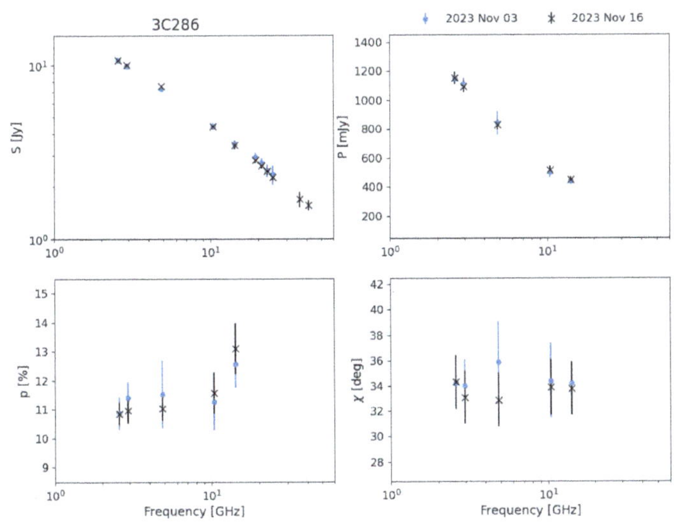

Figure 3. Radio observations of 3C 286 from November 2023. S is the flux density, P the polarized flux density, p the polarization percentage, and χ the polarization position angle. Measurements from November 3 are marked in blue and measurements from November 16 in black.

4. Chandra Imaging Observation

An observation of 3C 286 with the Chandra ACIS-S detector [88] was performed on 12 February 2023 with a duration of 35 ks. The data will be discussed in more detail by Yao et al. [79], but we use them here to search for additional sources in the vicinity of 3C 286 separately in the soft and hard X-ray band (Section 6).

5. Long-Term X-ray Light Curve

To construct the long-term X-ray light curve of 3C 286, data from previous X-ray missions were collected and converted into fluxes and then into Swift XRT count rates in the band 0.3–10 keV, assuming a constant spectral shape consisting of the single powerlaw description with Galactic absorption and Γ_X = 2.1. Count rate data are from Einstein [67], ROSAT [68], Chandra [69,79], and XMM-Newton [79], in addition to the Swift data itself.

We have always used the directly measured count rates for conversion into fluxes, as the previously published fluxes depend on the assumption of the spectral model and the amount of absorption. The single exception are the two Chandra observations, which are subject to a decrease in the ACIS low-energy quantum efficiency caused by increasing contamination on the Optical Blocking Filter, implying a decrease of the directly measured count rate. Instead, we used the measured powerlaw flux for conversion. Data from the very short 2 ks 2013 observation are from Massaro et al. [69]. For the Swift data of 2020 and 2023, the average count rate is shown. 3C 286 shows constant baseline emission between 1990 (ROSAT) and the 2020s (XMM-Newton), and evidence for flaring up to a factor of ~2 with Swift (Figure 4).

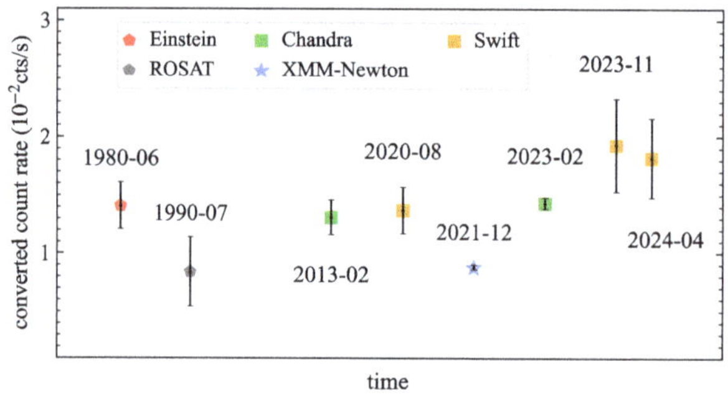

Figure 4. Long-term X-ray light curve of 3C 286 (displayed as Swift XRT count rate in the band 0.3–10 keV), observed with multiple missions between 1980 and 2024 (note that the time axis is not linear because of the widely spaced observations). All non-Swift data were first converted into fluxes in a homogeneous way and then into Swift XRT count rates (see text for details).

6. Discussion

We have presented new Swift optical, UV and X-ray observations, new radio data taken at the Effelsberg 100 m radio telescope, and the first deep imaging observations of 3C 286 in the X-ray regime (the latter will be further discussed by [79]). These data are used to address some of the open questions regarding the intriguing MWL properties of 3C 286.

6.1. *How Much Radio Variability?*

3C 286 is a well-known radio calibrator and the constancy of its flux density and polarization in the past has been well established [18–20,89–91]. At the same time, recent evidence for X-ray variability [29] and evidence for radio core variability [12] have raised the question whether 3C 286 is still constant in the radio regime (in low-resolution observations) and, especially, if the X-ray variability is driven by the disk or the jet. We have re-observed 3C 286 with the Effelsberg telescope in 2023 November quasi-simultaneous with Swift, in order to search for any possible radio variability.

Results confirm the constancy of the radio properties of 3C 286 at the Effelsberg telescope resolution. The measured flux densities were compared with past radio observations of 3C 286 at comparable resolution, [18,92–97] and values agree within the error bars. Further, the new observations show the well-known (small) increase in fractional polarization p with frequency [17,19,98]; also see Figure 3 of [99]. We measured a change of p from 11 to 13% as frequency increases. The polarization angle χ of 33–34° is consistent with previous measurements as well.

6.2. *X-ray Spectrum, Soft Excess, and Variability*

The X-ray spectral results are of interest for several reasons: disentangling the disk/jet contributions, constraining the amount of absorption along the line-of-sight, and addressing the question which component it is that varies: jet, disk, or absorber. The previous short X-ray observations of 3C 286 were well fit with a single powerlaw of photon index

$\Gamma_x \approx 2$ [29]. Excess absorption as seen in form of the DLA system was not required, which was a puzzle. However, the new X-ray models have shown that such excess absorption from *intervening* gas can be well accommodated when allowing for the presence of a soft excess component. With the new repeat Swift observations, we can now address which component it was that varied, and find that the data are consistent with a change in the strength of the soft excess, which represents the disk component. It is interpreted as the tail of the 'blue bump' seen in the optical–UV. Such a component is frequently observed in NLS1 galaxies [36].

NLS1 galaxies, the majority of them radio-quiet, are known to be highly variable in X-rays. It then depends on the variability mechanism(s) if and how the jet formation and emission at the launching region is affected. For instance, if the X-ray variability was due to partial-covering effects of an ionized disk wind, arose from hot spots in the disk, or was due to effects of relativistic reflection, the jet would or could remain unaffected. A number of follow-up observations are essential to constrain further the disk–jet links. For instance, deeper X-ray spectroscopy of 3C 286 will improve measurements of the soft excess, and long-term X-ray monitoring will quantify the amplitude and frequency of variability.

6.3. Extinction, Absorption, and the Very High Radio Loudness of 3C 286

Knowledge of the amount of extinction by dust along our line-of-sight is important for a variety of reasons: for instance, for deriving and modeling the intrinsic SED, for the optical spectroscopic identification of AGN class, for measuring SMBH mass and Eddington ratio, for constraining the mechanism of X-ray variability, and for determining the radio loudness. (3C 286 is the radio-loudest gamma-ray emitting NLS1 galaxy known. The question is raised whether extinction within 3C 286 itself or along our line-of-sight strongly dims the optical emission, such that, if uncorrected, the radio-loudness index R is highly boosted.) There are several ways to evaluate the amount of intrinsic or intervening extinction along our line-of-sight, employing (1) the BLR Balmer decrement, (2) the NLR Balmer decrement, (3) UV absorption lines (only if the absorber is dusty), (4) 21 cm absorption (only if dusty), and (5) X-ray absorption (only if dusty). We comment on each in turn (Table 5).

Table 5. Summary of different measurements of extinction and absorption towards 3C 286.

Waveband	Method	N_H 10^{20} cm^{-2}	Comment
radio	HI 21 cm abs. (z = 0.849)	<0.052	[100]
radio	HI 21 cm abs. (z = 0.692)		[58]
optical	BLR Balmer decrement	–	this paper
optical	NLR Balmer decrement		not yet available
UV	Ly α absorption (z = 0.692)	20	low dust content [55,101]
X-rays	neutral abs. (z = 0.692, fixed)	∼17.5	[79], this paper

(1) In the optical band, the most reliable of these is the BLR Balmer decrement, because dust is not expected to survive in the bulk of the BLR; therefore, the extinction must occur along the line of sight and will affect BLR and continuum emission in the same way. The Balmer decrement of the bluest AGN is close to the case B recombination value [102]. Missing Hα, we use the observed Hγ/Hβ flux ratio [29] of ∼0.36, which compares to the case B recombination value of 0.46 [103] and implies little or no extinction within the uncertainties that come with Hγ line measurement and decomposition. (2) The NLR Balmer decrement provides a measure of extinction as well, even though the method does not immediately imply the fraction of dust along our line-of-sight (affecting the continuum) since the dust can be intrinsic to the NLR clouds, not necessarily covering the continuum source. Since the NLR component of Hγ was not yet detected in the optical spectrum, this method has to await future higher S/N spectroscopy. (3) UV absorption lines provide a measure of the absorbing gas along the line-of-sight toward the UV continuum source. The absorber will only redden the continuum if it is dusty. Cohen et al. [55] estimated a column density

of the intervening Lyman α absorber of 3C 286 of 2×10^{21} cm^{-2}. However, its actual dust content was found to be low, about 5% of that of the Galactic-disk dust-to-gas ratio [101]. Therefore, little dust *extinction* is expected from this component. (4) The 21 cm absorption at the redshift of 3C 286 itself was found to be very low, $<0.052 \times 10^{20}$ cm^{-2} [100], providing a tight constraint on any *intrinsic* absorption directly along the line-of-sight to the radio continuum source. The DLA system at $z = 0.692$ is accompanied by HI 21 cm absorption at the same redshift. As mentioned above, it was argued that the dust content of this component is low. (5) X-ray spectroscopy is a potentially powerful way of measuring the amount of absorption along our line-of-sight, even though the dust content itself is usually not constrained (see high-column density dusty *ionized* absorbers for an exception [104]). Unfortunately, 3C 286 is weak in X-rays due to its high redshift, only medium-deep X-ray observations were obtained so far, and the soft excess and absorption are partially degenerate (Section 6.2). The best constraints so far come from the XMM-Newton spectrum of 3C 286 with absorption with a column density comparable to the DLA system (our Table 4 and [79]); intervening not intrinsic.

In summary, the absorption/extinction intrinsic to 3C 286 is very low, and the intervening absorber is known to have low dust content. This finding validates the method used to derive the high Eddington ratio. This finding also reconfirms that the exceptional radio loudness of 3C 286 is not an artifact of heavy extinction affecting the flux ratio $f_{5\,\mathrm{GHz}}/f_{4400\mathrm{A}}$ used to calculate the radio index $R_{5\,\mathrm{GHz}}$. The very high radio loudness is even more remarkable given that the jet of 3C 286 is not beamed and given that 3C 286 is a CSS so that the extended radio emission is rather compact.

6.4. Possible Detection of X-rays from the SW Lobe with Chandra

The SW lobe is spatially resolved from the core with Chandra, and three hard X-ray photons from its region were detected. Given the sharp PSF and low background of Chandra, this likely constitutes a detection, but deeper imaging is needed to confirm it. The lobe X-ray emission may arise from upscattered CMB photons (detected from other radio lobes as well [105]) and/or may be related to the gamma-ray emission of 3C 286. Assuming a flat powerlaw shape for this component with $\Gamma_x = 1.5$ results in an observed, absorbed (3–10 keV) flux of $\sim 3.1 \times 10^{-15}$ erg s^{-1}.

6.5. The Origin of the Gamma-Ray Emission of 3C 286

3C 286 is one of only a small number of radio steep-spectrum sources which have a Fermi gamma-ray detection. A variety of scenarios were discussed in the literature for CSS in general [106]. In application to 3C 286, An et al. [12] explored the possibility that the *inner* jet is misaligned with the outer jet, is pointing at us, and is relativistic, but found that that was not the case, with a viewing angle of the inner jet of 48°. Here, we discuss several scenarios for the gamma-ray emission of 3C 286 in turn.

Binary AGN, Binary SMBH, or Chance Projection of Multiple Sources

Pairs of AGN and binary SMBHs form in the course of galaxy mergers, and many systems have been identified in recent years [107,108]. If an AGN pair or binary SMBH was present in 3C 286, we could imagine one of them with NLS1–CSS properties, and the other one with blazar characteristics producing the detected gamma-rays. It is interesting to note that Zhang et al. [9] reported the presence of two core-like sources at the center of 3C 286 (C1 in the NE and C2 in the SW), speculating about these representing two separate nuclei. Their radio spectra and fluxes were found to be very similar. However, follow-up observations revealed SED differences below 8 GHz, where the NE source showed an inverted spectrum (based on non-simultaneous data). Therefore, An et al. [12] interpreted the NE component as the actual core and the SW component as a knot in the jet. Further, we do not find any other indication for the presence of a dual AGN; for instance, in the form of double-peaked optical emission lines.

One can think of two (or multiple) unrelated sources as well, at different redshifts but closely projected on the sky. For instance, one source could be a radio-quiet, variable NLS1 galaxy dominating the optical spectrum and UV–X-ray excess, and the other could be a gamma-ray emitting CSS. However, in that case, the intervening absorber at $z = 0.692$, seen in both the radio and UV, either implies that radio and UV arise from one and the same source, or that both are very close in projection such that both sightlines pass the intervening absorber. Available Hubble Space Telescope (HST) images did not yet reveal multiple optical sources [109]. Therefore, we conclude that there is no positive evidence for these scenarios at present.

Intervening Galaxy

Yao and Komossa [29] raised the question whether the known intervening galaxy at 2.5″ offset from 3C 286 could actually be an AGN and be the counterpart of the gamma-ray emission. However, this was rejected because the galaxy was not detected in the radio band, and is known to have low metallicity. With our deep Chandra observation, we have now checked for X-ray emission from this galaxy and do not detect any (Figure 5). We conclude that it is not the counterpart of the Fermi source.

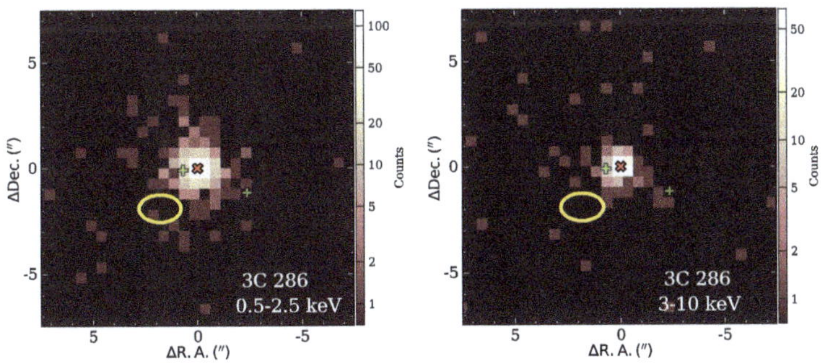

Figure 5. Chandra X-ray images of 3C 286 in the soft (0.5–2.5 keV; **left**) and hard (3–10 keV; **right**) X-ray band. Locations of several features of interest are marked: the red cross marks the X-ray center of 3C 286; the green plus symbols correspond to the two radio lobes in the East and South-West; and the yellow ellipse marks the position of the intervening low-surface-brightness galaxy.

Other Nearby, Unrelated Galaxies in the Field

Given the limited spatial resolution of Fermi, there is a residual small possibility that the gamma-ray emission does not arise from 3C 286 but rather from an unrelated foreground or background source. However, we find that all other X-ray sources detected (at $>3\sigma$) in the field with Swift and/or Chandra are much fainter in X-rays than 3C 286, by factors of >6–10. In particular, no other bright source near 3C 286 is detected at the spatial resolution of Chandra.

Microlensing

Next, we note that we cannot have a situation of microlensing, which temporarily intensified any intrinsically much fainter gamma-ray emission, because Fermi found persistent emission for multiple years.

Emission from Lobes

The radio lobes have been considered as potential sites for gamma-ray emission. For instance, it was suggested that inverse Compton (IC) processes involving local photons produce high-energy radiation [110,111] if electrons can be efficiently re-accelerated locally, for instance in hot spots, and then injected into the lobes. Upscattering of cosmic microwave background (CMB) photons is another mechanism [52,112,113]. Bremsstrahlung emission in the MeV–GeV regime from shocked plasma created from a cocoon expanding into the ISM was considered too [114]. In some of these scenarios, we expect enhanced off-nuclear X-ray emission at these sites and have searched for any such emission with Chandra. The SW lobe is spatially resolved, and three photons from its location were detected; too few to pin down their production mechanism.

It should be noted that some of the models above [110,111,114] are made for young CSS radio sources evolving into a dense medium at one-to-few kpc scales, while 3C 286 is already more evolved (see next paragraph), and no evidence for significant intrinsic absorption along our line-of-sight was found either.

Jet Deflection on Dense Gas

Alternatively, it was speculated that individual jet components are occasionally deflected and then beamed toward the observer as they encounter single dense blobs in the ISM along the path of the jet [115]. CSS are generally considered young AGN (review by [106]) with compact jets evolving through their dense ISM. However, whether this holds for 3C 286 in particular is less clear. Its radio emission is more extended than most CSS (up to a projected size of $\sim$20 kpc, beyond the confines of the host galaxy). Further, there is no evidence for absorbing material along our line-of-sight intrinsic to 3C 286 (our Section 6.3 [100]). The inner region remains spatially unresolved in X-rays, but a highly beamed component would likely produce radio variability, not observed.

Strong (and Variable) External Photon Field

Zhang et al. [52] highlighted the fact that gamma-ray emitting CSS, including 3C 286, have on average higher Eddington ratios than the rest of the CSS population. In particular, the NLS1 nature of 3C 286 explains the bright soft excess (disk) contribution to the SED and the high accretion near the Eddington rate, and its reprocessing on BLR and torus scales may provide an efficient external photon field for the inverse Compton processes. Based on the BLR radius–luminosity relation of Kaspi et al. [116], and the 5100 Å luminosity [29] of $\lambda\, L_\lambda(5100) = 1.6 \times 10^{45}$ erg s^{-1} (estimated from $L(H\beta)$) or up to $\lambda\, L_\lambda(5100) = 4.7 \times 10^{45}$ erg s^{-1} (the directly observed value, assuming the optical emission is dominated by the accretion disk), the BLR radius of 3C 286 is $R = 151$–318 lt-days. This approach is worthwhile to explore further, carrying out SED modeling using the most recent X-ray spectral constraints.

7. Summary and Conclusions

After a short review of the unique MWL properties of 3C 286, we have presented new observations from the radio to X-ray regime. Our main results can be summarized as follows:

- No radio variability is detected. The (2.6–41 GHz) flux density, polarization percentage, and polarization angle remain constant at Effelsberg telescope resolution when compared to past measurements, confirming the calibrator status of 3C 286.
- The X-ray spectrum of 3C 286 is well represented by a soft excess on top of a flat powerlaw. The long-term X-ray light curve confirms the evidence for X-ray variability by a factor $\sim$2, and spectral fits to Swift observations suggest that it is the soft (disk) component which varies; not unexpected in light of the NLS1 classification of 3C 286. Disk variability will not necessarily affect the inner jet, but it could.
- The amount of extinction/absorption along our line-of-sight was rigorously assessed. No reddening intrinsic to 3C 286 is found, and the excess X-ray absorption detected with XMM-Newton [79] is consistent with the intervening DLA system, which has low dust content as well. The lack of reddening validates the estimate of a high Eddington ratio, the NLS1 classification of 3C 286, and its high radio-loudness index. 3C 286 stands out as one of the radio-loudest NLS1 galaxies known.

- The Chandra image reveals tentative evidence for hard X-ray emission from the SW lobe.
- Different gamma-ray emission mechanisms were discussed. Several scenarios were found to be unlikely (no positive evidence for a dual AGN or binary SMBH so far, no alternative counterpart to the gamma-ray emission detected in X-rays, and no evidence that 3C 286 is a particularly *young* CSS with its jet still making its way out of the galaxy for the first time). It is speculated that the NLS1 character of the source and its high Eddington ratio, and efficient reprocessing of the disk photons in the BLR and torus, lead to efficient IC processes contributing to the observed gamma-ray emission.

A number of follow-up observations of 3C 286 suggest themselves. Deeper XMM-Newton observations will provide tight constraints on spectral variability, and a deeper Chandra observation is needed to confirm the presence of off-nuclear hard X-ray emission from the lobe(s). It will require the next generation of X-ray spectroscopic missions to detect individual metals in the intervening absorber at $z = 0.692$. Continuing Swift monitoring will allow us to quantify the frequency and amplitude of MWL variability. JWST imaging will reveal the properties of the intervening low-surface-brightness galaxy, which is the likely host of the DLA system, and will detect the host galaxy of 3C 286 itself for the first time with high significance (so far only marginally detected with HST [109]). VLBI at the highest resolution will determine the compactness of the core, so far always resolved in radio observations, and reveal variability of the inner jet if present.

3C 286, which was among the first quasars identified spectroscopically in the 1960s and among the first to be imaged with VLBI, continues to be an important cornerstone system for our understanding of the radio-loudest quasars known, the drivers behind NLS1 galaxies and their link with CSS sources, and the mechanism of gamma-ray emission in misaligned sources.

Author Contributions: Conceptualization and methodology, S.K.; data analysis, S.K., S.Y., D.G. and A.K.; writing—original draft preparation, S.K.; writing—review and editing, S.K., S.Y., D.G. and A.K. All authors have read and agreed to the published version of the manuscript.

Funding: This research received no external funding.

Data Availability Statement: All data are available upon reasonable request. The Swift data we proposed are available in the Swift archive at https://swift.gsfc.nasa.gov/archive/ [assessed on 1 July 2024].

Acknowledgments: We would like to thank the Swift team for carrying out the observations we proposed. SK acknowledges the hospitality of NAOC Beijing in October 2023. SY acknowledges support by an Alexander von Humboldt Foundation Fellowship between 2020 and 2022, when this project was started. We would like to thank our referee and T. Krichbaum for their useful comments on the manuscript. This work is partly based on data obtained with the 100 m telescope of the Max-Planck-Institut für Radioastronomie at Effelsberg. This work made use of data supplied by the UK Swift Science Data Centre at the University of Leicester [117]. This research made use of the XRT Data Analysis Software (XRTDAS) developed under the responsibility of the ASI Science Data Center (ASDC), Italy. This research made use of data obtained with the Chandra satellite and software provided by the Chandra X-ray Center (CXC). This work also made use of the NASA Astrophysics Data System Abstract Service (ADS) and the NASA/IPAC Extragalactic Database (NED), which is operated by the Jet Propulsion Laboratory, California Institute of Technology, under contract with the National Aeronautics and Space Administration.

Conflicts of Interest: The authors declare no conflicts of interest.

Appendix A

Appendix A.1. Flux Density Measurements

Here, we report the flux density and polarization measurements of 3C 286 obtained with Swift (Table A1) and the Effelsberg telescope (Table A2).

Table A1. Swift UVOT magnitudes of 3C 286, corrected for Galactic extinction. Uncertainties are ±0.1 mag in the optical filters (V, B, U) and ±0.05 mag in the UV filters in 2020 and 2024. They are larger in 2023 because of the satellite drift motion, with ±0.3 mag in the optical filters and ±0.1 mag in the UV filters.

Date	V	B	U	W1	M	W2
2020-08-16	17.30	17.67	16.33	15.99	15.79	16.00
2020-08-21	17.23	17.59	16.29	16.01	15.81	16.00
2023-11-04	17.42	17.45	16.31	16.14	15.93	16.14
2023-11-14	17.25	17.77	16.35	16.19	15.92	16.16
2023-11-29	...	17.55	16.58	16.17	15.89	16.11
2023-12-03	17.35	17.30	16.37	16.08	16.02	16.08
2024-04-18	17.34	17.77	16.55	16.19	16.07	16.12

Table A2. Radio flux density S and polarization percentage p of 3C 286 measured with the Effelsberg 100 m telescope on 2023 November 3 and 16.

ν_c GHz	S Jy 2023-11-03	2023-11-16	p 2023-11-03	2023-11-16
2.595	10.59 ± 0.41	10.65 ± 0.08	10.9 ± 0.6	10.8 ± 0.4
2.95	9.76 ± 0.28	9.97 ± 0.21	11.4 ± 0.5	11.0 ± 0.5
4.85	7.29 ± 0.22	7.52 ± 0.03	11.5 ± 1.2	11.0 ± 0.4
10.45	4.48 ± 0.18	4.44 ± 0.08	11.3 ± 1.0	11.6 ± 0.7
14.25	3.49 ± 0.17	3.43 ± 0.19	12.6 ± 0.8	13.1 ± 0.9
19.25	2.93 ± 0.19	2.83 ± 0.10		
22.85	2.46 ± 0.22	2.43 ± 0.11		
24.75	2.33 ± 0.29	2.25 ± 0.08		
36.25		1.68 ± 0.16		
41.25		1.55 ± 0.10		

Appendix A.2. Other X-ray Sources Detected in the Field with Swift

The individual Swift X-ray images between 2020 and 2023 were co-added and a source detection was performed on the merged data set. All other X-ray sources are much fainter than 3C 286 itself. In Table A3, all sources detected above 3σ are listed. With NED, optical–IR counterparts were searched within a region of 5″ radius. Two sources have WISE [118] counterparts and one coincides with a high-redshift SDSS quasar candidate.

Table A3. X-ray sources detected with Swift at $>3\sigma$ in the merged 2020–2023 data set, their average count rates CR in the 0.3–10 keV band, their detection significance σ, the nearest candidate optical–IR counterpart within 5″ including photometric redshifts if available, and the separation d of the nearest counterpart in arcminutes.

Source (1)	RA-J2000 (h, m, s) (2)	Decl-J2000 $(°, ′, ″)$ (3)	CR 10^{-3} cts/s (4)	σ (5)	Counterpart (6)	d (7)
X1	13 31 08.4	+30 30 36.0	15.7 ± 1.4	11.1	3C 286	0.056′
X2	13 30 56.0	+30 40 23.7	2.53 ± 0.60	4.2	WISE source	0.057′
X3	13 30 33.1	+30 33 29.2	1.90 ± 0.51	3.7	SDSS QSO candidate z = 1.745	0.069′
X4	13 30 57.2	+30 37 46.2	1.50 ± 0.46	3.2	WISE source	0.069′
X5	13 30 30.4	+30 33 54.4	1.53 ± 0.47	3.3	–	
X6	13 31 16.4	+30 35 52.1	1.50 ± 0.47	3.2	–	

Notes

1 3C 286 is actually a narrow-line type 1 quasar, not a NLS1 galaxy. However, we follow the approach common in the literature to refer to all narrow-line type 1 AGN (quasars and their lower luminosity equivalents of Seyfert galaxies) collectively as NLS1 galaxies.

2 Strong projection effects are unlikely to affect the Balmer-line profile, since 3C 286 is misaligned [28], i.e., it is not a blazar viewed near face-on. We further note that 3C 286, like many NLS1 galaxies [32,35,50], has broad wings in its Balmer lines, and its $H\beta_{broad}$ profile is alternatively well fit by a Lorentz profile (rather than double Gauß profiles) with FWHM($H\beta_{broad,L}$) = 1858 km/s.

3 At an association probability of 0.985 [60], 3C 286 is the most likely counterpart of the Fermi source, but there remains a small possibility of an alternative counterpart identification.

References

1. Wright, E.L. A Cosmology Calculator for the World Wide Web. *Publ. Astron. Soc. Pac.* **2006**, *118*, 1711–1715. [CrossRef]
2. Edge, D.O.; Shakeshaft, J.R.; McAdam, W.B.; Baldwin, J.E.; Archer, S. A survey of radio sources at a frequency of 159 Mc/s. *Mem. R. Astron. Soc.* **1959**, *68*, 37–60.
3. Matthews, T.A.; Sandage, A.R. Optical Identification of 3C 48, 3C 196, and 3C 286 with Stellar Objects. *Astrophys. J.* **1963**, *138*, 30. [CrossRef]
4. Wilkinson, P.N.; Readhead, A.C.S.; Anderson, B.; Purcell, G.H. VLBI observations of compact radio sources at 609 megahertz. *Astrophys. J.* **1979**, *232*, 365–381. [CrossRef]
5. Pearson, T.J.; Readhead, A.C.S.; Wilkinson, P.N. Maps of the quasars 3C 119,3C 286, 3C 345, 3C 454.3 and CTA 102 with a resolution of 5 milli-arcseconds at 1.67 GHz. *Astrophys. J.* **1980**, *236*, 714–723. [CrossRef]
6. Spencer, R.E.; McDowell, J.C.; Charlesworth, M.; Fanti, C.; Parma, P.; Peacock, J.A. MERLIN and VLA observations of compact steep-spectrum radio sources. *Mon. Not. R. Astron. Soc.* **1989**, *240*, 657–687. [CrossRef]
7. van Breugel, W.J.M.; Fanti, C.; Fanti, R.; Stanghellini, C.; Schilizzi, R.T.; Spencer, R.E. Compact Steep-Spectrum 3CR sources: VLA observations at 1.5, 15 and 22.5 GHz. *Astron. Astrophys.* **1992**, *256*, 56.
8. An, T.; Hong, X.Y.; Wang, W.H. The Kiloparsec-scale Structure of 3C 286. *Chin. J. Astron. Astrophys.* **2004**, *4*, 28–36. [CrossRef]
9. Zhang, F.J.; Spencer, R.E.; Schilizzi, R.T.; Fanti, C.; Baath, L.B.; Su, B.M. Are there two engines at the center of 3C 286? *Astron. Astrophys.* **1994**, *287*, 32–37.
10. Cotton, W.D.; Fanti, C.; Fanti, R.; Dallacasa, D.; Foley, A.R.; Schilizzi, R.T.; Spencer, R.E. VLBA polarimetric observations of 3C 286 at 5 GHz. *Astron. Astrophys.* **1997**, *325*, 479–483.
11. Kellermann, K.I.; Vermeulen, R.C.; Zensus, J.A.; Cohen, M.H. Sub-Milliarcsecond Imaging of Quasars and Active Galactic Nuclei. *Astron. J.* **1998**, *115*, 1295–1318. [CrossRef]
12. An, T.; Lao, B.Q.; Zhao, W.; Mohan, P.; Cheng, X.P.; Cui, Y.Z.; Zhang, Z.L. Parsec-scale jet properties of the gamma-ray quasar 3C 286. *Mon. Not. R. Astron. Soc.* **2017**, *466*, 952–959. [CrossRef]
13. Lister, M.L.; Aller, M.F.; Aller, H.D.; Hodge, M.A.; Homan, D.C.; Kovalev, Y.Y.; Pushkarev, A.B.; Savolainen, T. MOJAVE. XV. VLBA 15 GHz Total Intensity and Polarization Maps of 437 Parsec-scale AGN Jets from 1996 to 2017. *Astrophys. J. Suppl. Ser.* **2018**, *234*, 12. [CrossRef]
14. Peacock, J.A.; Wall, J.V. Bright extragalactic radio sources at 2.7 GHz-II. Observations with the Cambridge 5-km telescope. *Mon. Not. R. Astron. Soc.* **1982**, *198*, 843–860. [CrossRef]
15. Rudnick, L.; Jones, T.W. Rotation measures for compact variable radio sources. *Astron. J.* **1983**, *88*, 518–526. [CrossRef]
16. Akujor, C.E.; Garrington, S.T. Compact steep-spectrum sources–polarisation observations at 1.6, 4.9, 8.4 and 15 GHz. *Astron. Astrophys. Suppl. Ser.* **1995**, *112*, 235.
17. Nagai, H.; Nakanishi, K.; Paladino, R.; Hull, C.L.H.; Cortes, P.; Moellenbrock, G.; Fomalont, E.; Asada, K.; Hada, K. ALMA Science Verification Data: Millimeter Continuum Polarimetry of the Bright Radio Quasar 3C 286. *Astrophys. J.* **2016**, *824*, 132. [CrossRef]
18. Ott, M.; Witzel, A.; Quirrenbach, A.; Krichbaum, T.P.; Standke, K.J.; Schalinski, C.J.; Hummel, C.A. An updated list of radio flux density calibrators. *Astron. Astrophys.* **1994**, *284*, 331–339.
19. Perley, R.A.; Butler, B.J. Integrated Polarization Properties of 3C48, 3C138, 3C147, and 3C286. *Astrophys. J. Suppl. Ser.* **2013**, *206*, 16. [CrossRef]
20. Perley, R.A.; Butler, B.J. An Accurate Flux Density Scale from 1 to 50 GHz. *Astrophys. J. Suppl. Ser.* **2013**, *204*, 19. [CrossRef]
21. Burbidge, G.R.; Burbidge, E.M. Red-shifts of Quasi-stellar Objects and Related Extragalactic Systems. *Nature* **1969**, *222*, 735–741. [CrossRef]
22. Hirst, P.; Jackson, N.; Rawlings, S. Near-infrared spectroscopy of powerful compact steep-spectrum radio sources. *Mon. Not. R. Astron. Soc.* **2003**, *346*, 1009–1020. [CrossRef]
23. Schneider, D.P.; Hall, P.B.; Richards, G.T.; Strauss, M.A.; Vanden Berk, D.E.; Anderson, S.F.; Brandt, W.N.; Fan, X.; Jester, S.; Gray, J.; et al. The Sloan Digital Sky Survey Quasar Catalog. IV. Fifth Data Release. *Astron. J.* **2007**, *134*, 102–117. [CrossRef]
24. Schmidt, M. Spectrum of a Stellar Object Identified with the Radio Source 3c 286. *Astrophys. J.* **1962**, *136*, 684. [CrossRef]
25. Oke, J.B. The Redshift of the Quasi-Stellar Radio Source 3c 286. *Astrophys. J.* **1965**, *142*, 810. [CrossRef]
26. Dawson, K.S.; Schlegel, D.J.; Ahn, C.P.; Anderson, S.F.; Aubourg, É.; Bailey, S.; Barkhouser, R.H.; Bautista, J.E.; Beifiori, A.; Berlind, A.A.; et al. The Baryon Oscillation Spectroscopic Survey of SDSS-III. *Astron. J.* **2013**, *145*, 10. [CrossRef]
27. Yao, S. Ph.D. Thesis, National Astronomical Observatories of China, CAS, Beijing, China, 2016.
28. Berton, M.; Foschini, L.; Caccianiga, A.; Ciroi, S.; Congiu, E.; Cracco, V.; Frezzato, M.; La Mura, G.; Rafanelli, P. An orientation-based unification of young jetted active galactic nuclei: The case of 3C 286. *Front. Astron. Space Sci.* **2017**, *4*, 8. [CrossRef]

29. Yao, S.; Komossa, S. Spectroscopic classification, variability, and SED of the Fermi-detected CSS 3C 286: The radio-loudest NLS1 galaxy? *Mon. Not. R. Astron. Soc.* **2021**, *501*, 1384–1393. [CrossRef]
30. Osterbrock, D.E.; Pogge, R.W. The spectra of narrow-line Seyfert 1 galaxies. *Astrophys. J.* **1985**, *297*, 166–176. [CrossRef]
31. Goodrich, R.W. Spectropolarimetry of "Narrow-Line" Seyfert 1 Galaxies. *Astrophys. J.* **1989**, *342*, 224. [CrossRef]
32. Véron-Cetty, M.P.; Véron, P.; Gonçalves, A.C. A spectrophotometric atlas of Narrow-Line Seyfert 1 galaxies. *Astron. Astrophys.* **2001**, *372*, 730–754. [CrossRef]
33. Sulentic, J.W.; Zwitter, T.; Marziani, P.; Dultzin-Hacyan, D. Eigenvector 1: An Optimal Correlation Space for Active Galactic Nuclei. *Astrophys. J. Lett.* **2000**, *536*, L5–L9. [CrossRef] [PubMed]
34. Boroson, T.A. Black Hole Mass and Eddington Ratio as Drivers for the Observable Properties of Radio-loud and Radio-quiet QSOs. *Astrophys. J.* **2002**, *565*, 78–85. [CrossRef]
35. Xu, D.; Komossa, S.; Zhou, H.; Lu, H.; Li, C.; Grupe, D.; Wang, J.; Yuan, W. Correlation Analysis of a Large Sample of Narrow-line Seyfert 1 Galaxies: Linking Central Engine and Host Properties. *Astron. J.* **2012**, *143*, 83. [CrossRef]
36. Grupe, D.; Komossa, S.; Leighly, K.M.; Page, K.L. The Simultaneous Optical-to-X-ray Spectral Energy Distribution of Soft X-ray Selected Active Galactic Nuclei Observed by Swift. *Astrophys. J. Suppl. Ser.* **2010**, *187*, 64–106. [CrossRef]
37. Komossa, S. Narrow-line Seyfert 1 Galaxies. In Proceedings of the Revista Mexicana de Astronomia y Astrofisica Conference Series. *Revista Mexicana de Astronomia y Astrofisica Conference Series*, 2008, Volume 32, pp. 86–92. Available online: https://www.redalyc.org/pdf/571/57103230.pdf (accessed on 15 January 2024).
38. Gallo, L. X-ray perspective of Narrow-line Seyfert 1 galaxies. In Proceedings of the Revisiting Narrow-Line Seyfert 1 Galaxies and Their Place in the Universe, Padova, Italy, 9–13 April 2018; p. 34. [CrossRef]
39. Lister, M. Radio Properties of Narrow-Line Seyfert 1 Galaxies. In Proceedings of the Revisiting Narrow-Line Seyfert 1 Galaxies and Their Place in the Universe, Padova, Italy, 9–13 April 2018; p. 22. [CrossRef]
40. D'Ammando, F. Relativistic Jets in Gamma-Ray-Emitting Narrow-Line Seyfert 1 Galaxies. *Galaxies* **2019**, *7*, 87. [CrossRef]
41. Paliya, V.S. Gamma-ray emitting narrow-line Seyfert 1 galaxies: Past, present, and future. *J. Astrophys. Astron.* **2019**, *40*, 39. [CrossRef]
42. Komossa, S. Multi-wavelength properties of radio-loud Narrow-line Seyfert 1 galaxies. In Proceedings of the Revisiting Narrow-Line Seyfert 1 Galaxies and Their Place in the Universe, Padova, Italy, 9–13 April 2018; p. 15. [CrossRef]
43. Puchnarewicz, E.M.; Mason, K.O.; Cordova, F.A.; Kartje, J.; Brabduardi-A A Puchnarewicz, E.M.; Mason, K.O.; Cordova, F.A.; Kartje, J.; Branduardi-Raymont, G.; Mittaz, J.P.D.; et al. Optical properties of active galaxies with ultra-soft X-ray spectra. *Mon. Not. R. Astron. Soc.* **1992**, *256*, 589. [CrossRef]
44. Komossa, S.; Voges, W.; Xu, D.; Mathur, S.; Adorf, H.M.; Lemson, G.; Duschl, W.J.; Grupe, D. Radio-loud Narrow-Line Type 1 Quasars. *Astron. J.* **2006**, *132*, 531–545. [CrossRef]
45. Yuan, W.; Zhou, H.Y.; Komossa, S.; Dong, X.B.; Wang, T.G.; Lu, H.L.; Bai, J.M. A Population of Radio-Loud Narrow-Line Seyfert 1 Galaxies with Blazar-Like Properties? *Astrophys. J.* **2008**, *685*, 801–827. [CrossRef]
46. Abdo, A.A.; Ackermann, M.; Ajello, M.; Baldini, L.; Ballet, J.; Barbiellini, G.; Bastieri, D.; Bechtol, K.; Bellazzini, R.; Berenji, B.; et al. Radio-Loud Narrow-Line Seyfert 1 as a New Class of Gamma-Ray Active Galactic Nuclei. *Astrophys. J. Lett.* **2009**, *707*, L142–L147. [CrossRef]
47. Lähteenmäki, A.; Järvelä, E.; Ramakrishnan, V.; Tornikoski, M.; Tammi, J.; Vera, R.J.C.; Chamani, W. Radio jets and gamma-ray emission in radio-silent narrow-line Seyfert 1 galaxies. *Astron. Astrophys.* **2018**, *614*, L1. [CrossRef]
48. Berton, M.; Järvelä, E.; Crepaldi, L.; Lähteenmäki, A.; Tornikoski, M.; Congiu, E.; Kharb, P.; Terreran, G.; Vietri, A. Absorbed relativistic jets in radio-quiet narrow-line Seyfert 1 galaxies. *Astron. Astrophys.* **2020**, *636*, A64. [CrossRef]
49. Järvelä, E.; Savolainen, T.; Berton, M.; Lähteenmäki, A.; Kiehlmann, S.; Hovatta, T.; Varglund, I.; Readhead, A.C.S.; Tornikoski, M.; Max-Moerback, W.; et al. Unprecedented extreme high-frequency radio variability in early-stage active galactic nuclei. *arXiv* **2023**, arXiv:2312.02326. [CrossRef]
50. Sulentic, J.W.; Marziani, P.; Dultzin-Hacyan, D. Phenomenology of Broad Emission Lines in Active Galactic Nuclei. *Annu. Rev. Astron. Astrophys.* **2000**, *38*, 521–571. [CrossRef]
51. Marziani, P.; Dultzin, D.; Sulentic, J.W.; Del Olmo, A.; Negrete, C.A.; Martínez-Aldama, M.L.; D'Onofrio, M.; Bon, E.; Bon, N.; Stirpe, G.M. A main sequence for quasars. *Front. Astron. Space Sci.* **2018**, *5*, 6. [CrossRef]
52. Zhang, J.; Zhang, H.M.; Gan, Y.Y.; Yi, T.F.; Wang, J.F.; Liang, E.W. Jet Properties of Compact Steep-spectrum Sources and an Eddington-ratio-driven Unification Scheme of Jet Radiation in Active Galactic Nuclei. *Astrophys. J.* **2020**, *899*, 2. [CrossRef]
53. Kellermann, K.I.; Sramek, R.; Schmidt, M.; Shaffer, D.B.; Green, R. VLA Observations of Objects in the Palomar Bright Quasar Survey. *Astron. J.* **1989**, *98*, 1195. [CrossRef]
54. Paliya, V.S.; Stalin, C.S.; Domínguez, A.; Saikia, D.J. Narrow-line Seyfert 1 galaxies in Sloan Digital Sky Survey: A new optical spectroscopic catalogue. *Mon. Not. R. Astron. Soc.* **2024**, *527*, 7055–7069. [CrossRef]
55. Cohen, R.D.; Barlow, T.A.; Beaver, E.A.; Junkkarinen, V.T.; Lyons, R.W.; Smith, H.E. Conditions in the Z = 0.692 Absorber toward 3CR 286. *Astrophys. J.* **1994**, *421*, 453. [CrossRef]
56. Steidel, C.C.; Pettini, M.; Dickinson, M.; Persson, S.E. Imaging of Two Damped Lyman Alpha Absorbers at Intermediate Redshifts. *Astron. J.* **1994**, *108*, 2046. [CrossRef]
57. Le Brun, V.; Bergeron, J. Lyalpha absorbers at z <= 1: HST-CFHT imaging and spectroscopy in the field of 3C 286. *Astron. Astrophys.* **1998**, *332*, 814–824. [CrossRef]

58. Brown, R.L.; Roberts, M.S. 21-CENTIMETER Absorption at ≈ = 0.692 in the Quasar 3c 286. *Astrophys. J. Lett.* **1973**, *184*, L7. [CrossRef]
59. Ackermann, M.; Ajello, M.; Atwood, W.B.; Baldini, L.; Ballet, J.; Barbiellini, G.; Bastieri, D.; Becerra Gonzalez, J.; Bellazzini, R.; Bissaldi, E.; et al. The Third Catalog of Active Galactic Nuclei Detected by the Fermi Large Area Telescope. *Astrophys. J.* **2015**, *810*, 14. [CrossRef]
60. Ballet, J.; Bruel, P.; Burnett, T.H.; Lott, B.; The Fermi-LAT Collaboration. Fermi Large Area Telescope Fourth Source Catalog Data Release 4 (4FGL-DR4). *arXiv* **2023**, arXiv:2307.12546. [CrossRef]
61. Principe, G.; Di Venere, L.; Orienti, M.; Migliori, G.; D'Ammando, F.; Mazziotta, M.N.; Giroletti, M. Gamma-ray emission from young radio galaxies and quasars. *Mon. Not. R. Astron. Soc.* **2021**, *507*, 4564–4583. [CrossRef]
62. Fichtel, C.E.; Bertsch, D.L.; Chiang, J.; Dingus, B.L.; Esposito, J.A.; Fierro, J.M.; Hartman, R.C.; Hunter, S.D.; Kanbach, G.; Kniffen, D.A.; et al. The First Energetic Gamma-Ray Experiment Telescope (EGRET) Source Catalog. *Astrophys. J. Suppl. Ser.* **1994**, *94*, 551. [CrossRef]
63. Foschini, L.; Lister, M.L.; Andernach, H.; Ciroi, S.; Marziani, P.; Antón, S.; Berton, M.; Dalla Bontà, E.; Järvelä, E.; Marchã, M.J.M.; et al. A New Sample of Gamma-Ray Emitting Jetted Active Galactic Nuclei. *Universe* **2022**, *8*, 587. [CrossRef]
64. Foschini, L.; Berton, M.; Caccianiga, A.; Ciroi, S.; Cracco, V.; Peterson, B.M.; Angelakis, E.; Braito, V.; Fuhrmann, L.; Gallo, L.; et al. Properties of flat-spectrum radio-loud narrow-line Seyfert 1 galaxies. *Astron. Astrophys.* **2015**, *575*, A13. [CrossRef]
65. D'Ammando, F. Gamma-ray-emitting narrow-line Seyfert 1 galaxies: The Swift view. *Mon. Not. R. Astron. Soc.* **2020**, *496*, 2213–2229. [CrossRef]
66. Yao, S.; Komossa, S. Multiwavelength variability of γ-ray emitting narrow-line Seyfert 1 galaxies. *Mon. Not. R. Astron. Soc.* **2023**, *523*, 441–452. [CrossRef]
67. Tananbaum, H.; Wardle, J.F.C.; Zamorani, G.; Avni, Y. X-ray studies of quasars with the Einstein Observatory. III. The 3CR sample. *Astrophys. J.* **1983**, *268*, 60–67. [CrossRef]
68. Voges, W.; Aschenbach, B.; Boller, T.; Brauninger, H.; Briel, U.; Burkert, W.; Dennerl, K.; Englhauser, J.; Gruber, R.; Haberl, F.; et al. Rosat All-Sky Survey Faint Source Catalogue. *IAU Circ.* **2000**, *7432*, 3.
69. Massaro, F.; Harris, D.E.; Liuzzo, E.; Orienti, M.; Paladino, R.; Paggi, A.; Tremblay, G.R.; Wilkes, B.J.; Kuraszkiewicz, J.; Baum, S.A.; et al. The Chandra Survey of Extragalactic Sources in the 3CR Catalog: X-ray Emission from Nuclei, Jets, and Hotspots in the Chandra Archival Observations. *Astrophys. J. Suppl. Ser.* **2015**, *220*, 5. [CrossRef]
70. Gehrels, N.; Chincarini, G.; Giommi, P.; Mason, K.O.; Nousek, J.A.; Wells, A.A.; White, N.E.; Barthelmy, S.D.; Burrows, D.N.; Cominsky, L.R.; et al. The Swift Gamma-Ray Burst Mission. *Astrophys. J.* **2004**, *611*, 1005–1020. [CrossRef]
71. Poole, T.S.; Breeveld, A.A.; Page, M.J.; Landsman, W.; Holland, S.T.; Roming, P.; Kuin, N.P.M.; Brown, P.J.; Gronwall, C.; Hunsberger, S.; et al. Photometric calibration of the Swift ultraviolet/optical telescope. *Mon. Not. R. Astron. Soc.* **2008**, *383*, 627–645. [CrossRef]
72. Burrows, D.N.; Hill, J.E.; Nousek, J.A.; Kennea, J.A.; Wells, A.; Osborne, J.P.; Abbey, A.F.; Beardmore, A.; Mukerjee, K.; Short, A.D.T.; et al. The Swift X-ray Telescope. *Space Sci. Rev.* **2005**, *120*, 165–195. [CrossRef]
73. Hill, J.E.; Burrows, D.N.; Nousek, J.A.; Abbey, A.F.; Ambrosi, R.M.; Bräuninger, H.W.; Burkert, W.; Campana, S.; Cheruvu, C.; Cusumano, G.; et al. Readout modes and automated operation of the Swift X-ray Telescope. In *X-ray and Gamma-ray Instrumentation for Astronomy XIII*; Flanagan, K.A., Siegmund, O.H.W., Eds.; Society of Photo-Optical Instrumentation Engineers (SPIE) Conference Series; Society of Photo Optical: San Diego, CA, USA, 2004; Volume 5165, pp. 217–231. [CrossRef]
74. Arnaud, K.A. XSPEC: The First Ten Years. In *Astronomical Data Analysis Software and Systems V*; Jacoby, G.H., Barnes, J., Eds.; Astronomical Society of the Pacific Conference Series; Astronomical Society of the Pacific: San Francisco, CA, USA, 1996; Volume 101, p. 17.
75. Cash, W. Parameter estimation in astronomy through application of the likelihood ratio. *Astrophys. J.* **1979**, *228*, 939–947. [CrossRef]
76. Arnaud, K.; Gordon, C.; Dorman, B.; Rutkowski, K. XSPEC: An X-ray Spectral Fitting Package. 2024. Available online: https://heasarc.gsfc.nasa.gov/xanadu/xspec/manual/XspecManual.html (accessed on 15 June 2024).
77. HI4PI Collaboration; Ben Bekhti, N.; Flöer, L.; Keller, R.; Kerp, J.; Lenz, D.; Winkel, B.; Bailin, J.; Calabretta, M.R.; Dedes, L.; et al. HI4PI: A full-sky H I survey based on EBHIS and GASS. *Astron. Astrophys.* **2016**, *594*, A116. [CrossRef]
78. Wilms, J.; Allen, A.; McCray, R. On the Absorption of X-rays in the Interstellar Medium. *Astrophys. J.* **2000**, *542*, 914–924. [CrossRef]
79. Yao, S.; Komossa, S.; Kraus, A.; Grupe, D. First deep X-ray observations of 3C 286. *Mon. Not. R. Astron. Soc.* 2024, *submitted*.
80. Roming, P.W.A.; Kennedy, T.E.; Mason, K.O.; Nousek, J.A.; Ahr, L.; Bingham, R.E.; Broos, P.S.; Carter, M.J.; Hancock, B.K.; Huckle, H.E.; et al. The Swift Ultra-Violet/Optical Telescope. *Space Sci. Rev.* **2005**, *120*, 95–142. [CrossRef]
81. Cenko, B. Swift satellite in safe mode. *GRB Coord. Netw.* **2024**, *35953*, 1.
82. Breeveld, A.A.; Curran, P.A.; Hoversten, E.A.; Koch, S.; Landsman, W.; Marshall, F.E.; Page, M.J.; Poole, T.S.; Roming, P.; Smith, P.J.; et al. Further calibration of the Swift ultraviolet/optical telescope. *Mon. Not. R. Astron. Soc.* **2010**, *406*, 1687–1700. [CrossRef]
83. Cardelli, J.A.; Clayton, G.C.; Mathis, J.S. The Relationship between Infrared, Optical, and Ultraviolet Extinction. *Astrophys. J.* **1989**, *345*, 245. [CrossRef]

84. Komossa, S.; Grupe, D.; Kraus, A.; Gallo, L.C.; Gonzalez, A.G.; Parker, M.L.; Valtonen, M.J.; Hollett, A.R.; Bach, U.; Gómez, J.L.; et al. Project MOMO: Multiwavelength Observations and Modeling of OJ 287. *Universe* **2021**, *7*, 261. [CrossRef]
85. Komossa, S.; Kraus, A.; Grupe, D.; Gonzalez, A.G.; Gurwell, M.A.; Gallo, L.C.; Liu, F.K.; Myserlis, I.; Krichbaum, T.P.; Laine, S.; et al. MOMO. VI. Multifrequency Radio Variability of the Blazar OJ 287 from 2015 to 2022, Absence of Predicted 2021 Precursor-flare Activity, and a New Binary Interpretation of the 2016/2017 Outburst. *Astrophys. J.* **2023**, *944*, 177. [CrossRef]
86. Kraus, A.; Krichbaum, T.P.; Wegner, R.; Witzel, A.; Cimò, G.; Quirrenbach, A.; Britzen, S.; Fuhrmann, L.; Lobanov, A.P.; Naundorf, C.E.; et al. Intraday variability in compact extragalactic radio sources. II. Observations with the Effelsberg 100 m radio telescope. *Astron. Astrophys.* **2003**, *401*, 161–172. [CrossRef]
87. Turlo, Z.; Forkert, T.; Sieber, W.; Wilson, W. Calibration of the instrumental polarization of radio telescopes. *Astron. Astrophys.* **1985**, *142*, 181–188.
88. Weisskopf, M.C.; Brinkman, B.; Canizares, C.; Garmire, G.; Murray, S.; Van Speybroeck, L.P. An Overview of the Performance and Scientific Results from the Chandra X-ray Observatory. *Publ. Astron. Soc. Pac.* **2002**, *114*, 338108. [CrossRef]
89. Peng, B.; Kraus, A.; Krichbaum, T.P.; Witzel, A. Long-term monitoring of selected radio sources. *Astron. Astrophys. Suppl. Ser.* **2000**, *145*, 1–10. [CrossRef]
90. Agudo, I.; Thum, C.; Wiesemeyer, H.; Molina, S.N.; Casadio, C.; Gómez, J.L.; Emmanoulopoulos, D. 3C 286: A bright, compact, stable, and highly polarized calibrator for millimeter-wavelength observations. *Astron. Astrophys.* **2012**, *541*, A111. [CrossRef]
91. Perley, R.A.; Butler, B.J. An Accurate Flux Density Scale from 50 MHz to 50 GHz. *Astrophys. J. Suppl. Ser.* **2017**, *230*, 7. [CrossRef]
92. Kellermann, K.I.; Pauliny-Toth, I.I.K.; Williams, P.J.S. The Spectra of Radio Sources in the Revised 3c Catalogue. *Astrophys. J.* **1969**, *157*, 1. [CrossRef]
93. Kuehr, H.; Witzel, A.; Pauliny-Toth, I.I.K.; Nauber, U. A Catalogue of Extragalactic Radio Sources Having Flux Densities Greater than 1-JY at 5-GHZ. *Astron. Astrophys. Suppl. Ser.* **1981**, *45*, 367.
94. Gabányi, K.É.; Marchili, N.; Krichbaum, T.P.; Britzen, S.; Fuhrmann, L.; Witzel, A.; Zensus, J.A.; Müller, P.; Liu, X.; Song, H.G.; et al. The IDV source J 1128+5925, a new candidate for annual modulation? *Astron. Astrophys.* **2007**, *470*, 83–95. [CrossRef]
95. Massardi, M.; López-Caniego, M.; González-Nuevo, J.; Herranz, D.; de Zotti, G.; Sanz, J.L. Blind and non-blind source detection in WMAP 5-yr maps. *Mon. Not. R. Astron. Soc.* **2009**, *392*, 733–742. [CrossRef]
96. Mantovani, F.; Mack, K.H.; Montenegro-Montes, F.M.; Rossetti, A.; Kraus, A. Effelsberg 100-m polarimetric observations of a sample of compact steep-spectrum sources. *Astron. Astrophys.* **2009**, *502*, 61–65. [CrossRef]
97. Richards, J.L.; Max-Moerbeck, W.; Pavlidou, V.; King, O.G.; Pearson, T.J.; Readhead, A.C.S.; Reeves, R.; Shepherd, M.C.; Stevenson, M.A.; Weintraub, L.C.; et al. Blazars in the Fermi Era: The OVRO 40 m Telescope Monitoring Program. *Astrophys. J. Suppl. Ser.* **2011**, *194*, 29. [CrossRef]
98. Hull, C.L.H.; Girart, J.M.; Zhang, Q. 880 μm SMA Polarization Observations of the Quasar 3C 286. *Astrophys. J.* **2016**, *830*, 124. [CrossRef]
99. Pasetto, A.; Carrasco-González, C.; O'Sullivan, S.; Basu, A.; Bruni, G.; Kraus, A.; Curiel, S.; Mack, K.H. Broadband radio spectro-polarimetric observations of high-Faraday-rotation-measure AGN. *Astron. Astrophys.* **2018**, *613*, A74. [CrossRef]
100. Grasha, K.; Darling, J.; Bolatto, A.; Leroy, A.K.; Stocke, J.T. A Search for Intrinsic H I 21 cm and OH 18 cm Absorption toward Compact Radio Sources. *Astrophys. J. Suppl. Ser.* **2019**, *245*, 3. [CrossRef]
101. Meyer, D.M.; York, D.G. Observations of Zn ii, CR ii, Fe ii, and CA II in the Damped Lyman-Alpha Absorber at Z = 0.692 toward 3C 286. *Astrophys. J. Lett.* **1992**, *399*, L121. [CrossRef]
102. Gaskell, C.M. The case for cases B and C: Intrinsic hydrogen line ratios of the broad-line region of active galactic nuclei, reddenings, and accretion disc sizes. *Mon. Not. R. Astron. Soc.* **2017**, *467*, 226–238. [CrossRef]
103. Osterbrock, D.E. *Book: Astrophysics of Gaseous Nebulae and Active Galactic Nuclei;* University Science Books: Mill Valley, CA, USA, 1989.
104. Komossa, S.; Bade, N. Properties of dusty warm absorbers and the case of IRAS17020+4544. *Astron. Astrophys.* **1998**, *331*, L49–L52. [CrossRef]
105. Croston, J.H.; Birkinshaw, M.; Hardcastle, M.J.; Worrall, D.M. X-ray emission from the nuclei, lobes and hot-gas environments of two FR II radio galaxies. *Mon. Not. R. Astron. Soc.* **2004**, *353*, 879–889. [CrossRef]
106. O'Dea, C.P.; Saikia, D.J. Compact steep-spectrum and peaked-spectrum radio sources. *Annu. Rev. Astron. Astrophys.* **2021**, *29*, 3. [CrossRef]
107. Komossa, S.; Zensus, J.A. Compact object mergers: Observations of supermassive binary black holes and stellar tidal disruption events. In *Star Clusters and Black Holes in Galaxies across Cosmic Time;* Meiron, Y., Li, S., Liu, F.K., Spurzem, R., Eds.; Cambridge University Press: Cambridge, UK, 2016; Volume 312, pp. 13–25. [CrossRef]
108. De Rosa, A.; Vignali, C.; Bogdanović, T.; Capelo, P.R.; Charisi, M.; Dotti, M.; Husemann, B.; Lusso, E.; Mayer, L.; Paragi, Z.; et al. The quest for dual and binary supermassive black holes: A multi-messenger view. *New Astron. Rev.* **2019**, *86*, 101525. [CrossRef]
109. de Vries, W.H.; O'Dea, C.P.; Baum, S.A.; Sparks, W.B.; Biretta, J.; de Koff, S.; Golombek, D.; Lehnert, M.D.; Macchetto, F.; McCarthy, P.; et al. Hubble Space Telescope Imaging of Compact Steep-Spectrum Radio Sources. *Astrophys. J. Suppl. Ser.* **1997**, *110*, 191–211. [CrossRef]
110. Stawarz, Ł.; Ostorero, L.; Begelman, M.C.; Moderski, R.; Kataoka, J.; Wagner, S. Evolution of and High-Energy Emission from GHz-Peaked Spectrum Sources. *Astrophys. J.* **2008**, *680*, 911–925. [CrossRef]

111. Migliori, G.; Siemiginowska, A.; Kelly, B.C.; Stawarz, Ł.; Celotti, A.; Begelman, M.C. Jet Emission in Young Radio Sources: A Fermi Large Area Telescope Gamma-Ray View. *Astrophys. J.* **2014**, *780*, 165. [CrossRef]
112. Abdo, A.A.; Ackermann, M.; Ajello, M.; Atwood, W.B.; Baldini, L.; Ballet, J.; Barbiellini, G.; Bastieri, D.; Baughman, B.M.; Bechtol, K.; et al. Fermi Gamma-Ray Imaging of a Radio Galaxy. *Science* **2010**, *328*, 725. [CrossRef] [PubMed]
113. Yu, Y.W.; Zhang, H.M.; Gan, Y.Y.; Hu, X.K.; Wu, T.Z.; Zhang, J. Fermi Large Area Telescope Detection of Gamma-Rays from NGC 6251 Radio Lobe. *arXiv* **2024**, arXiv:2402.04591. [CrossRef]
114. Kino, M.; Ito, H.; Kawakatu, N.; Nagai, H. New prediction of extragalactic GeV γ-ray emission from radio lobes of young AGN jets. *Mon. Not. R. Astron. Soc.* **2009**, *395*, L43–L47. [CrossRef]
115. Bosch-Ramon, V.; Perucho, M.; Barkov, M.V. Clouds and red giants interacting with the base of AGN jets. *Astron. Astrophys.* **2012**, *539*, A69. [CrossRef]
116. Kaspi, S.; Maoz, D.; Netzer, H.; Peterson, B.M.; Vestergaard, M.; Jannuzi, B.T. The Relationship between Luminosity and Broad-Line Region Size in Active Galactic Nuclei. *Astrophys. J.* **2005**, *629*, 61–71. [CrossRef]
117. Evans, P.A.; Beardmore, A.P.; Page, K.L.; Tyler, L.G.; Osborne, J.P.; Goad, M.R.; O'Brien, P.T.; Vetere, L.; Racusin, J.; Morris, D.; et al. An online repository of Swift/XRT light curves of γ-ray bursts. *Astron. Astrophys.* **2007**, *469*, 379–385. [CrossRef]
118. Wright, E.L.; Eisenhardt, P.R.M.; Mainzer, A.K.; Ressler, M.E.; Cutri, R.M.; Jarrett, T.; Kirkpatrick, J.D.; Padgett, D.; McMillan, R.S.; Skrutskie, M.; et al. The Wide-field Infrared Survey Explorer (WISE): Mission Description and Initial On-orbit Performance. *Astron. J.* **2010**, *140*, 1868–1881. [CrossRef]

universe MDPI

Article

Changing-Look Narrow-Line Seyfert 1 Galaxies, their Detection with SVOM, and the Case of NGC 1566

D. W. Xu [1,2,*], S. Komossa [3,1,*], D. Grupe [4], J. Wang [5,1], L. P. Xin [1], X. H. Han [1], J. Y. Wei [1,2], J. Y. Bai [1], E. Bon [6] F. Cangemi [7,8], B. Cordier [9], M. Dennefeld [10], L. C. Gallo [11], W. Kollatschny [12], De-Feng Kong [5], M. W. Ochmann [12,13], Y. L. Qiu [1] and N. Schartel [14]

1 Key Laboratory of Space Astronomy and Technology, National Astronomical Observatories, Chinese Academy of Sciences, Beijing 100101, China
2 School of Astronomy and Space Science, University of Chinese Academy of Sciences, Beijing 100049, China
3 Max–Planck Institut für Radioastronomie, Auf dem Hügel 69, 53121 Bonn, Germany
4 Department of Physics, Geology, and Engineering Technology, Northern Kentucky University, 1 Nunn Drive, Highland Heights, KY 41099, USA
5 Guangxi Key Laboratory for Relativistic Astrophysics, School of Physical Science and Technology, Guangxi University, Nanning 530004, China
6 Astronomical Observatory Belgrade, Volgina 7, 11060 Belgrade, Serbia
7 Sorbonne Université, CNRS/IN2P3, Laboratoire de Physique Nucléaire et de Hautes Energies, LPNHE, 4 Place Jussieu, 75005 Paris, France
8 Université Paris-Cité, AstroParticules et Cosmologie, APC, 10 rue Alice Domon et Léonie Duquet, 75013 Paris, France
9 CEA-Saclay, IRFU/Departement d'Astrophysique, 91191 Gif-sur-Yvette, France
10 Institut d'Astrophysique de Paris, Sorbonne Université, CNRS, UMR 7095, 98 bis bd Arago, 75014 Paris, France
11 Department of Astronomy and Physics, Saint Mary's University, 923 Robie Street, Halifax, NS B3H 3C3, Canada
12 Institut für Astrophysik und Geophysik, Universität Göttingen, Friedrich-Hund Platz 1, 37077 Göttingen, Germany
13 Ruhr University Bochum, Faculty of Physics and Astronomy, Astronomical Institute (AIRUB), 44780 Bochum, Germany
14 ESA, European Space Astronomy Centre (ESAC), Villanueva de la Cañada, E-28692 Madrid, Spain
* Correspondence: dwxu@nao.cas.cn (D.W.X.); skomossa@mpifr.de (S.K.)

Abstract: We discuss applications of the study of the new and barely explored class of changing-look (CL) narrow-line Seyfert 1 (NLS1) galaxies and comment on their detection with the space mission SVOM (Space Variable Objects Monitor). We highlight the case of NGC 1566, which is outstanding in many respects, for instance as one of the nearest known CL AGN undergoing exceptional outbursts. Its NLS1 nature is discussed, and we take it as a nearby prototype for systems that could be discovered and studied in the near future, including with SVOM. Finally, we briefly examine the broader implications and applications of CL events in NLS1 galaxies and show that such systems, once discovered in larger numbers, will greatly advance our understanding of the physics of the environment of rapidly growing supermassive black holes. This White Paper is part of a sequence of publications which explore aspects of our understanding of (CL) NLS1 galaxy physics with future missions.

Keywords: active galactic nuclei; Seyfert galaxies; NGC 1566; broad-line region; supermassive black holes; space mission SVOM; accretion disks; optical spectra; X-ray spectra

1. Introduction

Highly variable and transient active galactic nuclei (AGN) provide us with important insights into the physics of the accretion of matter onto their central supermassive black holes (SMBHs). Of particular interest are changing-look (CL) AGN, which change their optical Seyfert-type (from type 1 to type 2, or vice versa) in response to high-amplitude continuum variability. First recognized in the 1970s, such systems have been more recently identified in larger numbers based on dedicated searches for either strong variability in the optical or X-ray continuum emission, or alternatively in the broad emission lines (see [1] for a review). Several models have been suggested to explain CL AGN. Some have explored extinction variability, for instance through dusty material crossing our line of sight or alternatively by dust destruction in response to an outburst [2]. The majority of theoretical models has focused on short-term changes in the accretion rate (on the time scale of months–decades), for instance due to disk instabilities or other mechanisms (e.g., the radiation pressure disk instability, the Hydrogen ionization instability, magnetic-pressure-supported disks, magnetic flux inversion, instabilities in warped disks at large radii, or shock formation in highly precessing disks) (e.g., [3–15]) or have explored the possibility of tidal disk interactions in binary SMBH systems [16].

The subgroup of AGN that has shown particularly high amplitudes of variability in the X-ray band are narrow-line Seyfert 1 (NLS1) galaxies (review by [17]). Therefore, the following question is raised: how common is the CL phenomenon among NLS1 galaxies? This subgroup of AGN is defined by their remarkable optical spectroscopic emission-line characteristics: narrow widths of their broad Balmer lines from the broad-line region (BLR) with FWHM(Hβ) < 2000 km s^{-1} and an emission-line intensity ratio of [O III]$\lambda5007/\mathrm{H}\beta_{\mathrm{total}} < 3$ [18,19]. The presence of Fe II complexes is often added as additional defining criterion [20]. A number of NLS1 galaxies also show strong high-ionization forbidden lines in their optical spectra.

NLS1 galaxies are of great interest because they are at one extreme end of the AGN correlation space [1], and trace AGN physics under extreme conditions (e.g., at low SMBH masses and high (near-Eddington) accretion rates). Sky surveys have successfully provided us with large samples of thousands of NLS1 galaxies. For instance, a large fraction of the soft X-ray AGN identified with the former space mission ROSAT turned out to be NLS1 galaxies [22,23]. More recently, large numbers of NLS1 galaxies were selected from the Sloan Digital Sky Survey (SDSS) [24–26].

NLS1 galaxies, as a class, are characterized by small SMBH masses and accretion rates near the Eddington limit ([27–29]; see the review in Komossa [30]). They therefore represent systems of rapidly growing SMBHs in the local universe. Given their near-Eddington accretion rates, they should be particularly prone to CL events under those mechanisms which operate close to Eddington. Here, we discuss implications and applications of the CL phenomenon in NLS1 galaxies, and discuss the detectability of such systems with the upcoming space mission SVOM. We highlight the example of the nearby CL NLS1 galaxy NGC 1566. We use a cosmology with $H_0 = 70$ km s^{-1} Mpc^{-1}, $\Omega_M = 0.3$ and $\Omega_\Lambda = 0.7$ and the cosmology calculator of Wright [31]. (In the case of NGC 1566, this corresponds to a Hubble-flow distance of 21.5 Mpc.)

2. CL Events in NLS1 Galaxies

Only few NLS1-type galaxies have so far been identified as CL AGN (Table 1) [2]. This Section provides a brief overview of their properties, and critically discusses the need to distinguish from other phenomena like type II SNe in dense media and (very rare) stellar tidal disruption events.

The AGN IC 3599 (Zwicky 159.034) harbored an extreme CL event, dramatically changing its optical emission-line spectrum following a high-amplitude soft X-ray (and optical) outburst with a peak luminosity of $L_x = 10^{44}$ erg s^{-1} [33,34]. Optical spectra taken during the first outburst revealed bright optical emission lines, including broad Balmer lines with a FWHM of 1200 km s^{-1} and high-ionization iron lines of similar width [33]. These then faded away strongly in subsequent years. In low-state, IC 3599 is classified as Sy 1.9 galaxy with faint broad-line emission only detectable in Hα and with a long-lived narrow-line region (NLR) including faint high-ionization iron lines [35]. While IC 3599 is sometimes referred to as NLS1 galaxy based on its strong variability and its FWHM(Hβ) $\sim$1200 km s^{-1}, it is not a typical NLS1 because no Fe II emission was detected at high-state, and because all permitted and forbidden lines had the same width at high-state more typical of a coronal-line region (CLR). However, in any case, IC 3599 can certainly be regarded as an extreme case of a CL AGN.

The high-amplitude outbursts of IC 3599 were suggested by Grupe et al. [5] to be due to possible thermal instabilities in the accretion disk, propagating through the inner region at the sound speed and causing an increase in the local accretion rate. Episodes of repeat flaring are then produced on the timescale of decades when the inner disk empties and refills. The two bright outbursts of IC 3599 were separated by 20 yrs. Other scenarios to produce repeated photoionizing outbursts, like an OJ 287-type binary SMBH scenario (where a secondary SMBH might interact in one way or another with the accretion disk around a primary SMBH), remain possible [36].

Table 1. CL events among candidate NLS1 galaxies. Column (1) provides the galaxy name, column (2) provides its redshift z, column (3) provides the total amplitude of variability in X-rays (or in the optical band if no X-ray observations were available), column (4) provides the FWHM of the broad Hβ emission line, and column (5) provides the variability time scale Δt_{var} between the optical spectroscopic change (from one type to another). Since often only two spectra exist instead of a dense monitoring, it is well possible that the AGN changed its type multiple times in between, but simply no spectra were taken to record this. Δt_{var} therefore usually is an upper limit. NGC 1566 changed within months from one type to the other; here, Δt_{var} of the fastest recorded change is listed. Column (6) lists further comments. For references on the source parameters, see Section 2.

Name (1)	z (2)	Amplitude of var. in X-rays (3)	FWHM(Hβ) in km s^{-1} (4)	Δt_{var} in yrs (5)	Comments (6)
IC 3599	0.021	>100	∼1200	0.75	two outbursts separated by 20 yrs no bona-fide NLS1, no Fe II
SDSS J123359.12 + 084211.5	0.256	∼2 (opt)	2430	11	change in both Balmer and Fe II emission
ZTF18aajupnt	0.037	10	940	16	spectral change from LINER to NLS1
J1406507–244250	0.046	∼1 (opt)	3000	18	little optical continuum variability
NGC 1566	0.005	30	1950	0.33	repeat Seyfert-type changes

SDSS J123359.12 + 084211.5 was identified as CL NLS1 [37] with changes in Balmer lines and particularly in Fe II emission. In the high state, its FWHM(Hβ) was 2430 km s^{-1} [38]. The continuum itself varied by a factor ∼2.

ZTF18aajupnt was classified as NLS1 in the high state, with a LINER spectrum in its low state [39]. Its high-state spectrum shows strong coronal lines. A luminous soft X-ray and mid-infrared flare was detected.

J1406507–244250 was identified from 6dFGS as a candidate CL NLS1 [40]. Its observed optical continuum is about constant, but its Hβ line was found to change its FWHM from 500 km s^{-1} to 3000 km s^{-1}, leading the authors to speculate about a possible NLS1 classification, taking the FWHM classification criterion loosely.

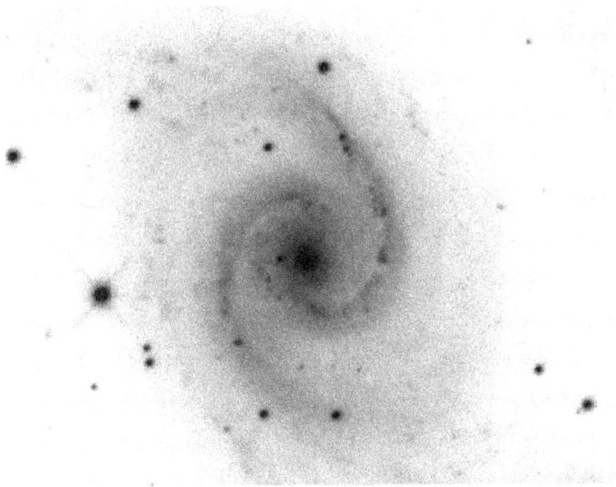

Figure 1. Optical image of NGC 1566 taken on 2023 October 19 with the LCOGT (Las Campanas Observatory Global Telescope) 1 m telescope at Siding Spring Observatory. North is up, east is left.

There are also CL impostors. Some other events with transient broad lines have been identified in AGN, but their nature remains uncertain. This is because two other transient phenomena are accompanied by the temporary emission of optical continuum and broad lines, therefore potentially mimicking a CL AGN. The first transients are supernovae (SN) of type IIn [41] which show temporary broad and narrow emission lines (which are believed to arise in interactions of the SN ejecta with dense circum-stellar matter), the other transients are optical stellar tidal disruption events (TDEs; [42]). These are often accompanied by broad emission lines of the Balmer series and/or He II [43,44] thought to arise from the temporary accretion disk [45]. Some TDEs also show transient high-ionization coronal lines [43,46,47]. If TDEs occur in quiescent, inactive host galaxies, then they can be identified from the host galaxy spectra showing inactivity before the transient event and/or long after the transient event. However, the situation is more complicated when the galaxy is an AGN, such that it is not immediately clear whether the change comes from an accretion disk (instability) or, less likely, from a (very rare) TDE or an SN IIn. The situation is also challenging when there is no pre-outburst spectrum available, or when there are only a few post-outburst spectra, so that long-lived AGN, rare TDE, or SN IIn scenarios cannot immediately be distinguished based on optical spectra alone (see the detailed discussion of Komossa et al. [43] and Komossa et al. [48]; and see Drake et al. [49], Blanchard et al. [50] and Zhang et al. [51] for examples in NLS1-like galaxies).

TDEs are distinguished by their supersoft X-ray emission [35,52,53], except for jetted TDEs which are associated with radio emission and have much harder X-ray spectra [54,55] and could therefore be directly triggered with the ECLAIRs instrument aboard SVOM [56]. SN IIn, on the other hand, are distinguished by the absence of (long-lived) bright X-ray emission, or the absence of X-rays altogether. Further, TDEs are very rare events with a rate of 10^{-4}–10^{-5} events per galaxy per year [57], and we therefore would not expect any significant number of these in the most nearby galaxies, if any at all. While TDEs themselves are another interesting source population for SVOM, these will not be discussed further here. Many will be detected with the X-ray satellite Einstein Probe [58], a Chinese mission with European participation which was launched in January 2024.

Why have so few CL events been observed so far in NLS1 galaxies? Balmer lines in NLS1 galaxies are bright because of high accretion rates near the Eddington limit in NLS1 galaxies as a class. Therefore, a *strong* change in BLR luminosity is required before a system is identified as CL event (instead, if the broad lines are already faint in the first place, their disappearance and therefore a change to the type 2 state would be much more easily detectable). Ideally, CL events should therefore require a threshold amplitude of variability in the Balmer lines and continuum flux of a factor of, e.g., >10, in order to identify CL systems homogeneously among AGN. Otherwise, AGN with faint (broad) emission lines would be preferentially classified as CL systems.

Finally, the particularly interesting case of CL events in NGC 1566 will be further discussed in the next Section, including new data taken in 2023 October–November.

3. The NLS1 Nature of NGC 1566 and the Feasibility of Detecting CL NLS1 Galaxies with SVOM

NGC 1566 (Figure 1) is the nearest known CL AGN, *repeatedly* changing its Seyfert type.[3] Here, we argue for its NLS1 classification. Initially known as a bright type 1 AGN [64], it became much fainter in subsequent years. Multiple spectroscopic observations between 1980 and 1982 found it to be a type 2 Seyfert in 1980, then gradually changing back to a type 1 Seyfert, and another CL event happened in 1985 [65,66]. Most of the changes occurred within only 4 months. The Balmer decrement of the broad lines did not change significantly during the CL event [65].

In 2018, an INTEGRAL detection of NGC 1566 at hard X-rays [67] triggered multi-wavelength (MWL) follow-up observations [68–71], revealing high-amplitude variability. These led to the identification of a new CL event in NGC 1566 [72–74].

The width of the broad Hβ line in NGC 1566 is relatively narrow. da Silva et al. [75] measured FWHM(Hβ) = 1950 $\pm$ 20 km s^{-1} and FWHM(Hα) = 1970 $\pm$ 10 km s^{-1} (see also [65,76]). Together, with a small ratio of [O III]λ5007/H$\beta_{\rm total}$ < 3, NGC 1566 fulfills the classification of an NLS1 galaxy. Fe II emission is detected in the type 1 and 1.5 spectral state [76]. This makes NGC 1566 the nearest of the few CL cases detected so far in NLS1 galaxies.

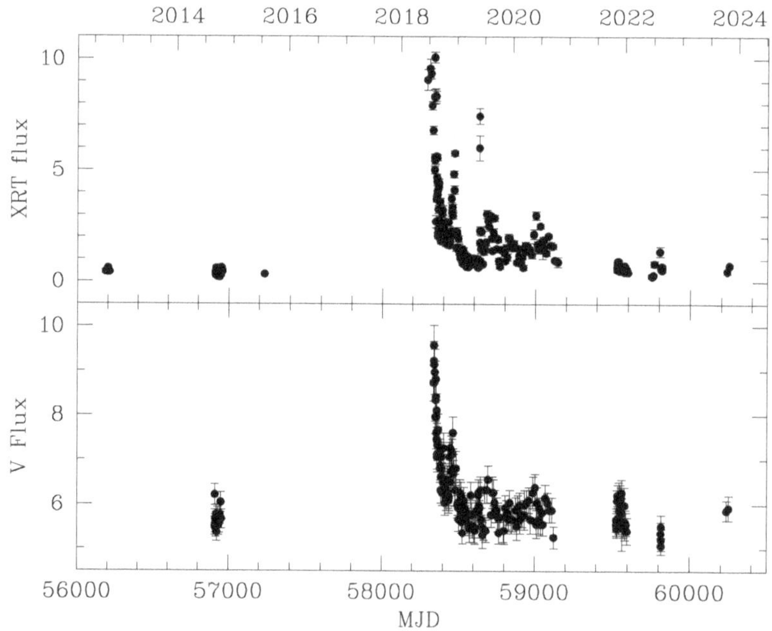

Figure 2. Swift lightcurve of NGC 1566. Upper panel: X-ray flux in units of 10^{-11} erg cm^{-2} s^{-1} (0.3–10 keV), lower panel: optical V flux in units of 10^{-11} erg cm^{-2} s^{-1}. The latest bright outburst occurred in 2018 and was detected in all Swift bands. The last two data points are from 2023 October 19 and November 2, and show NGC 1566 in quiescence.

The X-ray spectrum of NGC 1566 has a strong, hard X-ray component, enabling its detection in high-energy surveys in outburst, as was the case with INTEGRAL in 2018. Parker et al. [72] reported the presence of a typical type 1 X-ray spectrum at high state, extending to high energies and subject to ionized absorption and mild reflection, based on rapid X-ray follow-up observations carried out with XMM-Newton, NuSTAR, and the Neil Gehrels Swift observatory (Swift hereafter). An outflow with velocity of 500 km s^{-1}, possibly launched by the outburst, was detected with the reflection grating spectrometer (RGS) of XMM-Newton. An earlier hard X-ray flare is evident in 2010 in the Swift BAT lightcurve [77].

The Swift long-term optical and X-ray lightcurve is displayed in Figure 2 (V band and X-ray flux) and Figure 3 (X-ray spectral shape, parameterized by a power law). The majority of observations was proposed by (some of) us, with the addition of archival data including serendipitous observations in 2022 due to an SN program. Our most recent measurements of NGC 1566 of 2023 October 19 and November 2 show it to be in its quiescent state.

The Swift [78] data reduction was carried out following standard procedures. In brief, the UV-optical telescope (UVOT; [79]) observes in three optical and three UV filters. The UV and optical bands in NGC 1566 are closely correlated and we representatively show the V band fluxes here. The V filter central wavelength is λ = 5468 Å. For further analysis, the datasets of each V-band observation were first co-added. Source counts were then extracted in a region of circular size with an extraction radius of 3 arcseconds centered on NGC 1566. A nearby area of 20 arcseconds radius was used to extract the background region. The background-corrected counts were converted into VEGA magnitudes and flux densities using the latest calibration (the flux densities in each filter were determined with the UVOT F-tool *uvotsource* with the parameter *apercorr* set to *curveofgrowth* in order to compensate for the deviation from the calibrated circular 5 arcsecond radius source extraction region; the smaller 3 arcsecond extraction region was chosen here to minimize the contribution of the host galaxy). The V-band fluxes are reported as flux density multiplied by the central frequency of the V filter. The last two UVOT data points of 2023 are affected by a slight, uncorrected drift motion of the satellite such that the sources are no longer point-like. Therefore, an elliptical extraction region was chosen instead. Finally, the data were corrected for Galactic reddening based on the reddening curves of Cardelli et al. [80]. The X-ray data were obtained with the Swift X-ray telescope (XRT; [81]). To carry out the timing and spectral analysis, we selected source photons within a circular area with a radius of 20 detector pixels, where one pixel corresponds to a scale of 2.36 arcseconds. Background photons were extracted in a nearby circular region. X-ray spectra of the source and background in the energy band 0.3–10 keV were generated. The spectral analysis was then carried out with the package XSPEC [82].

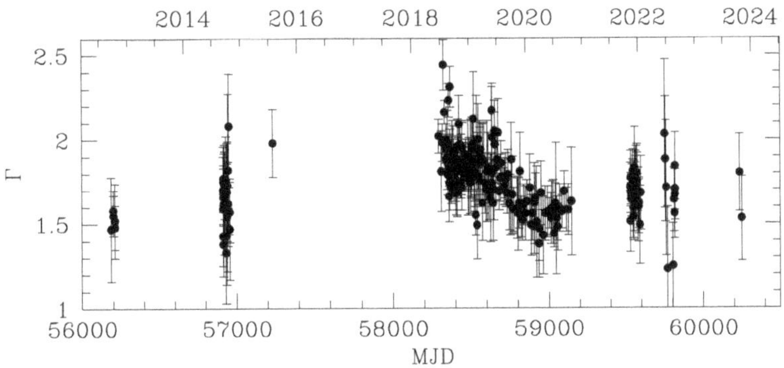

Figure 3. Variation in the X-ray photon index Γ_X of NGC 1566 from single-component powerlaw fits with absorption fixed at the Galactic value.

In order to measure its average X-ray spectral slope, we have merged all existing intermediate to low-state Swift observations of NGC 1566 into one single deep spectrum (amounting to a total exposure time of 0.26 Megaseconds) using the software tool of Evans et al. [83]. The spectrum was then fit with a single powerlaw subject to Galactic absorption ($N_H = 0.9 \times 10^{20}$ cm^{-2}). This results in an average powerlaw photon index of $\Gamma = 1.6$ (Figure 4) and demonstrates the rather hard X-ray spectrum of NGC 1566.

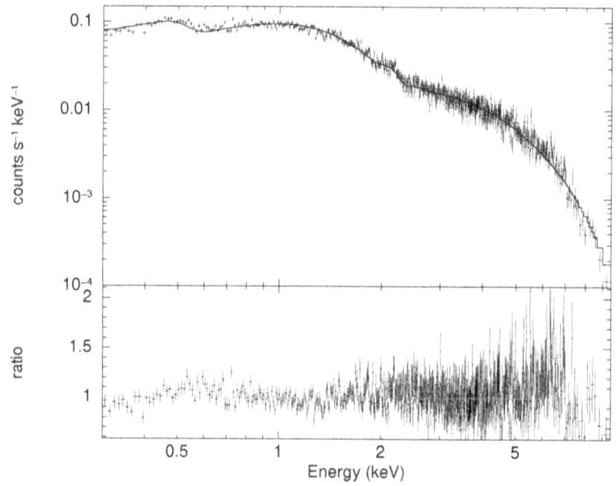

Figure 4. Composite Swift X-ray spectrum of NGC 1566 created from all intermediate to low-state observations taken in the PC (photon counting) mode, with a total exposure time of 264 ks, overbinned and fit with a single powerlaw spectrum.

CL systems like NGC 1566 will be important 'secondary science' targets for the SVOM mission. SVOM (Space Variable Objects Monitor) is a Sino-French space mission with international participation that is scheduled to launch in June 2024 [56] with four instruments onboard: two gamma-ray telescopes (ECLAIRs (4–150 keV) and GRM (15–5000 keV); [84–86]), one narrow-field X-ray telescope (MXT; 0.2–10 keV; [87,88]), and an optical telescope (VT; 400–650 nm and 650–1000 nm; [89]). The field of view of each detector is 2 sr (ECLAIRs and GRM), 64×64 arcmin^2 (MXT) and 26×26 arcmin^2 (VT), respectively. The satellite will be placed into a LEO (Low Earth Orbit) with an inclination of 30 degrees, an altitude of 625 km, and an orbital period of approximately 96 min. Its primary goal is the discovery and study of GRBs at all redshifts [56,90,91]. However, it will also enable a great wealth of other transient astrophysics, which is driven by proposals. The Target of Opportunity proposals are open to all scientists and are evaluated by the Principal Investigators. Here, we comment on the detection of CL NLS1 galaxies with SVOM, highlighting the galaxy NGC 1566 as a nearby prototype.

First, the initial trigger of an NGC 1566 like outburst would come from ECLAIRs. During the SWIFT BAT survey, NGC 1566 was detected at an average peak flux of 4 mCrab (or 10^{-10} erg cm^{-2} s^{-1}) in a one month interval during its 2010 outburst [77] in the BAT energy band of 14–195 keV with a powerlaw photon index $\Gamma = 2.2$. The sensitivity limit of ECLAIRs is 7.9×10^{-10} erg cm^{-2} s^{-1} for a 5 sigma detection in 1 ks [84,92,93]. During a new outburst, and at a flux level of 4 mCrab, NGC 1566 would be detected with ECLAIRs within 17 ks at 5 sigma. NGC 1566 would therefore be well above the detection threshold of ECLAIRs.

With the Swift XRT, NGC 1566 was not monitored during the 2010 outburst, but well during its 2018 outburst (Figure 2). Its (0.3–10 keV) peak flux was $f = 10^{-10}$ erg cm^{-2} s^{-1} [72]. A similar outburst would therefore be detected at 5 sigma with the SVOM MXT [87] with an integration time of 16 s. The detection limit of the VT is 22.5 mag in a 300 s observation, and systems like NGC 1566, or even much fainter, would be easily observable and followed-up with the VT, even in their low-emission state. Because of its two filters, VT also provides color information.

In general, AGN outbursts with an (isotropic) X-ray peak luminosity of 10^{43} erg s^{-1} will be detectable out to 200 Mpc with MXT in 10 ks.

4. Implications and Applications of Studying CL Events in NLS1 Galaxies

In addition to investigating the physics of the BLR in NLS1 galaxies and measuring the cross-band time delays, identification, and follow-ups of CL events in NLS1 galaxies have other important astrophysical applications.

First, the disk-instability models make CL NLS1 galaxies valuable targets in shedding light on the type and origin of the disk instability and the accretion disk physics in general under extreme physical conditions. By adopting values of the viscosity parameter α and the radiative efficiency parameter η of $\alpha = 0.1$ and $\eta = 0.1$, the timescales are estimated to be $\sim$10 and $\sim$6500 yrs for a $10^7\ M_\odot$ SMBH accreting at the Eddington limit in the thermal and viscous radial inflow scenarios (e.g., [94–96]), respectively. Additional mechanisms, such as elevation of the accretion disk by a magnetic field (e.g., [8,10]), or any of the other mechanisms discussed in the literature, may therefore be required to understand shorter timescales.

Second, the search for a population of type 2 NLS galaxies, if they exist [97], has been difficult to address, since the main NLS1 identification relies on the BLR properties (line width, Fe II strength, [O III]λ5007/Hβ_{total} ratio), unobservable in type 2 states. Few candidate type 2 NLS galaxies have therefore been identified so far, relying mostly on polarization observations (e.g., [98,99]). Once identified (no matter whether the type 2 state is long-lived and due to the unified model [100], or due to high-amplitude variability), then, in X-rays, type 2 states enable follow-up imaging spectroscopy. For example, the detection of X-ray lines from the circum-nuclear medium [101,102] and the determination of their excitation/ionization mechanism, or the detection of spatially resolved optical outflows, are of great interest for our understanding of the mechanisms of feeding and feedback, especially in these high-accretion systems which are rapidly growing their SMBHs.

Third, the optical stellar velocity dispersion measurements become more accessible in type 2 states, otherwise known to be particularly challenging in NLS1 galaxies with their bright continuum emission and strong Fe II complexes in their type 1 states.

Fourth, type 2 states will allow us to study the locus of NLS1 galaxies in diagnostic diagrams [103]. These are based on narrow emission-line ratios as diagnostic tools of the physics of the NLR and of the EUV emission of the accretion disk. This has so far rarely been performed because of the challenge of deblending broad and narrow lines when the broad-line FWHM is so small (<2000 km s^{-1}). Samples of at least 10–20 type 2 NLS galaxies will be required for a systematic study of their location in diagnostic diagrams. Along with comparative studies with other types of AGN, especially broad-line Seyfert 1 (BLS1) galaxies, this will provide valuable insights into understanding the characteristics of the NLR properties in NLS1 galaxies.

Fifth, as hypothesized 'young' AGN in the early stages of evolution, CL NLS1 galaxies provide an ideal platform for examining the (co)-evolution of AGN and their host galaxies [104]. This can be performed by directly investigating the properties of their host galaxies through their type 2 state. Such studies have been limited due to the strong nonstellar radiation from the luminous active nucleus, which typically overwhelms the starlight from the host galaxy.

Sixth, the stellar populations of the host galaxy can be easily measured in type 2 states. Studies of the stellar population of the host galaxy can give direct evidence if type 2 NLS galaxies possess young populations.

Finally, while the focus of this publication is on using the CL phenomenon for new applications to understand NLS1 physics, we also note that systematic comparisons between CL BLS1 and CL NLS1 galaxies will highlight possible systematic differences in their accretion disk properties (for instance, different types of disk instabilities operating at different $L/L_{\rm Edd}$).

5. Summary and Conclusions

The space mission SVOM, scheduled to be launched in 2024, will detect nearby CL AGN. Here, we have focused on NLS1 galaxies in particular, have discussed multiple new applications of observing NLS galaxies in their type 2 state, and have highlighted NGC 1566 as a nearby CL NLS1 prototype; it showed bright MWL outbursts in the past and is found to be in a quiescent state in our most recent Swift observations. New hard X-ray transients will first be triggered with ECLAIRs. The MXT with its soft X-ray sensitivity, and the highly sensitive VT, are then particularly well suited for follow-ups of all bright CL AGN; both NLS1 and BLS1 galaxies. In addition to serving as the discovery mission, SVOM can also be used to follow-up CL events first detected in other surveys and/or wavebands.

The identification of further CL AGN, and in particular more CL NLS1 galaxies, is also expected from large optical spectroscopic surveys with repeat spectroscopic coverage, like SDSS [105,106], the Large Sky Area Multi-Object Fiber Spectroscopic Telescope (LAMOST; [107]) or the Dark Energy Spectroscopic Instrument (DESI; [108]), and will continue to be extremely valuable. This approach, however, will generally miss the detection of new events as they happen, when they are brightest, so that MWL observations will not be triggered quasi-simultaneously. In contrast, the identification of transient events among AGN with SVOM will enable *rapid* follow-up observations including spectroscopy. SVOM will issue alerts on bright transients within $\sim$5 min [56].

In addition to their identification and X-ray and optical follow-ups with the instruments onboard SVOM to follow the lightcurve evolution, triggering deeper IR, optical and X-ray imaging spectroscopy will then allow important new applications, especially with respect to the study of type 2 NLS galaxies, their NLR properties, their circum-nuclear environment, and of mechanisms of feeding and feedback. This will provide new insights and constraints on the formation and evolution of SMBHs.

Author Contributions: D.W.X. and S.K. developed the initial idea and drafted the first version of the manuscript. All authors discussed about changing-look NLS1 galaxies and their detection with SVOM, commented on and contributed to the manuscript. All authors have read and agreed to the published version of the manuscript.

Funding: This research was funded by the National Natural Science Foundation of China of grant numbers 12273054 and 12173009. EB would like to acknowledge the support of the Serbian Ministry of Education, Science and Technological Development, through the contract number 451-03-66/2024-03/200002. MWO gratefully acknowledges the support of the German Aerospace Center (DLR) within the framework of the 'Verbundforschung Astronomie und Astrophysik' through grant 50OR2305 with funds from the German Federal Ministry for Economic Affairs and Climate Action (BMWK).

Data Availability Statement: All data are available upon reasonable request. The Swift data are available in the Swift archive at https://swift.gsfc.nasa.gov/archive/ (accessed on 1 January 2024).

Acknowledgments: We would like to thank the Swift team for carrying out the observations we proposed. Additional Swift observations were taken from the archive. We would like to thank Alexis Coleiro for very useful discussions, and our referees for their very useful comments. This work made use of data supplied by the UK Swift Science Data Centre at the University of Leicester. It is our pleasure to thank the organizers and participants of the conference on "Astroinformatics and Virtual Observatories" (Renhuai, October 2023) for an excellent meeting and discussions.

Conflicts of Interest: The authors declare no conflicts of interest.

Notes

1 In a different AGN classification scheme, they are referred to as 'population A' AGN, and the FWHM cut is set around 4000 km s^{-1} [21].

2 We note here that, throughout, we adopt the optical classification of CL events which has been common in recent years. An alternative scheme looks at X-ray spectral variability, and refers to CLs as events where the X-ray spectrum changes from highly absorbed as in type 2 AGN to barely or unabsorbed as in type 1 AGN [32]. We do not discuss possible such X-ray absorption events in NLS1 galaxies.

3 Even though this contribution is focused on NLS1 galaxies, we note in passing that some broad-line Seyfert 1 galaxies which changed their Seyfert subtype are at comparable distances to NGC 1566. One such example is NGC 4151. Traditionally known as an intermediate-type Seyfert galaxy (e.g., [59]), its broad Balmer lines were still detected [60,61] but became significantly fainter in 1984 [60–62] and were already back to their bright state in January 1985 [63].

References

1. Komossa, S.; Grupe, D. Extreme accretion events: TDEs and changing-look AGN. *Astron. Nachrichten* **2023**, *344*, e20230015. [CrossRef]
2. Oknyansky, V.L.; Malanchev, K.L.; Gaskell, C.M. Changing-look Narrow-Line Seyfert 1s? In Proceedings of the Revisiting Narrow-Line Seyfert 1 Galaxies and their Place in the Universe, Padova, Italy, 9–13 April 2018; p. 12. [CrossRef]
3. Nicastro, F. Broad Emission Line Regions in Active Galactic Nuclei: The Link with the Accretion Power. *Astrophys. J. Lett.* **2000**, *530*, L65–L68. [CrossRef]
4. Janiuk, A.; Czerny, B.; Siemiginowska, A. Radiation Pressure Instability Driven Variability in the Accreting Black Holes. *Astrophys. J.* **2002**, *576*, 908–922. [CrossRef]
5. Grupe, D.; Komossa, S.; Saxton, R. IC 3599 Did It Again: A Second Outburst of the X-Ray Transient Seyfert 1.9 Galaxy. *Astrophys. J. Lett.* **2015**, *803*, L28. [CrossRef]
6. Ross, N.P.; Ford, K.E.S.; Graham, M.; McKernan, B.; Stern, D.; Meisner, A.M.; Assef, R.J.; Dey, A.; Drake, A.J.; Jun, H.D.; et al. A new physical interpretation of optical and infrared variability in quasars. *Mon. Not. R. Astron. Soc.* **2018**, *480*, 4468–4479. [CrossRef]
7. Noda, H.; Done, C. Explaining changing-look AGN with state transition triggered by rapid mass accretion rate drop. *Mon. Not. R. Astron. Soc.* **2018**, *480*, 3898–3906. [CrossRef]
8. Dexter, J.; Begelman, M.C. Extreme AGN variability: Evidence of magnetically elevated accretion? *Mon. Not. R. Astron. Soc.* **2019**, *483*, L17–L21. [CrossRef]
9. Sniegowska, M.; Czerny, B.; Bon, E.; Bon, N. Possible mechanism for multiple changing-look phenomena in active galactic nuclei. *Astron. Astrophys.* **2020**, *641*, A167. [CrossRef]
10. Pan, X.; Li, S.L.; Cao, X. The Effects of Large-scale Magnetic Fields on the Model for Repeating Changing-look AGNs. *Astrophys. J.* **2021**, *910*, 97. [CrossRef]
11. Raj, A.; Nixon, C.J.; Doğan, S. Disk Tearing: Numerical Investigation of Warped Disk Instability. *Astrophys. J.* **2021**, *909*, 81. [CrossRef]
12. Feng, J.; Cao, X.; Li, J.w.; Gu, W.M. A Magnetic Disk-outflow Model for Changing Look Active Galactic Nuclei. *Astrophys. J.* **2021**, *916*, 61. [CrossRef]
13. Laha, S.; Meyer, E.; Roychowdhury, A.; Becerra Gonzalez, J.; Acosta-Pulido, J.A.; Thapa, A.; Ghosh, R.; Behar, E.; Gallo, L.C.; Kriss, G.A.; et al. A Radio, Optical, UV, and X-Ray View of the Enigmatic Changing-look Active Galactic Nucleus 1ES 1927 + 654 from Its Pre- to Postflare States. *Astrophys. J.* **2022**, *931*, 5. [CrossRef]
14. Kaaz, N.; Liska, M.T.P.; Jacquemin-Ide, J.; Andalman, Z.L.; Musoke, G.; Tchekhovskoy, A.; Porth, O. Nozzle Shocks, Disk Tearing, and Streamers Drive Rapid Accretion in 3D GRMHD Simulations of Warped Thin Disks. *Astrophys. J.* **2023**, *955*, 72. [CrossRef]
15. Cao, X.; You, B.; Wei, X. An accretion disc with magnetic outflows triggered by a sudden mass accretion event in changing-look active galactic nucleus 1ES 1927 + 654. *Mon. Not. R. Astron. Soc.* **2023**, *526*, 2331–2340. [CrossRef]
16. Wang, J.M.; Bon, E. Changing-look active galactic nuclei: Close binaries of supermassive black holes in action. *Astron. Astrophys.* **2020**, *643*, L9. [CrossRef]
17. Gallo, L. X-ray perspective of Narrow-line Seyfert 1 galaxies. In Proceedings of the Revisiting Narrow-Line Seyfert 1 Galaxies and their Place in the Universe, Padova, Italy, 9–13 April 2018; p. 34. [CrossRef]
18. Osterbrock, D.E.; Pogge, R.W. The spectra of narrow-line Seyfert 1 galaxies. *Astrophys. J.* **1985**, *297*, 166–176. [CrossRef]
19. Goodrich, R.W. Spectropolarimetry of "Narrow-Line" Seyfert 1 Galaxies. *Astrophys. J.* **1989**, *342*, 224. [CrossRef]
20. Véron-Cetty, M.P.; Véron, P.; Gonçalves, A.C. A spectrophotometric atlas of Narrow-Line Seyfert 1 galaxies. *Astron. Astrophys.* **2001**, *372*, 730–754. [CrossRef]
21. Marziani, P.; Dultzin, D.; Sulentic, J.W.; Del Olmo, A.; Negrete, C.A.; Martínez-Aldama, M.L.; D'Onofrio, M.; Bon, E.; Bon, N.; Stirpe, G.M. A main sequence for quasars. *Front. Astron. Space Sci.* **2018**, *5*, 6. [CrossRef]
22. Grupe, D.; Beuermann, K.; Mannheim, K.; Thomas, H.C. New bright soft X-ray selected ROSAT AGN. II. Optical emission line properties. *Astron. Astrophys.* **1999**, *350*, 805–815. [CrossRef]

23. Xu, D.W.; Komossa, S.; Wei, J.Y.; Qian, Y.; Zheng, X.Z. An Active Galactic Nucleus Sample with High X-Ray-to-Optical Flux Ratio from RASS. II. Optical Emission Line Properties of Seyfert 1-Type Active Galactic Nuclei. *Astrophys. J.* **2003**, *590*, 73–85. [CrossRef]
24. Zhou, H.; Wang, T.; Yuan, W.; Lu, H.; Dong, X.; Wang, J.; Lu, Y. A Comprehensive Study of 2000 Narrow Line Seyfert 1 Galaxies from the Sloan Digital Sky Survey. I. The Sample. *Astrophys. J. Suppl. Ser.* **2006**, *166*, 128–153. [CrossRef]
25. Rakshit, S.; Stalin, C.S.; Kotilainen, J. Spectral Properties of Quasars from Sloan Digital Sky Survey Data Release 14: The Catalog. *Astrophys. J. Suppl. Ser.* **2020**, *249*, 17. [CrossRef]
26. Paliya, V.S.; Stalin, C.S.; Domínguez, A.; Saikia, D.J. Narrow-line Seyfert 1 galaxies in Sloan Digital Sky Survey: A new optical spectroscopic catalogue. *Mon. Not. R. Astron. Soc.* **2024**, *527*, 7055–7069. [CrossRef]
27. Boroson, T.A. Black Hole Mass and Eddington Ratio as Drivers for the Observable Properties of Radio-loud and Radio-quiet QSOs. *Astrophys. J.* **2002**, *565*, 78–85. [CrossRef]
28. Grupe, D. A Complete Sample of Soft X-Ray-selected AGNs. II. Statistical Analysis. *Astron. J.* **2004**, *127*, 1799–1810. [CrossRef]
29. Xu, D.; Komossa, S.; Zhou, H.; Lu, H.; Li, C.; Grupe, D.; Wang, J.; Yuan, W. Correlation Analysis of a Large Sample of Narrow-line Seyfert 1 Galaxies: Linking Central Engine and Host Properties. *Astron. J.* **2012**, *143*, 83. [CrossRef]
30. Komossa, S. Multi-wavelength properties of radio-loud Narrow-line Seyfert 1 galaxies. In Proceedings of the Revisiting Narrow-Line Seyfert 1 Galaxies and their Place in the Universe, Padova, Italy, 9–13 April 2018; p. 15. [CrossRef]
31. Wright, E.L. A Cosmology Calculator for the World Wide Web. *Publ. Astron. Soc. Pac.* **2006**, *118*, 1711–1715. [CrossRef]
32. Matt, G.; Guainazzi, M.; Maiolino, R. Changing look: From Compton-thick to Compton-thin, or the rebirth of fossil active galactic nuclei. *Mon. Not. R. Astron. Soc.* **2003**, *342*, 422–426. [CrossRef]
33. Brandt, W.N.; Pounds, K.A.; Fink, H. The unusual X-ray and optical properties of the ultrasoft active galactic nucleus Zwicky 159.034 (RE J1237 + 264). *Mon. Not. R. Astron. Soc.* **1995**, *273*, L47–L52. [CrossRef]
34. Grupe, D.; Beuermann, K.; Mannheim, K.; Bade, N.; Thomas, H.C.; de Martino, D.; Schwope, A. X-ray outburst of the peculiar Seyfert galaxy IC 3599. *Astron. Astrophys.* **1995**, *299*, L5. [CrossRef]
35. Komossa, S.; Bade, N. The giant X-ray outbursts in NGC 5905 and IC 3599: Follow-up observations and outburst scenarios. *Astron. Astrophys.* **1999**, *343*, 775–787.
36. Komossa, S.; Grupe, D.; Saxton, R.; Gallo, L. Seyfert galaxies with Swift: Giant flares, rapid drops, and other surprises. In Proceedings of the Proceedings of Swift: 10 Years of Discovery (SWIFT 10), Rome, Italy, 2–5 December 2014; p. 143. [CrossRef]
37. MacLeod, C.L.; Green, P.J.; Anderson, S.F.; Bruce, A.; Eracleous, M.; Graham, M.; Homan, D.; Lawrence, A.; LeBleu, A.; Ross, N.P.; et al. Changing-look Quasar Candidates: First Results from Follow-up Spectroscopy of Highly Optically Variable Quasars. *Astrophys. J.* **2019**, *874*, 8. [CrossRef]
38. Liu, H.Y.; Liu, W.J.; Dong, X.B.; Zhou, H.; Wang, T.; Lu, H.; Yuan, W. A Comprehensive and Uniform Sample of Broad-line Active Galactic Nuclei from the SDSS DR7. *Astrophys. J. Suppl. Ser.* **2019**, *243*, 21. [CrossRef]
39. Frederick, S.; Gezari, S.; Graham, M.J.; Cenko, S.B.; van Velzen, S.; Stern, D.; Blagorodnova, N.; Kulkarni, S.R.; Yan, L.; De, K.; et al. A New Class of Changing-look LINERs. *Astrophys. J.* **2019**, *883*, 31. [CrossRef]
40. Hon, W.J.; Wolf, C.; Onken, C.A.; Webster, R.; Auchettl, K. SkyMapper colours of Seyfert galaxies and changing-look AGN - II. Newly discovered changing-look AGN. *Mon. Not. R. Astron. Soc.* **2022**, *511*, 54–70. [CrossRef]
41. Filippenko, A.V. Optical Spectra of Supernovae. *Annu. Rev. Astron. Astrophys.* **1997**, *35*, 309–355. [CrossRef]
42. Rees, M.J. Tidal disruption of stars by black holes of 10^6–10^8 solar masses in nearby galaxies. *Nature* **1988**, *333*, 523–528. [CrossRef]
43. Komossa, S.; Zhou, H.; Wang, T.; Ajello, M.; Ge, J.; Greiner, J.; Lu, H.; Salvato, M.; Saxton, R.; Shan, H.; et al. Discovery of Superstrong, Fading, Iron Line Emission and Double-peaked Balmer Lines of the Galaxy SDSS J095209.56 + 214313.3: The Light Echo of a Huge Flare. *Astrophys. J. Lett.* **2008**, *678*, L13. [CrossRef]
44. Onori, F.; Cannizzaro, G.; Jonker, P.G.; Kim, M.; Nicholl, M.; Mattila, S.; Reynolds, T.M.; Fraser, M.; Wevers, T.; Brocato, E.; et al. The nuclear transient AT 2017gge: A tidal disruption event in a dusty and gas-rich environment and the awakening of a dormant SMBH. *Mon. Not. R. Astron. Soc.* **2022**, *517*, 76–98. [CrossRef]
45. Liu, F.K.; Zhou, Z.Q.; Cao, R.; Ho, L.C.; Komossa, S. Disc origin of broad optical emission lines of the TDE candidate PTF09djl. *Mon. Not. R. Astron. Soc.* **2017**, *472*, L99–L103. [CrossRef]
46. Wang, T.G.; Zhou, H.Y.; Komossa, S.; Wang, H.Y.; Yuan, W.; Yang, C. Extreme Coronal Line Emitters: Tidal Disruption of Stars by Massive Black Holes in Galactic Nuclei? *Astrophys. J.* **2012**, *749*, 115. [CrossRef]
47. Short, P.; Lawrence, A.; Nicholl, M.; Ward, M.; Reynolds, T.M.; Mattila, S.; Yin, C.; Arcavi, I.; Carnall, A.; Charalampopoulos, P.; et al. Delayed appearance and evolution of coronal lines in the TDE AT2019qiz. *Mon. Not. R. Astron. Soc.* **2023**, *525*, 1568–1587. [CrossRef]
48. Komossa, S.; Zhou, H.; Rau, A.; Dopita, M.; Gal-Yam, A.; Greiner, J.; Zuther, J.; Salvato, M.; Xu, D.; Lu, H.; et al. NTT, Spitzer, and Chandra Spectroscopy of SDSSJ095209.56 + 214313.3: The Most Luminous Coronal-line Supernova Ever Observed, or a Stellar Tidal Disruption Event? *Astrophys. J.* **2009**, *701*, 105–121. [CrossRef]
49. Drake, A.J.; Djorgovski, S.G.; Mahabal, A.; Anderson, J.; Roy, R.; Mohan, V.; Ravindranath, S.; Frail, D.; Gezari, S.; Neill, J.D.; et al. The Discovery and Nature of the Optical Transient CSS100217:102913 + 404220. *Astrophys. J.* **2011**, *735*, 106. [CrossRef]
50. Blanchard, P.K.; Nicholl, M.; Berger, E.; Guillochon, J.; Margutti, R.; Chornock, R.; Alexander, K.D.; Leja, J.; Drout, M.R. PS16dtm: A Tidal Disruption Event in a Narrow-line Seyfert 1 Galaxy. *Astrophys. J.* **2017**, *843*, 106. [CrossRef]

51. Zhang, W.J.; Shu, X.W.; Sheng, Z.F.; Sun, L.M.; Dou, L.M.; Jiang, N.; Wang, J.G.; Hu, X.Y.; Wang, Y.B.; Wang, T.G. Discovery of late-time X-ray flare and anomalous emission line enhancement after the nuclear optical outburst in a narrow-line Seyfert 1 Galaxy. *Astron. Astrophys.* **2022**, *660*, A119. [CrossRef]
52. Li, D.; Saxton, R.D.; Yuan, W.; Sun, L.; Liu, H.Y.; Jiang, N.; Cheng, H.; Zhou, H.; Komossa, S.; Jin, C. Multiwavelength Study of an X-Ray Tidal Disruption Event Candidate in NGC 5092. *Astrophys. J.* **2020**, *891*, 121. [CrossRef]
53. Hampel, J.; Komossa, S.; Greiner, J.; Reiprich, T.H.; Freyberg, M.; Erben, T. A New X-Ray Tidal Disruption Event Candidate with Fast Variability. *Res. Astron. Astrophys.* **2022**, *22*, 055004. [CrossRef]
54. Burrows, D.N.; Kennea, J.A.; Ghisellini, G.; Mangano, V.; Zhang, B.; Page, K.L.; Eracleous, M.; Romano, P.; Sakamoto, T.; Falcone, A.D.; et al. Relativistic jet activity from the tidal disruption of a star by a massive black hole. *Nature* **2011**, *476*, 421–424. [CrossRef]
55. Bloom, J.S.; Giannios, D.; Metzger, B.D.; Cenko, S.B.; Perley, D.A.; Butler, N.R.; Tanvir, N.R.; Levan, A.J.; O'Brien, P.T.; Strubbe, L.E.; et al. A Possible Relativistic Jetted Outburst from a Massive Black Hole Fed by a Tidally Disrupted Star. *Science* **2011**, *333*, 203, [CrossRef] [PubMed]
56. Wei, J.; Cordier, B.; Antier, S.; Antilogus, P.; Atteia, J.L.; Bajat, A.; Basa, S.; Beckmann, V.; Bernardini, M.G.; Boissier, S.; et al. The Deep and Transient Universe in the SVOM Era: New Challenges and Opportunities - Scientific prospects of the SVOM mission. *arXiv* **2016**, arXiv:1610.06892. [CrossRef]
57. Wang, J.; Merritt, D. Revised Rates of Stellar Disruption in Galactic Nuclei. *Astrophys. J.* **2004**, *600*, 149–161. [CrossRef]
58. Yuan, W.; Komossa, S.; Zhang, C.; Feng, H.; Ling, Z.X.; Zhao, D.H.; Zhang, S.N.; Osborne, J.P.; O'Brien, P.; Willingale, R.; et al. Detecting tidal disruption events of massive black holes in normal galaxies with the Einstein Probe. Proc. Star Clust. Black Holes Galaxies Across Cosm. *Time* **2016**, *312*, 68–70. [CrossRef]
59. Seyfert, C.K. Nuclear Emission in Spiral Nebulae. *Astrophys. J.* **1943**, *97*, 28. [CrossRef]
60. Penston, M.V.; Perez, E. An evolutionary link between Seyfert I and II galaxies. *Mon. Not. R. Astron. Soc.* **1984**, *211*, 33P–39P. [CrossRef]
61. Kielkopf, J.; Brashear, R.; Lattis, J. High-resolution observations of H alpha in NGC 4151. *Astrophys. J.* **1985**, *299*, 865–872. [CrossRef]
62. Lyutyj, V.M.; Oknyanskij, V.L.; Chuvaev, K.K. NGC 4151: Sy in a deep photometric minimum. *Pisma V Astron. Zhurnal* **1984**, *10*, 803–807.
63. Peterson, B.M. NGC 4151. *IAU Circ.* **1985**, *4036*, 1.
64. Shobbrook, R.R. Southern groups and clusters of galaxies. I, Spectra and radial velocities of nineteen southern galaxies. *Mon. Not. R. Astron. Soc.* **1966**, *131*, 293. [CrossRef]
65. Alloin, D.; Pelat, D.; Phillips, M.; Whittle, M. Recent spectral variations in the active nucleus of NGC 1566. *Astrophys. J.* **1985**, *288*, 205–220. [CrossRef]
66. Alloin, D.; Pelat, D.; Phillips, M.M.; Fosbury, R.A.E.; Freeman, K. Recurrent Outbursts in the Broad-Line Region of NGC 1566. *Astrophys. J.* **1986**, *308*, 23. [CrossRef]
67. Ducci, L.; Siegert, T.; Diehl, R.; Sanchez-Fernandez, C.; Ferrigno, C.; Savchenko, V.; Bozzo, E. INTEGRAL detection of hard X-ray emission from NGC 1566. *Astron. Telegr.* **2018**, *11754*, 1.
68. Ferrigno, C.; Siegert, T.; Sanchez-Fernandez, C.; Kuulkers, E.; Ducci, L.; Savchenko, V.; Bozzo, E. Swift follow-up observations determine enhanced nuclear activity from NGC 1566. *Astron. Telegr.* **2018**, *11783*, 1.
69. Grupe, D.; Komossa, S.; Schartel, N. Swift observations in all filters of the Seyfert galaxy NGC 1566 in outburst. *Astron. Telegr.* **2018**, *11903*, 1.
70. Dai, X.; Stanek, K.Z.; Kochanek, C.S.; Shappee, B.J.; ASAS-SN Collaboration. ASAS-SN Light Curve Reveals Dramatic Variability of Seyfert 1.5 AGN in NGC 1566. *Astron. Telegr.* **2018**, *11893*, 1.
71. Oknyansky, V.L.; Lipunov, V.M.; Gorbovskoy, E.S.; Winkler, H.; van Wyk, F.; Tsygankov, S.; Buckley, D.A.H. New changing look case in NGC 1566. *Astron. Telegr.* **2018**, *11915*, 1. [CrossRef]
72. Parker, M.L.; Schartel, N.; Grupe, D.; Komossa, S.; Harrison, F.; Kollatschny, W.; Mikula, R.; Santos-Lleó, M.; Tomás, L. X-ray spectra reveal the reawakening of the repeat changing-look AGN NGC 1566. *Mon. Not. R. Astron. Soc.* **2019**, *483*, L88–L92. [CrossRef]
73. Oknyansky, V.L.; Winkler, H.; Tsygankov, S.S.; Lipunov, V.M.; Gorbovskoy, E.S.; van Wyk, F.; Buckley, D.A.H.; Tyurina, N.V. New changing look case in NGC 1566. *Mon. Not. R. Astron. Soc.* **2019**, *483*, 558–564. [CrossRef]
74. Ochmann, M.W.; Kollatschny, W.; Zetzl, M. Spectral changes and BLR kinematics of eruptive changing-look AGN. *Contrib. Astron. Obs. Skaln. Pleso* **2020**, *50*, 318–327. [CrossRef]
75. da Silva, P.; Steiner, J.E.; Menezes, R.B. NGC 1566: Analysis of the nuclear region from optical and near-infrared Integral Field Unit spectroscopy. *Mon. Not. R. Astron. Soc.* **2017**, *470*, 3850–3876. [CrossRef]
76. Ochmann, M.; Kollatschny, W.; Probst, M.; Romero-Colmenero, E.; Buckley, D.; Chelouche, D.; Chini, R.; Grupe, D.; Haas, M.; Kaspi, S.; et al. The transient event in NGC 1566 from 2017 to 2019. I. *Astron. Astrophys.* **2023**, *submitted*.
77. Oh, K.; Koss, M.; Markwardt, C.B.; Schawinski, K.; Baumgartner, W.H.; Barthelmy, S.D.; Cenko, S.B.; Gehrels, N.; Mushotzky, R.; Petulante, A.; et al. The 105-Month Swift-BAT All-sky Hard X-Ray Survey. *Astrophys. J. Suppl. Ser.* **2018**, *235*, 4. [CrossRef]
78. Gehrels, N.; Chincarini, G.; Giommi, P.; Mason, K.O.; Nousek, J.A.; Wells, A.A.; White, N.E.; Barthelmy, S.D.; Burrows, D.N.; Cominsky, L.R.; et al. The Swift Gamma-Ray Burst Mission. *Astrophys. J.* **2004**, *611*, 1005–1020. [CrossRef]

79. Roming, P.W.A.; Kennedy, T.E.; Mason, K.O.; Nousek, J.A.; Ahr, L.; Bingham, R.E.; Broos, P.S.; Carter, M.J.; Hancock, B.K.; Huckle, H.E.; et al. The Swift Ultra-Violet/Optical Telescope. *Space Sci. Rev.* **2005**, *120*, 95–142. [CrossRef]
80. Cardelli, J.A.; Clayton, G.C.; Mathis, J.S. The Relationship between Infrared, Optical, and Ultraviolet Extinction. *Astrophys. J.* **1989**, *345*, 245. [CrossRef]
81. Burrows, D.N.; Hill, J.E.; Nousek, J.A.; Kennea, J.A.; Wells, A.; Osborne, J.P.; Abbey, A.F.; Beardmore, A.; Mukerjee, K.; Short, A.D.T.; et al. The Swift X-Ray Telescope. *Space Sci. Rev.* **2005**, *120*, 165–195. [CrossRef]
82. Arnaud, K.A. XSPEC: The First Ten Years. In *Astronomical Data Analysis Software and Systems*; Jacoby, G.H., Barnes, J., Eds.; Astronomical Society of the Pacific Conference Series; ASP: London, UK, 1996; Volume 101, p. 17.
83. Evans, P.A.; Beardmore, A.P.; Page, K.L.; Tyler, L.G.; Osborne, J.P.; Goad, M.R.; O'Brien, P.T.; Vetere, L.; Racusin, J.; Morris, D.; et al. An online repository of Swift/XRT light curves of γ-ray bursts. *Astron. Astrophys.* **2007**, *469*, 379–385. [CrossRef]
84. Godet, O.; Nasser, G.; Atteia, J.; Cordier, B.; Mandrou, P.; Barret, D.; Triou, H.; Pons, R.; Amoros, C.; Bordon, S.; et al. The x-/gamma-ray camera ECLAIRs for the gamma-ray burst mission SVOM. In *Space Telescopes and Instrumentation 2014: Ultraviolet to Gamma Ray*; Takahashi, T., den Herder, J.W.A., Bautz, M., Eds.; Society of Photo-Optical Instrumentation Engineers (SPIE) Conference Series; SPIE: Bellingham, WA, USA, 2014; Volume 9144, p. 914424. [CrossRef]
85. Godet, O.; Atteia, J.L.; Amoros, C.; Roger, P.; Bouchet, L.; Dezalay, J.P.; Yassine, M.; Arcier, B.; Bordon, S.; Lacombe, K.; et al. On-ground calibration highlights for the SVOM/ECLAIRs camera. In *Space Telescopes and Instrumentation 2022: Ultraviolet to Gamma Ray*; den Herder, J.W.A., Nikzad, S., Nakazawa, K., Eds.; Society of Photo-Optical Instrumentation Engineers (SPIE) Conference Series; SPIE: Bellingham, WA, USA, 2022; Volume 12181, p. 1218150. [CrossRef]
86. He, J.; Sun, J.C.; Wen, X.; Dong, Y.W.; Zhang, J.; Li, L.; Liu, J.T.; Liu, S.; Liu, X.; Liu, Y.; et al. In-orbit background simulation study of SVOM/GRM. *Astrophys. Space Sci.* **2020**, *365*, 167. [CrossRef]
87. Götz, D.; Osborne, J.; Cordier, B.; Paul, J.; Evans, P.; Beardmore, A.; Martindale, A.; Willingale, R.; O'Brien, P.; Basa, S.; et al. The microchannel x-ray telescope for the gamma-ray burst mission SVOM. In *Space Telescopes and Instrumentation 2014: Ultraviolet to Gamma Ray*; Takahashi, T., den Herder, J.W.A., Bautz, M., Eds.; Society of Photo-Optical Instrumentation Engineers (SPIE) Conference Series; SPIE: Bellingham, WA, USA, 2014; Volume 9144, p. 914423. [CrossRef]
88. Götz, D.; Boutelier, M.; Burwitz, V.; Chipaux, R.; Cordier, B.; Feldman, C.; Ferrando, P.; Fort, A.; Gonzalez, F.; Gros, A.; et al. The scientific performance of the microchannel X-ray telescope on board the SVOM mission. *Exp. Astron.* **2023**, *55*, 487–519. [CrossRef]
89. Wang, J.; Qiu, Y.L.; Wei, J.Y. A pilot study of catching high-z GRBs and exploring circumburst environment in the forthcoming SVOM era. *Res. Astron. Astrophys.* **2020**, *20*, 124. [CrossRef]
90. Dagoneau, N.; Schanne, S.; Atteia, J.L.; Götz, D.; Cordier, B. Ultra-Long Gamma-Ray Bursts detection with SVOM/ECLAIRs. *Exp. Astron.* **2020**, *50*, 91–123. [CrossRef]
91. Yu, S.j.; Gonzalez, F.; Wei, J.y.; Zhang, S.n.; Cordier, B. SVOM: A Joint Gamma-ray Burst Detection Mission. *Chin. Astron. Astrophys.* **2020**, *44*, 269–282. [CrossRef]
92. Paul, J.; Wei, J.; Basa, S.; Zhang, S.N. The Chinese-French SVOM mission for gamma-ray burst studies. *Comptes Rendus Phys.* **2011**, *12*, 298–308. [CrossRef]
93. Cordier, B.; Wei, J.; Atteia, J.L.; Basa, S.; Claret, A.; Daigne, F.; Deng, J.; Dong, Y.; Godet, O.; Goldwurm, A.; et al. The SVOM gamma-ray burst mission. *arXiv* **2015**, arXiv:1512.03323. [CrossRef]
94. Siemiginowska, A.; Czerny, B.; Kostyunin, V. Evolution of an Accretion Disk in an Active Galactic Nucleus. *Astrophys. J.* **1996**, *458*, 491. [CrossRef]
95. LaMassa, S.M.; Cales, S.; Moran, E.C.; Myers, A.D.; Richards, G.T.; Eracleous, M.; Heckman, T.M.; Gallo, L.; Urry, C.M. The Discovery of the First "Changing Look" Quasar: New Insights Into the Physics and Phenomenology of Active Galactic Nucleus. *Astrophys. J.* **2015**, *800*, 144. [CrossRef]
96. Lawrence, A. Quasar viscosity crisis. *Nat. Astron.* **2018**, *2*, 102–103. [CrossRef]
97. Dewangan, G.C.; Griffiths, R.E. Type 2 Counterparts of Narrow-Line Seyfert 1 Galaxies. *Astrophys. J. Lett.* **2005**, *625*, L31–L34. [CrossRef]
98. Pan, X.; Lu, H.; Komossa, S.; Xu, D.; Yuan, W.; Sun, L.; Smith, P.S.; Zhang, S.; Jiang, P.; Yang, C.; et al. A Deeply Buried Narrow-line Seyfert 1 Nucleus Uncovered in Scattered Light. *Astrophys. J.* **2019**, *870*, 75. [CrossRef]
99. Pan, X.; Zhou, H.; Yang, C.; Sun, L.; Smith, P.S.; Ji, T.; Jiang, N.; Jiang, P.; Liu, W.; Lu, H.; et al. Mrk 1239: A Type-2 Counterpart of Narrow-line Seyfert-1? *Astrophys. J.* **2021**, *912*, 118. [CrossRef]
100. Antonucci, R. Unified models for active galactic nuclei and quasars. *Annu. Rev. Astron. Astrophys.* **1993**, *31*, 473–521. [CrossRef]
101. Tomás, L.; Matzeu, G.A.; Jiménez Bailón, E.; Kalfountzou, E.; Santos-Lleó, M.; Parker, M.L.; Ballo, L.; Loiseau, N.; Ehle, M.; Rodríguez-Pascual, P.; et al. The changing-look AGN NGC 1566 in quiescence with XMM-Newton: A nuclear starburst and an AGN competing in power? *Mon. Not. R. Astron. Soc.* **2022**, *514*, 403–415. [CrossRef]
102. Buhariwalla, M.Z.; Gallo, L.C.; Mao, J.; Komossa, S.; Jiang, J.; Gonzalez, A.; Grupe, D. The collisional and photoionized plasma in the polarized NLS1 galaxy Mrk 1239. *Mon. Not. R. Astron. Soc.* **2023**, *521*, 2378–2390. [CrossRef]
103. Veilleux, S.; Osterbrock, D.E. Spectral Classification of Emission-Line Galaxies. *Astrophys. J. Suppl. Ser.* **1987**, *63*, 295. [CrossRef]
104. Heckman, T.M.; Best, P.N. The Coevolution of Galaxies and Supermassive Black Holes: Insights from Surveys of the Contemporary Universe. *Annu. Rev. Astron. Astrophys.* **2014**, *52*, 589–660. [CrossRef]

105. MacLeod, C.L.; Ross, N.P.; Lawrence, A.; Goad, M.; Horne, K.; Burgett, W.; Chambers, K.C.; Flewelling, H.; Hodapp, K.; Kaiser, N.; et al. A systematic search for changing-look quasars in SDSS. *Mon. Not. R. Astron. Soc.* **2016**, *457*, 389–404. [CrossRef]
106. Runco, J.N.; Cosens, M.; Bennert, V.N.; Scott, B.; Komossa, S.; Malkan, M.A.; Lazarova, M.S.; Auger, M.W.; Treu, T.; Park, D. Broad Hβ Emission-line Variability in a Sample of 102 Local Active Galaxies. *Astrophys. J.* **2016**, *821*, 33. [CrossRef]
107. Yang, Q.; Wu, X.B.; Fan, X.; Jiang, L.; McGreer, I.; Shangguan, J.; Yao, S.; Wang, B.; Joshi, R.; Green, R.; et al. Discovery of 21 New Changing-look AGNs in the Northern Sky. *Astrophys. J.* **2018**, *862*, 109. [CrossRef]
108. Guo, W.J.; Zou, H.; Fawcett, V.A.; Canning, R.; Juneau, S.; Davis, T.M.; Alexander, D.M.; Jiang, L.; Aguilar, J.N.; Ahlen, S.; et al. Changing-look Active Galactic Nuclei from the Dark Energy Spectroscopic Instrument. I. Sample from the Early Data. *arXiv* **2023**, arXiv:2307.08289. [CrossRef]

universe

MDPI

Article

Multi-Epoch Optical Spectroscopy Variability of the Changing-Look AGN Mrk 883

Erika Benítez [1,*], Castalia Alenka Negrete [1], Héctor Ibarra-Medel [1], Irene Cruz-González [1] and José Miguel Rodríguez-Espinosa [2,3]

1 Instituto de Astronomía, Universidad Nacional Autónoma de México, AP 70-264, Mexico City 04510, Mexico; alenka@astro.unam.mx (C.A.N.); hibarram@astro.unam.mx (H.I.-M.)
2 Instituto de Astrofísica de Andalucía, E-18008 Granada, Spain
3 Departamento de Astrofísica, Universidad de La Laguna, E-38206 La Laguna, Spain
* Correspondence: erika@astro.unam.mx; Tel.: +52-5556223986

Abstract: In this work, we present multi-epoch optical spectra of the Seyfert 1.9 galaxy Mrk 883. Data were obtained with the Gran Telescopio Canarias and the *MEGARA* Integral Field Unit mode, archival data from the SDSS-IV MaNGA Survey and the SDSS-I Legacy Survey, and new spectroscopic observations obtained at San Pedro Mártir Observatory. We report the appearance of the broad component of Hβ emission line, showing a maximum FWHM $\sim$ 5927 $\pm$ 481 km s^{-1} in the MaNGA spectra, finding evidence for a change from Seyfert 1.9 (23 June 2003) to Seyfert 1.8 (18 May 2018). The observed changing-look variation from Sy1.9 to Sy1.8 has a timescale Δt$\sim$15 y. In addition, we observe profile and flux broad emission line variability from 2018 to 2023, and a wind component in [OIII]5007 Å, with a maximum FWHM = 1758 $\pm$ 178 km s^{-1}, detected on 15 April 2023. In all epochs, variability of the broad lines was found to be disconnected from the optical continuum emission, which shows little or no variations. These results suggest that an ionized-driven wind in the polar direction could be a possible scenario to explain the observed changing-look variations.

Keywords: active galactic nuclei; Seyfert galaxies: variability; optical spectroscopy; AGN winds

1. Introduction

The so-called unified model for AGNs [1] has been for a long time the paradigm used to explain the observed phenomenology of the different classes of the jetted and non-jetted AGN family [2]. The model is mainly based on the assumption that a ubiquitous clumpy parsec-scale dusty torus presenting different viewing angles can explain the division into Type 1 (unobscured) and Type 2 (obscured) AGNs (for a review, see [3]).

AGNs are also known for showing multi-wavelength variability with time scales from hours to years (e.g., [4]). Some AGNs display more dramatic variations, including spectral-type variations, e.g., changing from Type 1 to Type 2 classes in time-scales $\sim$years. AGNs that display this type of variation are identified as 'Changing-look' (CL). They have been observed in the optical and the X-ray bands, and $\sim$100 CL AGNs have been identified by multi-epoch spectroscopy [5]. CL AGNs are known to present a temporary appearance/disappearance of the broad emission line (BLR) components with timescales from years to decades, and these kinds of coherent changes are found to be difficult to explain—see, e.g., [6,7]. The spectral change variations can be from Type 1 to Type 2, progressing also through the intermediate types, e.g., from Type 1 to 1.8, and 1.9.

The Seyfert galaxy Mrk 883 is a nearby AGN at z = 0.038. Classified as Sy1.9 by [8], it was identified by [9] as an object that shows spectral-type variations. In particular, broad emission line variations were detected in October 1993, producing a change from Sy1.9 to Sy1.8 (or even Sy1.5), with little change in the optical continuum emission. The author demonstrates that the detected line flux changes may not be caused by changes in the reddening of the BLR, i.e., not due to an obscuring material, suggesting that other mechanisms are involved in the appearance/disappearance of the broad component (BC) in Hβ (Hβ_{BC}). Later, analyzing the archival SDSS spectrum of Mrk 883 obtained on 23 June 2003, it was found that the central nucleus has an optical spectrum of Sy1.9 (see [10]). These authors also report that Mrk 883 is a double-peaked narrow emission line AGN (DPAGN), showing one Gaussian component in the locus of Sy galaxies in the Baldwin, Peterson, Terlevich (BPT) diagnostic diagrams [11], and the second Gaussian component is found in the so-called composite region (AGN+SB). This object was also included in a study of X-ray variability of Sy1.8/Sy1.9 galaxies by [12]. Analyzing four observations obtained with XMM–Newton between 2006 and 2010, they reported flux variations of 28% in both the soft and hard X-ray bands with a timescale of $\Delta t \sim 4$ y, and found that Mrk 883 is a Compton-thin (unobscured) AGN.

Mrk 883 also shows disturbed morphology in the optical images of the SDSS [13]. A gaseous ring and tidal tales can be identified, supporting evidence of past merger activity. A double nucleus in its center can be seen—see Figure 1. Our *MEGARA* data presented in this work allowed us to establish that the two nuclear sources (see Section 7), one at the center of Mrk 883 and the other one at the southwest, have a projected separation of 2.2″ or ∼1.7 kpc.

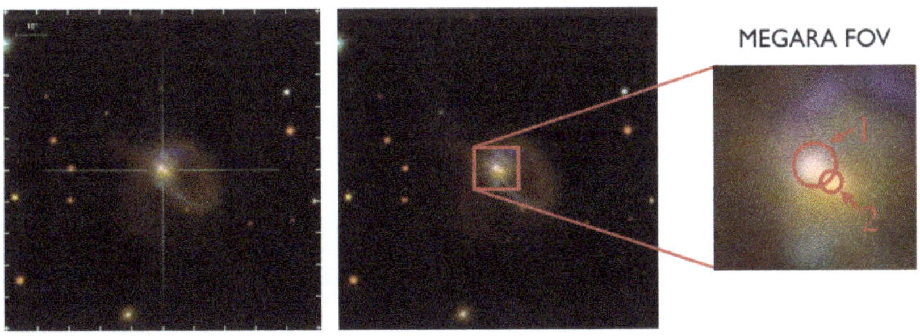

Figure 1. (**Left panel**): SDSS RGB color composed image (g-r-i bands), centered at Mrk 883, obtained at MJD 52813 (23 June 2003) showing the two nuclear sources. Mrk 883 is located at the central coordinates. A second nucleus is located in the southwest. (**Central panel**): The same as in the left image, but showing in a red box the *MEGARA* FOV. (**Right panel**): In the *MEGARA* FOV image, the red circles show the two apertures used in this work. Aperture 1 marks the locus of the aperture with a size of 3″. The selection matches the same aperture used in the SDSS fiber aperture position. A second aperture of 1.5″ was selected at the position of the second nucleus.

In this work, we present the results of a new variability study of Mrk 883, through the analysis of optical spectroscopic multi-epoch data obtained with the Gran Telescopio Canarias (GTC) and *MEGARA* in IFU mode, and also analyze the archived 3D spectrum of the central source obtained with the MaNGA survey. We also include the analysis of two long-slit spectra, one from the SDSS database, and a recent spectrum obtained at the Observatorio Astronómico Nacional at San Pedro Mártir (OAN-SPM) in Baja California Mexico, with the 2.1 m telescope and the Boller & Chivens spectrograph. We find that the *MEGARA* spectrum clearly shows the appearance of a broad Hβ emission line component with extended wings. Therefore, we report Seyfert 1.9 to Seyfert 1.8 or CL variability in Mrk 883 on timescales of ∼15 years. The variations in the broad components occurred with little changes in the optical continuum emission. We discuss later a possible mechanism that could be triggering the CL variations observed at different epochs. The cosmology adopted in this work is $H_0 = 69.6$ km/s^{-1} Mpc^{-1}, $\Omega_m = 0.286$, and $\Omega_\Lambda = 0.714$ [14]. Therefore, 1″ corresponds to 756 pc.

2. Data Acquisition and Observations

We analyzed the Integral Field Spectroscopy (IFS) data obtained in the Mapping Nearby Galaxies at the Apache Point Observatory survey (MaNGA [15]). Also, we analyzed the single-fiber spectra obtained by the Sloan Digital Sky Survey Data Release 7 (SDSS [13,16,17]). The IFS MaNGA survey aims to obtain systematic observations of 11,273 targets and is part of the SDSS phase IV [18]. MaNGA observations are carried out using five IFUs: 19, 37, 61, 91, and 127 fiber IFU bundles, with a 2″ fiber diameter size [19]. For MaNGA data, the eBOSS spectrograph has a spectral coverage range of 3600 Å to 10,300 Å, and a spectral resolution of R $\approx$ 2000 [20].

MaNGA observations of MrK 883 were obtained with the 61-fiber IFU on 18 May 2018 (MJD 58256). The single-fiber spectrum of MrK 883 was drawn from the SDSS database. The observation was obtained on 23 June 2003 (MJD 52813). The SDSS uses the Sloan spectrograph, which is the previous version of the MaNGA eBOSS spectrograph with spectral coverage from 3800 Å to 9200 Å and a spectral resolution of R $\approx$ 2000 [20]. The SDSS single-fiber observation uses a fiber diameter size of 3″.

Additionally, we obtained new IFS observations of Mrk 883 taken with the Multi-Espectrógrafo de Alta Resolución para Astronomía *MEGARA* [21]), located at the Gran Telescopio Canarias (GTC). The observations were obtained as a part of a project dedicated to studying DPAGN at kpc scales to establish the extension and origin of the ionized gas, along with its kinematics (GTC3-21AMEX; PI Benítez). We used the Large Compact Bundle (LCB) mode of *MEGARA* that provides an IFS observation with a field-of-view (FOV) of 12.5″ × 11.3″. We use the Volume Phase Holographic (VPH) gratings VPH675 LR-R and VPH521 MR-G. The VPH675 provides a spectral resolution of R = 5900 and a spectral range of 6096–7303 Å. The VPH521 has a spectral range of 4963–5445 Å, with a spectral resolution R = 12,000. The observations were performed in two observing blocks (OB). The first OB was obtained on 30 March 2021, during a photometric night with a bright moon and an atmospheric seeing of 1.3″. During this OB, three exposures of 663 s for each VPH were carried out. The second OB was obtained on 3 April 2021 on a photometric night with a bright moon and an atmospheric seeing of 1.0″. In this OB, three exposures of 663 s each with the two VPHs were obtained.

Recently, new spectra of Mrk 883 were obtained using the 2.1 m telescope and the Boller & Chivens spectrograph at OAN-SPM (hereafter, these data are denoted as SPM). SPM data were obtained from a long-term follow-up AGN variability program (PI Negrete). Long-slit spectroscopy observations were performed using 600 lines/mm grating and with a slit width of 150 microns. Three exposures of 1800 s were obtained to achieve a minimum signal-to-noise ratio (S/N) $\sim$ 30. We use two configurations: the blue arm, which covers the spectral range of 3300–5600 Å with a spectral resolution of R = 1000, and the red arm, which covers the spectral range of 5100–7500 Å with a spectral resolution of R = 1300. The observations were taken during gray nights and with photometric conditions on 14 and 17 April 2023.

3. Data Reduction

The SPM spectra were reduced using the standard long-slit IRAF packages. Therefore, the final spectra were processed through bias and flat field corrections, and then wavelength calibration, aperture extractions, and, finally, flux calibration.

For the 3D spectra, the reduction procedure is based on the one described in [22]. The objective is to reduce and reconstruct the calibrated *MEGARA* data cube. The procedure is summarized as follows:

(a) 2D reduction: We use the official *MEGARA* Data Reduction Pipeline (DRP) to perform the basic spectroscopic reduction. The DRP executes the bias subtraction, fiber extraction, flat-fielding, fiber flexure correction, and the wavelength and flux calibration of the GTC raw data. In this stage, we correct the spectra for the heliocentric velocity;

(b) 3D reduction: With the extracted wavelength and flux-calibrated multi-fiber spectra, we reconstruct the data cube. We use the 3D spatial re-sampling methodology described in [22]. The spatial re-sampling takes into account the *MEGARA* fiber size of 0.6″ to define a data cube of 38 × 40 spaxels with a size of 0.35″. In this stage, we correct the spatial position of the target from the atmospherically differential refraction. In addition, we combine the Observed Blocks (OBs) of the VPH521 MR-G and the VPH675 LR-R. We homogenize the spectral resolution to the lowest VPH resolution: R = 5900, that is, the resolution of VPH675. The final spectral sampling was fixed to a linear sampling of 0.3 Å per pixel. We homogenize the spectral resolution to R = 5900 to apply a full spectral fitting in Section 7;

(c) We perform a second-order spectrophotometric calibration to correct for any flux fluctuation on the surface brightness of the target. The method consists of comparing a set of photometric observations to the synthetic photometry maps from our data cube and measuring any flux deviations. However, the spectral coverage of the VPH521 and VPH675 did not match the full spectral coverage of the standard broadband photometric filters. Therefore, we use the MaNGA IFS observation of Mrk 883 to obtain synthetic photometric maps within the spectral regions of the VPH521 and VPH675. With these maps, we calculate a second-order flux correction map, and use it to obtain the final flux-corrected data cube;

(d) With the final data cube, we now proceed to calculate accurate astrometry. We use both synthetic SDSS r-band images from the MaNGA [15,19] and our final *MEGARA* data cube to match the astrometry and also match it with the SDSS spectrum. To do so, we calculate the centroids of the two nuclei within the MaNGA image and compare them with the centroids of the same nuclei within the *MEGARA* FOV. Therefore, we obtained the World Coordinate System (WCS) for our final *MEGARA* data cube to match the WCS of MaNGA.

4. Methodology

We extracted the integrated spectra of the *MEGARA* and MaNGA data cubes using the same aperture of the SDSS fiber to match the archival spectrum of Mrk 883 obtained by the SDSS in MJD 53226. The same aperture was chosen for extracting the spectra of the SPM observations.

The position and size of the SDSS fiber aperture appear in Table 1, and the same for MaNGA, *MEGARA* and SPM data. Also, in Figure 2, we present the emission line profiles in the Hβ (left) and Hα (right) regions obtained from *MEGARA*, MaNGA, SPM, and SDSS observations. Using the continuum around 5100 Å, we have estimated S/N ratios of 71, 100, 47, and 41, respectively. We applied the redshift correction using the value provided by the SDSS (z $\sim$ 0.04).

Similarly, we extracted the spectra for the second aperture for the *MEGARA* data (see Section 7, where we compared both Apertures 1 and 2). The S/N of the continuum around 5100 Å is 47.

Table 1. Astrometric position of the SDSS fiber: RA and DEC in (J2000). The aperture radius is in arcsec.

Instrument	RA	DEC	Aperture Radius	Date	S/N
SDSS	16:29:52.886	+24:26:38.40	1.5	23 June 2003	41
MaNGA	16:29:52.886	+24:26:38.40	1.5	18 May 2018	100
MEGARA	16:29:52.886	+24:26:38.40	1.5	1 Jan 2021	71
MEGARA	16:29:52.794	+24:26:37.25	0.75	1 Jan 2021	47
SPM	16:29:52.886	+24:26:38.40	1.5	15 April 2023	47

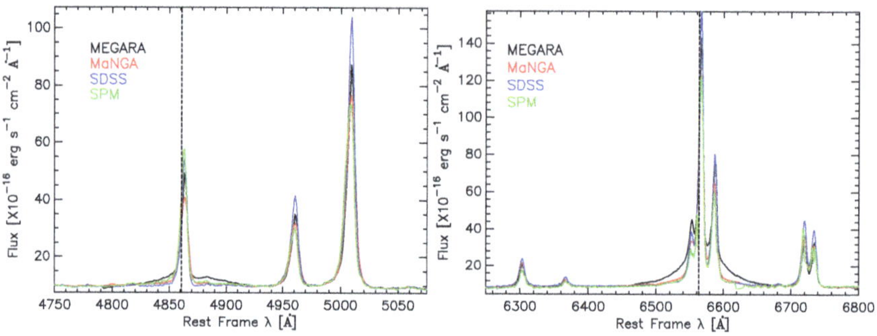

Figure 2. (**Left**): Mrk 883 spectra obtained with SPM (green), SDSS (blue), MaNGA (red), and *MEGARA* (black) are shown in the Hβ and [OIII]4959,5007 regions. (**Right**): The same but in the Hα, [NII]$\lambda\lambda$6548,6583, and [OI]$\lambda\lambda$6300,6364 regions. The abscissa is the rest-frame wavelength. The ordinate is the flux. The dashed vertical lines are in the Hβ and Hα rest-frame positions, respectively.

The best way to quantify the amount of spectral variability in the same object that has multi-epoch spectra is by making the decomposition of the most prominent emission lines. For this purpose, we used the task *Specfit*, which is part of the IRAF astronomical package [23]. *Specfit* allow us to model, at the same time, the underlying continuum, the emission lines, and, if necessary, absorption lines. For the line fitting, we split the spectral range into two regions: 4850 Å and 5050 Å for Hβ, and 6200 Å and 6420 Å for Hα. The Hα region includes the emissions of [NII]$\lambda\lambda$6548,6583, [SII]$\lambda\lambda$6716,6731, and [OI]$\lambda\lambda$6300,6364 doublets. The [OIII]$\lambda\lambda$4959,5007 emission is included in the Hβ region. The first step is to set the underlying continuum, modeled using a power law. The emission at the locus of nucleus 1 is dominated by the AGN and completely overshadows the host continuum (see Section 7). We found that the power law index was practically zero for the four spectra. Then, we considered Gaussian profiles to model all the emission line components. Each Gaussian uses three input parameters: the central wavelength, the line intensity, and the FWHM. Once we had the initial guesses (number of components and initial parameters), we minimized the χ^2 to obtain the best fit using the Marquardt algorithm for 5–10 iterations.

We started with the line fitting in the Hα spectral region because it has the most prominent lines. For all the spectra, we modeled Hα, the doublets of [NII], [SII], and [OI] narrow components (NC) constraining the FWHM, to be equal for all the lines. We assumed that they originate in the same clouds and therefore share the same kinematics. The intensities of [NII]λ6548 and [OI]λ6364 were linked to [NII]λ6583 and [OI]λ6300, respectively, taking into account the theoretical 3:1 ratio. The lambda shift of the three doublets was also linked to the same. We included an Hβ_{BC} in the fitting of the four spectra. Since the SDSS spectrum was previously modeled [10], we first reproduced their line fitting, confirming that two Gaussian components are needed to account for the narrow emission lines. We then fit the remaining SPM, MaNGA, and *MEGARA* spectra, imposing the same considerations. The spectral fitting of the Hα region is shown in the right panels of Figure 3.

For the Hβ spectral range, we fit the same two narrow components for the [OIII] doublet and the Hβ_{NC}, trying to set very similar initial conditions in the FWHM and shifts, as found in the Hα region. The residuals obtained in the fitting around [OIII]λ4959 indicated that an additional blueshifted Gaussian component is needed to obtain the best fit. Also, we do not find a broad component for Hβ. However, we have to include an Hβ_{BC} for the SPM, *MEGARA*, and MaNGA spectra (see left panels in Figure 3). Thus, these results are in agreement with the previous fittings carried out on the SDSS spectrum by [10].

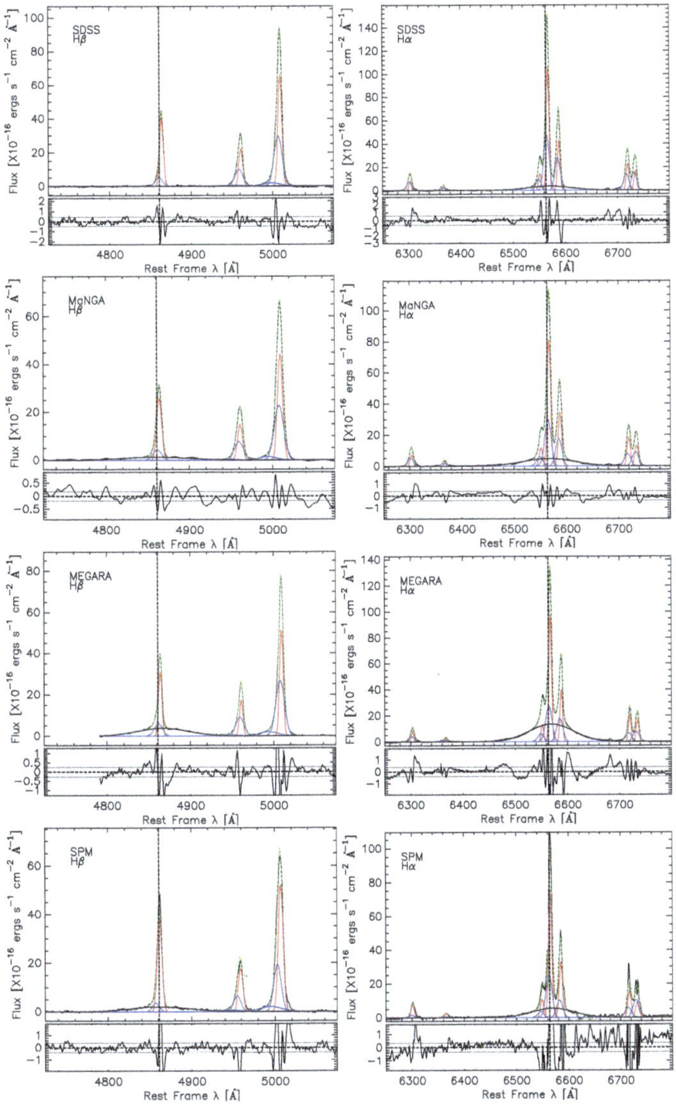

Figure 3. (**Left**): Hβ and [OIII] line-fitting continuum subtracted for the SDSS, MaNGA, SPM, and *MEGARA* spectra. (**Right**): The same but in the Hα [NII], [SII], and [OI] regions. The upper panels show the fit, while the lower panels report the residuals. The dashed horizontal black line is at zero level, and the solid lines mark $\pm 2\sigma$. For both spectral ranges, the thin black line is the observed spectra, the green line is the fitted model, and the dashed vertical lines are the Hβ and Hα rest-frames, considering the z reported by the SDSS (z = 0.03787). The thick black line is the Balmer broad component. The red and blue lines are the blue and red components for each narrow emission line. In the Hβ range, the thick blue line is an extra blueshifted component. The abscissa is the rest-frame wavelength. The ordinate is the flux.

5. Results

The emission line fluxes, FWHM, and EW(Hα) measurements obtained from the line modeling are presented in Table 2. For the line doublets of [OIII], [NII], and [OI], we report the fluxes only for the strongest line (see details in Section 4). We report the FHWM only for the Balmer components. The blue and red FWHMs of the forbidden lines are the same for $H\beta_{NC}$ and $H\alpha_{NC}$ components, except for the extra blueshifted component needed to fit the [OIII] 5007 Å. The EWs of $H\alpha_{NC}$ are reported at the bottom of the table. We find consistency in the velocity separation of the blue and red narrow components, with average values of $\Delta v(H\beta)$ 119 ± 8 km s^{-1} and $\Delta v(H\alpha)$ 107 ± 10 km s^{-1}, in agreement with [10] for the SDSS spectrum. The average shift velocities with respect to the rest frame are 109 ± 12 km s^{-1}, 113 ± 12 km s^{-1}, and −45 ± 8 km s^{-1} for the broad, red, and blue components, and −689 ± 5 km s^{-1} extra blueshifted component for [OIII]. This component is denoted as the wind component.

Table 2. Mrk 883 line fluxes in units of 10^{-16} erg s^{-1} cm^{-2} Å^{-1}, FWHM of the broad and narrow components in km s^{-1}, and $H\alpha_{NC}$ EW in Å.

Line	SDSS	MaNGA	*MEGARA*	SPM
Flux				
$H\alpha_{BC}$	355.2 ± 10.7	570.7 ± 8.2	1406.4 ± 13.3	534.0 ± 29.6
$H\alpha_{NC}$ B	600.0 ± 6.0	446.4 ± 20.1	410.0 ±7.8	370.1 ± 56.6
$H\alpha_{NC}$ R	671.6 ± 5.7	652.4 ± 20.1	594.6 ± 7.3	599.0 ± 11.4
[NII]λ6583 B	350.0 ± 7.0	263.1 ± 10.5	268.6 ± 7.8	151.8 ± 20.4
[NII]λ6583 R	272.2 ± 5.8	281.7 ± 10.1	243.5 ± 5.6	258.7 ± 8.5
[SII]λ6716 B	183.9 ± 5.2	120.3 ± 6.4	106.2 ± 4.5	90.0 ± 13.4
[SII]λ6716 R	150.1 ± 4.3	151.1 ± 5.6	129.2 ± 3.5	123.6 ± 12.5
[SII]λ6731 B	194.3 ± 5.3	136.3 ± 5.9	119.8 ± 4.6	158.4 ± 15.2
[SII]λ6731 R	109.1 ± 4.4	109.5 ± 5.5	107.3 ± 3.6	72.1 ± 17.0
[OI]λ6300 B	94.3 ± 3.8	77.3 ± 4.1	62.1 ± 3.4	23.4 ± 10.0
[OI]λ6300 R	49.7 ± 2.9	56.6 ± 3.6	48.5 ± 2.5	59.0 ± 7.1
$H\beta_{BC}$	…	134.2 ± 12.3	281.4 ± 8.8	174.1 ± 16.3
$H\beta_{NC}$ B	59.5 ± 19.8	57.6 ± 5.6	73.3 ± 5.7	38.6 ± 6.5
$H\beta_{NC}$ R	231.2 ± 4.9	193.3 ± 5.1	157.6 ± 4.9	270.0 ± 10.0
[OIII]λ5007 B	357.2 ± 5.1	317.2 ± 12.1	340.0 ± 12.1	214.0 ± 10.0
[OIII]λ5007 R	390.1 ± 0.8	340.3 ± 13.5	276.4 ± 13.3	380.0 ± 10.0
[OIII]$\lambda5007_W$	56.0 ± 19.8	36.8 ± 5.6	53.9 ± 5.7	66.0 ± 6.5
FWHM				
$H\alpha_{BC}$	4336 ± 172	5181 ± 112	4473 ± 88	4199 ± 121
$H\alpha_{NC}$ B	538 ± 6	636 ± 11	642 ± 12	629 ± 92
$H\alpha_{NC}$ R	270 ± 2	345 ± 4	262 ± 2	338 ± 7
$H\beta_{BC}$	…	5927 ± 481	4353 ± 117	5076 ± 373
$H\beta_{NC}$ B	672 ± 144	777 ± 78	712 ± 71	621 ± 25
$H\beta_{NC}$ R	332 ± 21	431 ± 6	297 ± 6	408 ± 6
[OIII]$\lambda5007_W$	1586 ± 159	1425 ± 110	1599 ± 224	1758 ± 178
EW				
$H\alpha_{NC}$ B	64.49 ± 0.69	49.55 ± 1.20	45.30 ± 0.94	43.20 ± 6.67
$H\alpha_{NC}$ R	72.18 ± 0.66	72.41 ± 2.29	65.70 ± 0.88	69.92 ± 6.98
Derived quantities				
$\log L_{bol}$	44.23 ± 0.20	44.25 ± 0.19	44.26 ± 0.18	44.24 ± 0.20
$\log M_{BH}$	7.93 ± 0.20	8.07 ± 0.19	7.92 ± 0.20	7.80 ± 0.20
$\log L_{Edd}$	46.03 ± 0.20	46.17 ± 0.19	46.03 ± 0.08	45.90 ± 0.20
$\log L_{bol}/L_{Edd}$	0.02 ± 0.01	0.01 ± 0.01	0.02 ± 0.01	0.02 ± 0.02
Balmer decrement	…	4.3±0.5	5.0 ±0.2	3.1 ±0.5

For each spectrum, we computed the bolometric luminosity (L_{bol}) using the continuum luminosity at 5100 Å and considering a bolometric correction of 10.33 [24], the black hole mass (M_{BH}, following [25] using the FWHM($H\beta_{BC}$)), the Eddington luminosity (L_{Edd}), and the Eddington ratio (L_{bol}/L_{Edd}), defined as the ratio of the bolometric and Eddington luminosities. For the SDSS spectrum, we calculated the FWHM($H\beta_{BC}$) using Equation (3) in [26]. We find average values using the four epochs of $\log L_{bol}$ = 44.25 ± 0.19, $\log M_{BH} \doteq$ 7.93 ± 0.20, $\log L_{Edd}$ = 46.03 ± 0.17, and $\log L_{bol}/L_{Edd}$ = 0.02 ± 0.02, with a standard deviation of 0.01,

0.11, 0.01, and 0.005, respectively. Note that, in Sy1.8 and Sy1.9, black hole mass estimates can be affected by possible obscuration of the continuum and the BLR. We report the estimations of the Balmer decrement (BD) and find that it also shows variability. This is an expected result in AGN with a low Eddington ratio (see Figure 2 in [27]).

Diagnostic Diagrams

The emission line profile modeling, previously shown, allows us to explore the ionized emission properties of Mrk 883 at different epochs. For instance, the line ratios were used to obtain BPT diagnostic diagrams. The regions in the BPT diagram are delimited by the [28] demarcation line (red line in Figure 4): above that line, the objects have an ionization that can be associated with an AGN, and the [29] demarcation line (black line in Figure 4): below that region, the ionization can be associated to an SF region. Objects that fall in between SF and AGN regions are usually considered composite. In this region, we can have objects with emissions produced by aged stars that are difficult to identify. As such, these objects can be misidentified from being aged stars as a LINER-like object [30]. The area below the [31] line denotes the LINER region (blue line in Figure 4). Therefore, we have also obtained the EWHα vs. [NII]/Hα diagram (WHAN [30]) using the line ratios obtained in this work. The WHAN diagram is designed to disentangle the ionization source produced by Hot Low-mass evolved stars (HOLMES [30,32]) from AGNs and SF regions. [30] separate the WHAN diagram into four regions: the SF region—log [NII]/H$\alpha < 0.4$ and EW(Hα) > 3 Å; the fake AGN region—EW(Hα) < 3 Å; the weak AGN region—log [NII]/H$\alpha > -0.4$ and 3 Å $<$ EW(Hα) < 6 Å; and the strong AGN emission—log [NII]/H$\alpha > -0.4$ and EW(Hα) > 6 Å. The BPT diagrams and the WHAN diagram obtained for Mrk 883 are shown in Figure 4.

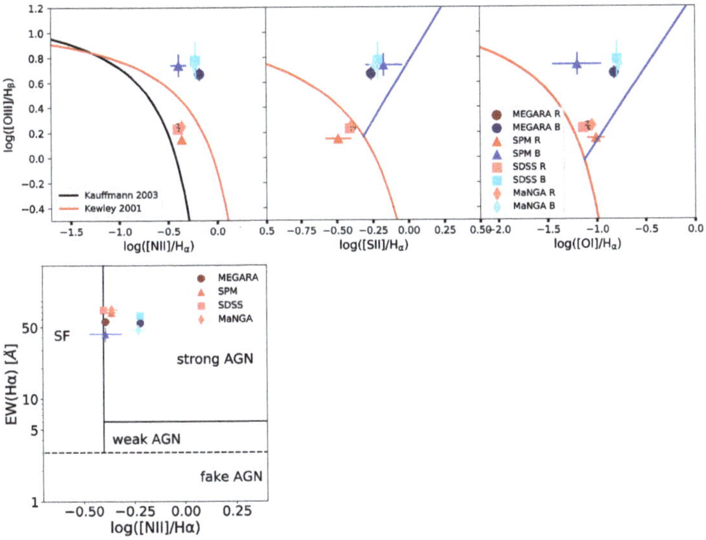

Figure 4. (**Upper panels**): BPT diagrams [11] obtained with the line ratios of Mrk 883. Different colors and symbols were used to illustrate the two Gaussian components needed to fit the narrow emission lines obtained during the multi-epoch observations. The line ratios obtained with the red Gaussian components from SDSS, MaNGA, *MEGARA*, and SPM spectra appear in the locus of composite objects in two diagrams and as AGN in the third one. Note that the *MEGARA* points are behind the pink SDSS and MaNGA points. In the case of the blue Gaussian components, these appear in the AGN locus. For more details, see the text. (**Lower panel**): The WHAN diagram (same colors and symbols used in the upper panels) shows that the blue and red Gaussian components from SDSS, MaNGA, *MEGARA*, and SPM appear, within the errors, in the strong AGN locus.

6. Variability Analysis

Flux variations in the Balmer broad components were detected, using the SDSS spectrum as a reference, in the MaNGA, *MEGARA*, and SPM spectra. Using the spectral decomposition described in Section 4, we have isolated the Hα and Hβ broad emission lines to compare the four activity states. This was performed through the subtraction of the modeled narrow components. In the upper panels of Figure 5, we show the profile variations in Hβ_{BC} and Hα_{BC} in the four epochs (with the NLR subtracted). The minimum intensity and width of broad lines were detected in the SDSS spectrum, and the maximum flux was detected in the *MEGARA* spectrum. However, it is worth noting that the maximum FWHM for the Hα and Hβ broad components were found in MaNGA data. Also, we found that the MaNGA and SPM have activity states in between the maximum and minimum activity states.

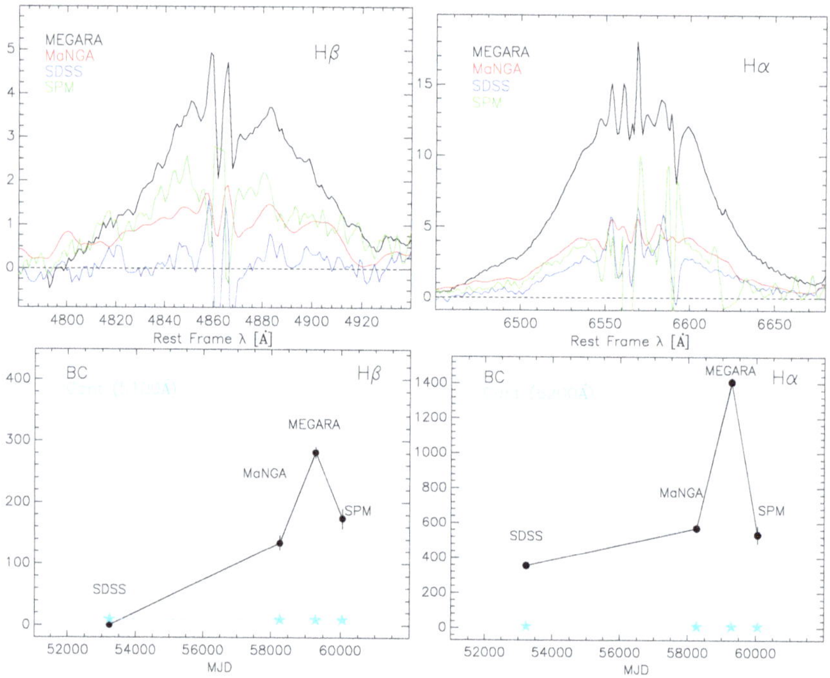

Figure 5. (**Upper panels**): Profile variations in Hβ_{BC} and Hα_{BC} are observed in four different epochs. The SDSS spectrum shows no broad Hβ component. The horizontal axis shows the MJD. The vertical shows the flux in units 10^{-16} erg s^{-1} cm^{-1} Å^{-1}. The appearance of the Hβ_{BC} is evident in the MaNGA spectrum. The maximum intensity of Hβ_{BC} and Hα_{BC} was observed three years later in the *MEGARA* spectrum. The SPM spectrum shows that the intensity of Hβ_{BC} and Hα_{BC} is declining after its maximum. (**Lower panels**): The optical continuum flux measured at 5000 Å and in 6200 Å (turquoise-filled stars) shows little or no variation during the four epochs, while flux variations (same units as in upper panels) are evident in the Hβ and Hα broad components.

A more detailed variability analysis is shown in the lower panels of Figure 5. The flux variations for the four stages of the Hα and Hβ broad components as a function of time are presented. The lower left panel shows the flux variations in Hβ and the flux continuum values for each epoch, measured around 5100 Å using a window of 50 Å. It is important to mention that the continuum flux at 5100 Å remains almost constant at around $9.1 \pm 0.2 \times 10^{-16}$ erg s^{-1} cm^{-2}Å^{-1}. Similar variations are seen in Hα and the flux continuum around 6200 Å measured using a window of 50 AA, shown in the lower right panel. For both Hα and Hβ broad components, we can observe clear flux variations, where *MEGARA* spectrum shows the maximum flux values in the Hα_{BC} and Hβ_{BC}.

7. Spatially Resolved *MEGARA* Data

Until this point, we have discussed only the data analysis of the nucleus of Mrk 883 (Aperture 1). Now, we will explore the spatially resolved broad emission line components of Hβ and Hα spectral regions in both nuclei (see Figure 1). We model the emission line profiles spaxel by spaxel using BADASS3D (Bayesian AGN Decomposition Analysis for SDSS Spectra, [33]) software tool. BADASS3D allow us to perform a Bayesian spectral decomposition of the host galaxy contribution and a multi-component emission line fit of the emission lines at the same time for IFS data. With this tool, we performed a modeling of the broad components of the Hα and Hβ spectral regions plus the host galaxy for each of the 38 × 40 spaxels that comprise the *MEGARA* FOV. To minimize the host-AGN degeneracy on the spectral fitting, we applied BADASS3D to fit the full spectral range from 4963 to 7300 once, masking the gap between the two VPHs. Therefore, it was necessary to homogenize the spectral resolution of R = 5900. The emission line modeling is not as detailed as the SPECFIT modeling that we performed in the previous section, but allows us to obtain the automatic modeling of the spectral lines along all the FOV of *MEGARA* with good accuracy (see [33] for more details). Therefore, we show the line modeling results from both nuclei (see Figure 6).

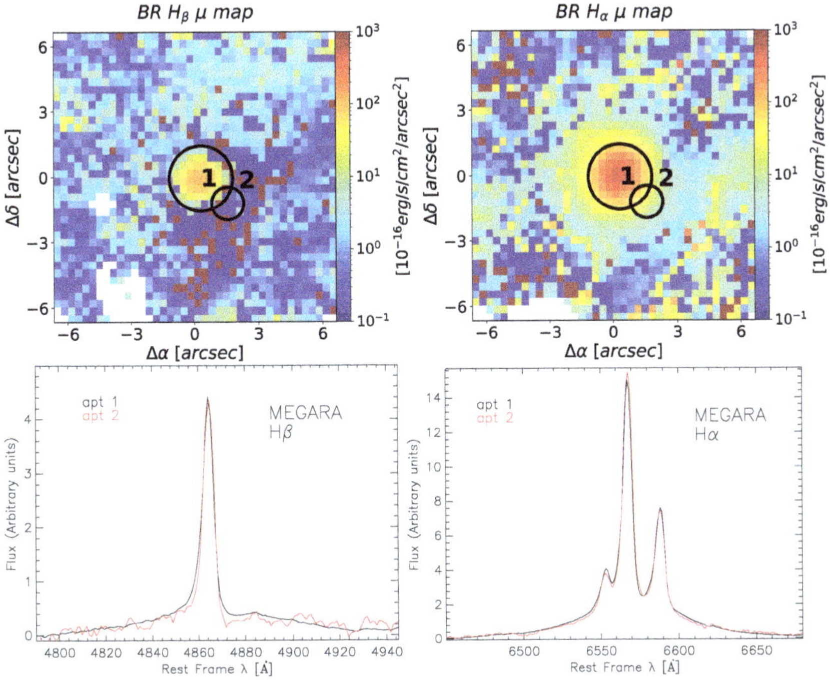

Figure 6. (**Upper panel**): Spatially resolved images showing the total flux of the broad Hβ and Hα components. (**Lower panel**) Nucleus 2 (Aperture 2) spectrum was analyzed with the code BADASS3. Since both nuclei are separated by ∼1.7 kpc and nucleus 2 shows a very similar spectrum to nucleus 1, the analyses provide similar results. This is due to ionized gas originating from the AGN in Mrk 883 that contaminates (at kpc scales) the emission from nucleus 2.

In Figure 6, we show the locus occupied by both nuclei. The *MEGARA* data were used to estimate the projected separation in the sky between both nuclei (∼1.7 kpc) and we also found that the spatial distribution of the flux of the broad component is always located at nucleus 1 (Aperture 1). Our data did not show that nucleus 2 is producing AGN activity. Instead, the spatial analysis shows that the emission line spectrum of this source is the result of ionized gas produced by nucleus 1 that is spread by the PSF to nucleus 2. This explains why the integrated spectra from aperture 2 show a very similar spectrum and, therefore, this spectrum was not analyzed. The contamination of the spectra from nuclei 1 to 2 does not allow for disentangling the true nature of nucleus 2.

8. Discussion

The multi-epoch spectral observations show that flux variations are related to spectroscopic changes. Since the $H\beta_{BC}$ is not observed in the SDSS spectrum, we found, in agreement with previous results by [10], that the SDSS spectrum is consistent with a Sy1.9 galaxy. Fifteen years later, we observed the appearance of the $H\beta_{BC}$ in the MaNGA spectra, consistent with a Sy1.8 galaxy. Therefore, our multi-epoch observations show that Mrk 883 is a CL AGN. Since the CL behavior was previously observed by [9], where he reported spectral-type variations from Sy1.9 (May 1979) to Sy1.8 (October 1993), we have found an object that presents recurrent CL variability in timescales $\sim$15 years in the optical bands. Since we do not have data between SDSS (2003) and MaNGA (2018) spectra, the recurrent variability timescale could be shorter. This could be better established if yearly optical spectral monitoring of the source is performed. Our multi-epoch spectra allow us to observe the evolution of the Balmer line profiles during the last five years. The variations include profile and flux variations, with an almost non-varying optical continuum.

Since the observed optical continuum variation in Mrk 883 is small, the appearance of the broad component in $H\beta$ may not be caused by changes in the accretion rate, i.e., changes may be due to obscuration by outflows. Some models suggest even accelerating outflows [6,34,35] as the origin of the CL variations. Accretion disk instabilities can also explain the CL variations, e.g., [36,37]. For example, radiation pressure instabilities can provide a viable mechanism to explain repeated outbursts on timescales of a few years (see [38]). Some studies have also found that CL AGNs tend to have low Eddington ratios, e.g., [37]. Our data confirm that Mrk 883 has low Eddington ratio values. For this kind of CL AGN, the disk-wind broad-line-region model is found to be a plausible explanation of the CL phenomenon [5,37]. Recently, [7] suggested that CL AGNs could be produced by non-steady outflows even when the luminosity of the central source remains constant throughout variations. Thus, dynamical processes occurring outside the accretion disk can be the origin of the variations. Therefore, based on our observations, it could be possible that an ionized-driven wind originating outside the accretion disk escaping in the polar direction could be producing the CL behavior in Mrk 883.

In addition, we report, for the first time, the presence of a wind component with a maximum FWHM = 1758 ± 178 km s^{-1} in the [OIII]5007 Å line in Mrk 883. It is worth noting that the wind component is narrower when the flux and width of the broad components increase. Also, we find that, in the spatially resolved WHAN diagram, Mrk 883 is a strong AGN. On the other hand, due to the contamination produced by the closeness of the two nuclei ($\sim$1.7 kpc), we cannot establish the true nature of the second one. The study of nature of the second nuclei and the origin of the wind component will be addressed in a forthcoming paper.

Author Contributions: Conceptualization, E.B. and C.A.N.; methodology, E.B., H.I.-M. and C.A.N.; software, C.A.N. and H.I.-M.; validation, I.C.-G. and J.M.R.-E.; formal analysis, E.B., H.I.-M. and C.A.N.; investigation, E.B.; resources, E.B., H.I.-M., C.A.N., I.C.-G. and J.M.R.-E.; data curation, H.I.-M.; writing—original draft preparation, E.B., H.I.-M. and C.A.N.; writing—review and editing, E.B., H.I.-M. and I.C.-G.; visualization, H.I.-M.; supervision, E.B.; project administration, I.C.-G.; funding acquisition, I.C.-G. All authors have read and agreed to the published version of the manuscript.

Funding: All authors acknowledge support from DGAPA-UNAM grant IN119123. E.B. and I.C.-G. thank the support from CONAHCyT grant CF-2023-G-100. C.A.N. thanks support from CONAHCyT project Paradigmas y Controversias de la Ciencia 2022-320020. C.A.N. and H.I.-M. thank support from DGAPA-UNAM grant IN111422. H.I.-M. thanks support from DGAPA-UNAM grant IN106823.

J.M.R.-E. acknowledges the Spanish State Research Agency under grant number AYA2017-84061-P. J.M.R.-E. also acknowledges the Canarian Government under the project PROID2021010077, and is indebted to the Severo Ochoa Programme at the IAC. Funding for the SDSS and SDSS-II has been provided by the Alfred P. Sloan Foundation, the Participating Institutions, the National Science Foundation, the U.S. Department of Energy, the National Aeronautics and Space Administration, the Japanese Monbukagakusho, the Max Planck Society, and the Higher Education Funding Council for England. The SDSS website is www.sdss4.org. Funding for the Sloan Digital Sky Survey IV has been provided by the Alfred P. Sloan Foundation, the U.S. Department of Energy Office of Science, and the Participating Institutions.

Data Availability Statement: The GTC/*MEGARA* data are available in the public archive at https://gtc.sdc.cab.inta-csic.es/gtc/index.jsp (accessed on 21 December 2023). SDSS data are available at https://data.sdss.org/sas/ (accessed on 21 December 2023). The SPM data underlying this article will be shared on reasonable request to the corresponding author.

Acknowledgments: This work is based on observations made with the Gran Telescopio Canarias (GTC), installed at the Spanish Observatorio del Roque de los Muchachos of the Instituto de Astrofísica de Canarias, on the island of La Palma. This work is based on data obtained with *MEGARA* instrument, funded by European Regional Development Funds (ERDF), through Programa Operativo Canarias FEDER 2014-2020. SDSS acknowledges support and resources from the Center for High-Performance Computing at the University of Utah. The SDSS website is www.sdss4.org. This work is based on observations at the Observatorio Astronómico Nacional on the Sierra San Pedro Mártir (OAN-SPM), Baja California, Mexico. IRAF was distributed by the National Optical Astronomy Observatory, managed by the Association of Universities for Research in Astronomy (AURA) under a cooperative agreement with the National Science Foundation. This research has made use of the NASA/IPAC Extragalactic Database (NED), which is operated by the Jet Propulsion Laboratory, California Institute of Technology, under contract with the National Aeronautics and Space Administration.

Conflicts of Interest: The authors declare no conflicts of interest.

References

1. Antonucci, R. Unified models for active galactic nuclei and quasars. *Annu. Rev. Astron. Astrophys.* **1993**, *31*, 473. [CrossRef]
2. Padovani, P.; Alexander, D.M.; Assef, R.J.; De Marco, B.; Giommi, P.; Hickox, R.C.; Richards, G.T.; Smolčić, V.; Hatziminaoglou, E.; Mainieri, V.; et al. Active Galactic Nuclei: What's in a name? *Astron. Astrophys. Rev.* **2017**, *25*, 2. [CrossRef]
3. Hickox, R.C.; Alexander, D.M. Obscured Active Galactic Nuclei. *Annu. Rev. Astron. Astrophys.* **2018**, *56*, 625–671. [CrossRef]
4. Ulrich, M.-H.; Maraschi, L.; Urry, C.M. Variability of Active Galactic Nuclei. *Annu. Rev. Astron. Astrophys.* **1997**, *35*, 445. [CrossRef]
5. Wang, J.; Zheng, W.K.; Xu, D.W.; Brink, T.G.; Filippenko, A.V.; Gao, C.; Sun, S.S.; Wei, J.Y. B3 0749+460A: A New Repeat "Changing-look" Active Galactic Nucleus Associated with X-Ray Spectral Slope Variations. *Res. Astron. Astrophys.* **2022**, *22*, 015011. [CrossRef]
6. Ricci, C.; Trackhtenbrot, B. Changing-look Active Galactic Nuclei. *Nat. Astron.* **2023**, *7*, 1282. [CrossRef]
7. Wada, K.; Kudoh, Y.; Nagao, T. Multiphase gas nature in the sub-pc region of the active galactic nuclei—II. Possible origins of the changing-state AGNs. *Mon. Not. R. Astron. Soc.* **2023**, *526*, 2717. [CrossRef]
8. Osterbrock, D.E.; Dahari, O. Spectra of Seyfert galaxies and Seyfert galaxy candidates. *Astrophys. J.* **1983**, *273*, 478. [CrossRef]
9. Goodrich, R.W. Dust in the Broad Line Region of Seyfert Galaxies. *Astrophys. J.* **1995**, *440*, 141. [CrossRef]
10. Benítez, E.; Méndez-Abreu, J.; Fuentes-Carrera, I.; Cruz-González, I.; Martínez, B.; López-Martin, L.; Jiménez-Bailón, E.; Chavushyan, V.; León-Tavares, J. Characterization of a Sample of Intermediate-type Active Galactic Nuclei. I. Spectroscopic properties and serendipitous discovery of new dual AGNs. *Astrophys. J.* **2013**, *763*, 36. [CrossRef]
11. Baldwin, J.A.; Phillips, M.M.; Terlevich, R. Classification parameters of the emission-line spectra of extragalactic objects. *Publ. Astron. Soc. Pac.* **1981**, *93*, 5. [CrossRef]
12. Hernández-García, L.; Masegosa, J.; González-Martín, O.; Márquez, I.; Guainazzi, M.; Panessa, F. X-ray variability of Seyfert 1.8/1.9 galaxies. *Astron. Astrophys.* **2017**, *602*, A65. [CrossRef]
13. Abazajian, K.N.; Adelman-McCarthy, J.K.; Agüeros, M.A.; Allam, S.S.; Prieto, C.A.; An, D.; Anderson, K.S.; Anderson, S.F.; Annis, J.; Bahcall, N.A.; et al. The Seventh Data Release of the Sloan Digital Sky Survey. *Astrophys. J. Suppl. Ser.* **2009**, *182*, 2. [CrossRef]
14. Bennett, C.L.; Larson, D.; Weiland, J.L.; Hinshaw, G. The 1% Concordance Hubble Constant. *Astrophys. J.* **2014**, *794*, 135. [CrossRef]
15. Bundy, K.; Bershady, M.A.; Law, D.R.; Yan, R.; Drory, N.; MacDonald, N.; Wake, D.A.; Cherinka, B.; Sánchez-Gallego, J.R.; Weijmans, A.M.; et al. Overview of the SDSS-IV MaNGA Survey: Mapping Nearby Galaxies at Apache Point Observatory. *Astrophys. J.* **2015**, *798*, 7. [CrossRef]

16. York, D.G.; Adelman, J.; Anderson, J.E., Jr.; Anderson, S.F.; Annis, J.; Bahcall, N.A.; Bakken, J.A.; Barkhouser, R.; Bastian, S.; Berman, E.; et al. The Sloan Digital Sky Survey: Technical Summary. *Astron. J.* **2000**, *120*, 1579. [CrossRef]
17. Strauss, M.A.; Weinberg, D.H.; Lupton, R.H.; Narayanan, V.K.; Annis, J.; Bernardi, M.; Blanton, M.; Burles, S.; Connolly, A.J.; Dalcanton, J.; et al. Spectroscopic Target Selection in the Sloan Digital Sky Survey: The Main Galaxy Sample. *Astron. J.* **2002**, *124*, 1810. [CrossRef]
18. Abdurro'uf, N.; Accetta, K.; Aerts, C.; Aguirre, V.S.; Ahumada, R.; Ajgaonkar, N.; Ak, N.F.; Alam, S.; Prieto, C.A.; Almeida, A.; et al. The Seventeenth Data Release of the Sloan Digital Sky Surveys: Complete Release of MaNGA, MaStar, and APOGEE-2 Data. *Astrophys. J. Suppl. Ser.* **2022**, *259*, 35. [CrossRef]
19. Drory, N.; MacDonald, N.; Bershady, M.A.; Bundy, K.; Gunn, J.; Law, D.R.; Smith, M.; Stoll, R.; Tremonti, C.A.; Wake, D.A.; et al. The MaNGA Integral Field Unit Fiber Feed System for the Sloan 2.5 m Telescope. *Astron. J.* **2015**, *149*, 2. [CrossRef]
20. Smee, S.A.; Gunn, J.E.; Uomoto, A.; Roe, N.; Schlegel, D.; Rockosi, C.M.; Carr, M.A.; Leger, F.; Dawson, K.S.; Olmstead, M.D.; et al. The Multi-object, Fiber-fed Spectrographs for the Sloan Digital Sky Survey and the Baryon Oscillation Spectroscopic Survey. *Astron. J.* **2013**, *146*, 32. [CrossRef]
21. Carrasco, E.; de Paz, A.G.; Gallego, J.; Iglesias-Páramo, J.; Cedazo, R.; Vargas, M.G.; Arrillaga, X.; Avilés, J.L.; Bouquin, A.; Carbajo, J.; et al. *MEGARA*, R = 6000–20000 IFU MOS GTC. *Proc. SPIE* **2018**, *10702*, 1070216.
22. Benítez, E.; Ibarra-Medel, H.J.; Negrete, C.A.; Cruz-González, I.; Rodríguez-Espinosa, J.M.; Liu, X.; Shen, Y. 3D spectroscopy with GTC-*MEGARA* of the Triple AGN Candidate in SDSS J10270057+174900. *Astrophys. J.* **2023**, *952*, 45. [CrossRef]
23. Kriss, G. Fitting Models to UV and Optical Spectral Data. In *Astronomical Data Analysis Software and Systems III*; ASP Conference Series; Astronomical Society of the Pacific: San Francisco, CA, USA, 1994; Volume 61, p. 437.
24. Richards, G.T.; Lacy, M.; Storrie-Lombardi, L.J.; Hall, P.B.; Gallagher, S.C.; Hines, D.C.; Fan, X.; Papovich, C.; Berk, D.E.V.; Trammell, G.B.; et al. Spectral Energy Distributions and Multiwavelength Selection of Type 1 Quasars. *Astrophys. J. Suppl. Ser.* **2006**, *166*, 470. [CrossRef]
25. Vestergaard, M.; Peterson, B. Determining central black hole masses in distant active galaxies and quasars II. Improved optical and UV scaling relationships. *Astrophys. J.* **2006**, *641*, 689. [CrossRef]
26. Shen, J.; Vanden Berk, D.E.; Schneider, D.P.; Hall, P.B. The black hole-bulge relationship in luminous broad-line active galactic nuclei and galaxies. *Astron. J.* **2008**, *135*, 928. [CrossRef]
27. Wu, J.; Wu, Q.; Xue, H.; Lei, W.; Lyu, B. Steep Balmer Decrement in Weak AGNs May Not Be Caused by Dust Extinction: Clues from Low-luminosity AGNs and Changing-look AGNs. *Astrophys. J.* **2023**, *950*, 106. [CrossRef]
28. Kewley, L.J.; Dopita, M.A.; Sutherland, R.S.; Heisler, C.A.; Trevena, J. Theoretical Modeling of Starburst Galaxies. *Astrophys. J.* **2001**, *556*, 121. [CrossRef]
29. Kauffmann, G.; Heckman, T.M.; White, S.D.M.; Charlot, S.; Tremonti, C.; Brinchmann, J.; Bruzual, G.; Peng, E.W.; Seibert, M.; Bernardi, M.; et al. Stellar masses and star formation histories for 105 galaxies from the Sloan Digital Sky Survey. *Mon. Not. R. Astron. Soc.* **2003**, *341*, 33. [CrossRef]
30. Cid Fernandes, R.; Stasińska, G.; Mateus, A.; Vale Asari, N. A comprehensive classification of galaxies in the Sloan Digital Sky Survey: How to tell true from fake AGN? *Mon. Not. R. Astron. Soc.* **2011**, *413*, 1687. [CrossRef]
31. Kewley, L.J.; Groves, B.; Kauffmann, G.; Heckman, T. The host galaxies and classification of active galactic nuclei. *Mon. Not. R. Astron. Soc.* **2006**, *372*, 961. [CrossRef]
32. Stasińska, G.; Mateus, A.; Sodré, L.; Szczerba, R. What drives the Balmer extinction sequence in spiral galaxies? Clues from the Sloan Digital Sky Survey. *Astron. Astrophys.* **2004**, *420*, 475. [CrossRef]
33. Sexton, R.O.; Matzko, W.; Darden, N.; Canalizo, G.; Gorjian, V. Bayesian AGN Decomposition Analysis for SDSS spectra: A correlation analysis of [OIII]λ5007 outflow kinematics with AGN and host galaxy properties. *Mon. Not. R. Astron. Soc.* **2021**, *500*, 2871. [CrossRef]
34. Shapovalova, A.I.; Popovic, L.C.; Burenkov, A.N.; Chavushyan, V.H.; Ilić, D.; Kovačević, A.; Bochkarev, N.G.; León-Tavares, J. Long-term variability of the optical spectra of NGC 4151. II. Evolution of the broad Hα and Hβ emission-line profiles. *Astron. Astrophys.* **2010**, *509*, A106. [CrossRef]
35. La Massa, S.; Cales, S.; Moran, E.C.; Myers, A.D.; Richards, G.T.; Eracleous, M.; Heckman, T.M.; Gallo, L.; Urry, C.M. The Discovery of the First "Changing Look" Quasar: New Insights Into the Physics and Phenomenology of Active Galactic Nucleus. *Astrophys. J.* **2015**, *800*, 144. [CrossRef]
36. Noda, H.; Done, C. Explaining changing-look AGN with state transition triggered by rapid mass accretion rate drop. *Mon. Not. R. Astron. Soc.* **2018**, *480*, 3898. [CrossRef]
37. MacLeod, C.M.; Green, P.J.; Anderson, S.F.; Bruce, A.; Eracleous, M.; Graham, M.; Homan, D.; Lawrence, A.; LeBleu, A.; Ross, N.P.; et al. Changing-look Quasar Candidates: First Results from Follow-up Spectroscopy of Highly Optically Variable Quasars. *Astrophys. J.* **2019**, *874*, 8. [CrossRef]
38. Sniegowska, M.; Czerny, B.; Bon, E.; Bon, N. Possible mechanism for multiple changing-look phenomena in active galactic nuclei. *Astron. Astrophys.* **2020**, *641*, A167. [CrossRef]

universe MDPI

Review

Spectral Variability Studies in Active Galactic Nuclei: Exploring Continuum and Emission Line Regions in the Age of LSST and JWST

Swayamtrupta Panda [1,*,†], Paola Marziani [2], Bożena Czerny [3], Alberto Rodríguez-Ardila [1,4] and Francisco Pozo Nuñez [5]

1 Laboratório Nacional de Astrofísica, Rua dos Estados Unidos, 154 Bairro nas Nações, Itajubá 37504-364, MG, Brazil; aardila@lna.br
2 INAF-Astronomical Observatory of Padova, Vicolo dell'Osservatorio 5, 35122 Padua, Italy; paola.marziani@inaf.it
3 Center for Theoretical Physics, Polish Academy of Sciences, Al. Lotników 32/46, 02-668 Warsaw, Poland; bcz@cft.edu.pl
4 Divisão de Astrofísica, Instituto Nacional de Pesquisas Espaciais, Avenida dos Astronautas 1758, São José dos Campos 12227-010, SP, Brazil
5 Astroinformatics, Heidelberg Institute for Theoretical Studies, Schloss-Wolfsbrunnenweg 35, 69118 Heidelberg, Germany; francisco.pozonunez@h-its.org
* Correspondence: spanda@lna.br
† CNPq Fellow.

Abstract: The investigation of emission line regions within active galaxies (AGNs) has a rich and extensive history, now extending to the use of AGNs and quasars as "standardizable" cosmological indicators, shedding light on the evolution of our universe. As we enter the era of advanced observatories, such as the successful launch of the JWST and the forthcoming Vera C. Rubin Observatory's Legacy Survey of Space and Time (LSST), the landscape of AGN exploration across cosmic epochs is poised for exciting advancements. In this work, we delve into recent developments in AGN variability research, anticipating the substantial influx of data facilitated by LSST. The article highlights recent strides made by the AGN Polish Consortium in their contributions to LSST. The piece emphasizes the role of quasars in cosmology, dissecting the intricacies of their calibration as standard candles. The primary focus centers on the relationship between the broad-line region size and luminosity, showcasing recent breakthroughs that enhance our comprehension of this correlation. These breakthroughs encompass a range of perspectives, including spectroscopic analyses, photoionization modeling, and collaborative investigations with other cosmological tools. The study further touches on select studies, underlining how the synergy of theoretical insights and advancements in observational capabilities has yielded deeper insights into these captivating cosmic entities.

Keywords: galaxies: active; (galaxies): quasars: emission lines; (galaxies): quasars: supermassive black holes; accretion, accretion disks; radiation mechanisms: thermal; radiative transfer; methods: numerical; methods: observational; techniques: photometric; techniques: spectroscopic; (cosmology): distance scale

1. Introduction

Active Galactic Nuclei (AGNs) are among the most luminous and powerful sources of energyin our universe. AGNs are dynamic and powerful phenomena arising from the interaction between supermassive black holes and their surrounding environment [1]. The intense radiation originating from their very centers, how that interacts with the surrounding media, and their presence across a wide range of redshifts, make them crucial objects of study in understanding the cosmos at various scales and epochs [2–58]. They are characterized by the presence of a supermassive black hole (SMBH), accreting material from their surroundings, which leads to intense emission across a wide range of wavelengths. AGNs play a vital part in shaping the evolution of galaxies in the universe. As material falls in due to the gravitational pull of the black hole (BH), it forms an accretion disk (AD)—a flattened structure of matter spiraling inwards as it loses angular momentum. The disk becomes significantly hot, of the order of $\sim 10^5$ K, and emits radiation peaking in the X-ray and UV wavelengths. The accretion disk emission becomes intercepted by the gas clouds that, in turn, become ionized and emit the broad emission lines. This region is known as the broad line region (BLR) since the lines are characterized by the "broadness" of their line profiles. The broadening of the lines is due to a Keplerian distribution with velocities of the order of thousands to a few tens of thousand kilometers per second. Beyond the BLR lies the Narrow Line Region (NLR), another region of ionized gas emitting narrower emission lines. The NLR is characterized by lower ionization levels and extends farther out from the central BH [27,29].

The distinguishing emission features in the BLR and NLR led to further categorization of the AGNs, especially through detailed studies of the nearby "Seyfert" galaxies [2]. Two distinct categories of Seyfert galaxies were eventually established based on the characteristics of their emission lines in the spectrum: those galaxies that exhibit permitted emission lines that are notably broader than their forbidden emission lines are classified as "Type 1". In contrast, galaxies displaying permitted and forbidden lines with similar widths are labeled as "Type 2" [8,9]. Over the subsequent years, the number of recognized Seyfert galaxies surged, and successive investigations unveiled the remarkable features of these celestial objects: (1) they possess remarkably compact and luminous nuclear regions, and (2) their spectrum contains a variable ultraviolet and optical continuum that complements their unusually strong and broad emission lines (see [11] for an early review). AGNs are further classified into two well-known categories based on the strength of their radio emission: "radio-loud" (RL) and "radio-quiet" (RQ) AGN, a classification initially applied in the context of quasars (historically identified by sources with B-band magnitude $M_B < -23$) (see, for example, [59–74]). This categorization was based on two primary criteria: (1) radio flux density or luminosity (as demonstrated in [60]), and (2) the ratio of radio-to-optical flux density or luminosity (as illustrated in [75]). A critical parameter in this classification is the "radio loudness", characterized by the radio-to-optical flux density ratio, commonly referred to as the Kellerman index (as detailed in [61]). This index facilitated the grouping of radio-detected sources into these traditional categories, a practice that continues in the literature, albeit primarily applicable to Type-1 AGN [76,77]. However, recent studies by Padovani [78] and Padovani et al. [29] have emphasized the need for a more precise distinction. They propose that the two classes represent fundamentally distinct entities, with RL AGN emitting a substantial portion of their energy non-thermally, often in conjunction with powerful relativistic jets. In contrast, the multi-wavelength emission of RQ AGN is primarily dominated by thermal emission linked to the accretion disk. As a result, the primary physical discrepancy between these two categories is the presence or absence of strong relativistic jets. Hence, these sources should be viewed more appropriately in terms of "jetted" and "non-jetted" rather than adhering to the traditional RL-RQ dichotomy [29]. In this review, we will focus on the "non-jetted" sources and their salient properties.

The 1990s marked the rise of comprehensive surveys, such as the Bright Quasar Survey [13], enabling detailed statistical analyses of these sources, e.g., [79]. This opened up the way to group Type-1 AGNs with the estimation of the width of their Hβ emission line profile. This gave rise to a new dichotomy in these sources—Population A (Pop. A) and Population B (Pop. B). The notion of these two quasar populations was introduced as an analogy to the simplified classification of stars, e.g., [17]. Stars were cataloged into seven principal spectral types (OBAFGKM) in the HR Diagram. Population A sources included local Narrow-Line Seyfert 1 galaxies (NLS1s) as well as more massive, highly accreting objects that are predominantly classified as radio-quiet or non-jetted sources, e.g., [29,80,81]. These sources demonstrate FWHM(Hβ) estimates of $\leq$4000 km s^{-1}. Notably, the Hβ profile shape for Population A sources is typically Lorentzian-like, e.g., [82–84]. On the other hand,

Population B sources are noted to exhibit broader FWHM(Hβ) values ($\geq$4000 km s^{-1}), and are frequently found to be "jetted" objects, e.g., [29,85]. Their Hβ profiles are better described by Gaussian functions, and for those with even higher FWHMs, double Gaussian profiles characteristic of disk-like structures are observed in Balmer lines (see, e.g., [86]). Addressing the significance of the strict FWHM(Hβ) threshold, studies by, e.g., [17,84,87] indicate that AGN properties (such as the outflow intensity, accretion rate and metal content) exhibit notable shifts around this broader linewidth cutoff (at around 4000 km s^{-1}), mirroring the earlier division seen in Seyfert galaxies.

In this contribution, we group the studies based on the sub-structures that they are concerned with, i.e., Section 2 deals with our efforts to understand and utilize the information from the AD and the BLR, while Section 4 primarily highlights our efforts to probe the NLR using high transmission spectroscopy in the optical and near-infrared (NIR) regimes. AGNs offer insights into the universe's structure, evolution, and fundamental properties. They can serve as distance indicators, probe large-scale structures, and contribute to our understanding of dark matter, dark energy, and galaxy formation. We discuss how AGNs that are and will be probed at high redshifts bridge the studies of the local and early universe. We discuss some of these aspects in Section 3. The review concentrates on results from recently published works led by me and my collaborators. Hence, we do not intend to provide an exhaustive coverage of the literature in this review. We highlight the essential references throughout the text in this work to guide the readers for further reading.

We highlight some recent progress in the studies of accretion disk structure and the development of a database of AGN SEDs, especially for the sources accreting at or above the Eddington limit (see Section 2). We discuss the effectiveness of the BLR radius vs. AGN luminosity relation(s) in providing us with an independent way to estimate the luminosity distances to these cosmic sources, help build the Hubble diagram, and eventually test and retrieve cosmological parameters of our universe (see Section 3.1). Next, we propose a novel BH mass scaling relation traced by the intensities of high-ionization, forbidden emission lines (known as coronal lines) that we have discovered (see Section 4). These coronal lines, a bulk of them produced in the NLR, are shown to evolve with the change in the underlying accretion disk structure, which is the primary source of the photons that produce emission of such lines (see Section 4.1). We discuss our new-found BH mass scaling relation and how they can benefit from the up-and-coming JWST spectroscopic datasets. Wherever necessary, we provide notes for future studies in these aforementioned sections. We then showcase our efforts in preparation for the upcoming Vera Rubin Observatory's Legacy Survey for Space and Time (LSST) in the context of the accretion disk and BLR time delay measurements using photometric monitoring and scrutinizing the survey strategies that are under discussion (see Sections 5.1 and 5.2). We conclude with a summary of our work presented in this review with some notes for the future.

2. Understanding Accretion Disk and BLR Structure and Emission

In their comprehensive review, Sulentic and Marziani [88] raised a pivotal question: "Do Population A and B represent two distinct quasar populations, or are they merely extremes on the same continuum? Could they be interconnected through a gradual transition in accretion mode?" Figure 1 offers insights into this inquiry, suggesting that the latter scenario is operative. The figure underscores the transition in accretion disk structure from Population A to Population B, or vice versa, further elucidated in subsequent sections[1].

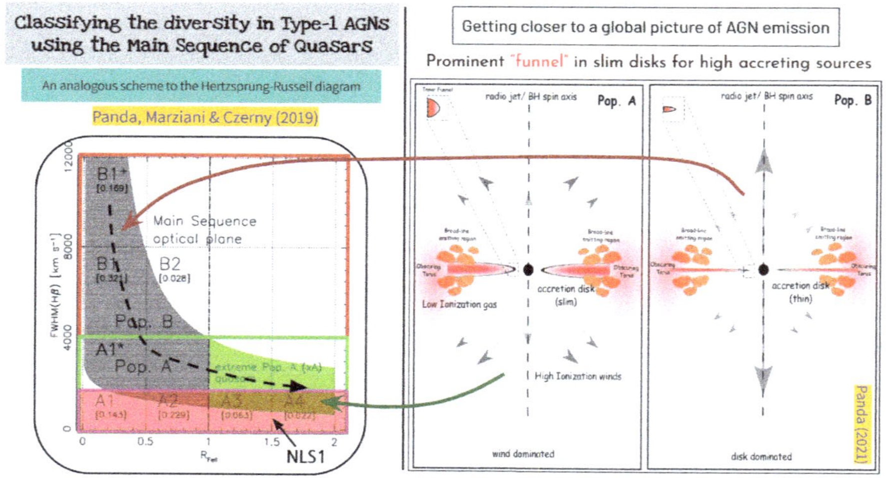

Figure 1. (**Left panel**): Schematic view of the main sequence of quasars in the optical regime. The figure demonstrates the spectral types (e.g., A1, A2, B1, B2, B1+) based on the full-width at half maximum (FWHM) of the broad component of Hβ versus the strength of the Fe II flux in the optical (4434–4684 Å) to the Hβ flux. Objects with FWHM(Hβ) $\lesssim$ 2000 km s^{-1} are highlighted as Narrow-line Seyfert 1 (NLS1) galaxies, while those with FWHM(Hβ) $\lesssim$ 4000 km s^{-1} belong to the Population A (Pop. A) and their counterparts with FWHM(Hβ) $\gtrsim$ 4000 km s^{-1} are marked as Population B (Pop. B). Abridged figure based on Panda et al. [87]. (**Right panel**): Schematic view of important components in the vicinity of supermassive black holes (SMBH) in AGNs. This illustration highlights the distinguishing characteristics of two different AGN types: Pop. A (left) and Pop. B (right). In both panels, an SMBH is located at the center, with a radio jet aligned along the BH's spin axis. Notably, the two panels contrast in terms of the accretion disk (AD) structure: the left panel showcases a slim AD, while the right panel features a thin disk conforming to the standard disk model (Shakura and Sunyaev [89]). Within this depiction, the BLR is divided into two regions: a high-ionization emitting region and a low-ionization emitting region. Additionally, the illustration includes the presence of a dusty, obscuring torus, with a covering factor nearly matching that of the low-ionization part of the BLR (Gaskell [90]). It's worth noting the modifications introduced by the distinct accretion disk structures in the inset panels (see the top left corner of each panel). Specifically, the funnel-like shape within the innermost region of the slim accretion disk is notably more prominent in comparison to the scenario presented by the thin disk case. Abridged figure based on Panda [91].

The reverberation mapping (RM) technique uses the notion of the extra time taken by the continuum photon—one that is intercepted by the broad line region clouds first before reaching us, and another that directly comes to us [12,16,18,23,90,92–110]. This extra time is then translated to how far the BLR clouds are located from the continuum emitting source, i.e., R_{BLR}. An important outcome of the investigations into RM is the establishment of an empirical power-law relationship between the R_{BLR} and the AGN's continuum luminosity, often expressed as $R_{BLR} \approx c\tau \propto L_{5100}{}^{\alpha}$. Here, τ represents the delay in time taken to respond to the variation in continuum for a given emission line, for example, Hβ. The recent developments in RM have led to the expansion of the empirical R_{BLR}-L_{5100} observational dataset [16,23], with a substantial increase in the total count, now exceeding 200 sources [111]. This growth has been primarily driven by the inclusion of sources monitored under the SEAMBH project [25,26,28,100,106,112], as well as contributions from the SDSS-RM campaigns [30,111,113] and OzDES survey (see, e.g., [109]). However, this expansion has introduced a new challenge: an inherent dispersion in the R_{BLR}-L_{5100} relation that has become apparent after incorporating these new sources (as depicted in Figure 1 of Panda and Marziani [114]). Interestingly, some of these new sources have contributed to an increase in the dispersion for the relation, which correlates with the Eddington ratio. Studies have found that the sources that demonstrate larger deviation from the empirical R_{BLR}-L_{5100} relation are found to have higher Eddington ratios [31,114,115].

Martínez-Aldama et al. [115] investigated the connection of the increased scatter in the standard R_{BLR}-L_{5100} relation with the Eddington ratio (L_{bol}/L_{Edd}) for the first time. However, the Eddington ratio is a parameter estimated from the luminosity itself and thus creates circular reasoning. To alleviate this circularity, Du and Wang [31] found that a separate observable parameter—the Fe II emission strength in the optical regime, denoted as R_{FeII}—can be used to correct for the added dispersion and the relation could be adjusted back with a slope close to 0.5. The introduction of the R_{FeII} term is supported by earlier studies which have shown its close connection to the L_{bol}/L_{Edd} [17,31,84,87,116,117]. Eventually, the introduction of the R_{FeII} in the R_{BLR}-L_{5100} relation compensates for the shorter time lags where sources with smaller R_{BLR} sizes are the ones with strong Fe II emission.

Addressing the shorter time lags observed in AGNs necessitates advancements in our photoionization modeling setup. This involves accounting for the Eddington ratio's influence on the continuum emission from the accretion disk and the resultant alterations in the net vertical distribution of the disk, particularly in locations near the innermost stable circular orbit (ISCO) of an SMBH. Wang et al. [118] developed steady-state, analytical solutions for "slim" accretion disks, encompassing a broad range in accretion rates, from sub-Eddington to super-Eddington. In their models, they note the emergence of a funnel-like structure at the onset of the accretion disk, close to the SMBH. They attribute this to high accretion rates obtained from solutions of slim accretion disks. These solutions deviate [118–120] from the standard accretion disk scenario of Shakura and Sunyaev [89]. We present a schematic view in the right panel of Figure 1 which depicts the scenario. The highlighted structural changes to the disk not only impact the net anisotropic emission but also highlight the inclination effects due to the axisymmetric nature of these systems. Consequently, at high accretion rates, it is justified to account for viewing-angle-dependent anisotropy while modeling the continuum emission from the accretion disks. This scenario can explain well the dichotomy in the location and subsequent emission lines produced from the low- and high-ionization regions. For example, as shown in Figure 1, the gas clouds emitting low-ionization lines are situated close to the disk plane, and effectively see a diminished disk luminosity relative to an observer with a line of sight nearly along the axis of rotation of the SMBH. This lowering in the flux incident on the BLR clouds in this region allows the clouds to encroach closer to the SMBH. This reduction in the location of the emitting regions aligns with observed estimates from RM campaigns (see [121] for additional details).

In our modeling approach [114,122], we construct a database of composite SEDs comprising two main elements: a slim accretion disk (in red) and a hot corona (in green) in each panel of Figure 2. The two components are adjusted through the spectral index (α_{ox}), which represents the ratio of the flux at 2 keV to the flux at 2500 Å [123]. The value for the spectral index is borrowed from the observed SEDs from Ferland et al. [124] where we match our model's composite SEDs viewed at 10° with these observed SEDs. We identify the best-fitting composite modeled SED for the corresponding cases of observed SED as a function of the dimensionless accretion rate (or $\dot{\mathcal{M}}$, [100,125]). Figure 2 (upper panels) demonstrates the best-fit modeled SEDs with increasing $\dot{\mathcal{M}}$. We also explore the relation of the anisotropic emission on the viewing angle in these models. While we assume the hot corona originates from an isotropic radiation source and thus remains unaffected by changes in the viewing angle, the disk emission does vary with the viewing angle. This is highlighted in the right panels of Figure 2. The funnel-like structure gives a preferred direction for continuum radiation from the disk, more importantly for the photons that have origins near to the SMBH.

With the introduction of the anisotropic emission from a slim disk, we connect the BLR and the accretion structure where the high-ionization lines could be emitted within the funnel or, at the very least, in a location where the ionized gas receives the bulk of the continuum. Conversely, low-ionization lines might be exposed to a weaker continuum, necessitating the ionized clouds to move in to account for the diminished flux of the photons. A comprehensive assessment of outcomes from our photoionization modeling will be presented in an upcoming work.

3. AGNs for Cosmology

AGNs play a significant role in cosmology by serving as valuable tools for various aspects of studying the universe's large-scale structure, evolution, and fundamental properties. AGNs with known luminosities, such as quasars, can serve as distance indicators in the universe. Since their intrinsic brightness is relatively consistent, their observed brightness can be used to estimate distances, aiding in measuring cosmological distances and understanding the expansion of the universe. In the next section, we discuss some salient details of this technique and highlight its shortcomings and how we as a community have devised ways to overcome these challenges to reinstate the use of quasars as standardizable distance indicators.

3.1. Effectiveness of the Radius-Luminosity Relation(s)

An important outcome of the investigations into reverberation mapping is the establishment of an empirical relationship between the radius of the BLR measured from the time delay (R_{BLR}) and the AGN's luminosity, often expressed as $R_{BLR} \approx c\tau \propto L_{5100}{}^{\alpha}$. Here, τ represents the delay in light travel time responding to the variation in the continuum of a given emission line, typically[2]. Notably, Bentz et al. [23] conducted a study on 41 AGNs covering a luminosity range spanning four orders of magnitude and determined a best-fit power-law slope value of $\alpha = 0.533^{+0.035}_{-0.033}$. This result is remarkably similar to $\alpha = 0.5$ (from theoretical expectations), which is necessary to preserve spectral similarity [126–128]. By combining this R_{BLR}-L_{5100} relation with the line profile widths of broad emission lines estimated from spectroscopy made at single or multiple epochs, one can gauge M_{BH}, rendering the relation particularly valuable for extensive surveys across wide redshift ranges [19,129].

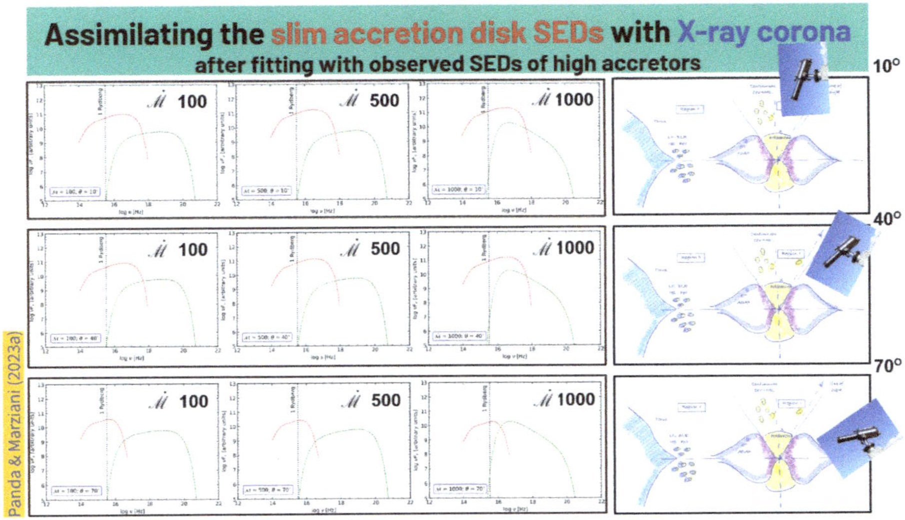

Figure 2. A collection of composite spectral energy distributions (SEDs) that combines a slim accretion disk (in red) and a high-energy, hot corona (in green). (From left to right:) The SEDs are shown as a function of increasing $\dot{\mathcal{M}}$ (or dimensionless accretion rate): (**left panels**) 100, (**middle panels**) 500, and (**right panels**) 1000. (From top to bottom:) Each $\dot{\mathcal{M}}$ case is accompanied by three distinct viewing angles (relative to an observer at a distance), denoted as i: (**upper panels**) 10°, (**middle panels**) 40°, and (**bottom panels**) 70°. The SEDs shown here assume a BH mass = 10^8 solar masses. The hot coronal emission is scaled to the slim disk. This adjustment employs the spectral index α_{ox} that is obtained by comparing the modeled SEDs at the lowest inclination angle (i.e., 10°) to the corresponding SEDs made directly from multi-wavelength observations. These observations are systematically arranged based on their $\dot{\mathcal{M}}$ values, as depicted in Figure 1 of Ferland et al. [124]. On the far right, a schematic view is provided, offering a conceptual representation of the inner region of a highly accreting SMBH. This representation is inspired by Wang et al. [118], although it is presented in a condensed format and not to scale. Abridged figure based on Panda and Marziani [122].

Dedicated monitoring has significantly expanded the population in the R_{BLR}-L_{5100} observational domain, surpassing a count of 100 (see [111,115,130] for more details). Notably, this was achieved through contributions from projects such as the SEAMBH project (Super-Eddington Accreting Massive Black Holes; [25,26,28,100,106,112]), the SDSS-RM campaigns [30,113], and individual photometric reverberation campaigns [131–133]. However, these developments have introduced a new issue, i.e., the increased scatter in the R_{BLR}-L_{5100} relation arising after the inclusion of these new sources. The left panel of Figure 2 in Panda and Marziani [114], which is a modified version from Martínez-Aldama et al. [115] and Panda et al. [130], illustrates the R_{BLR}-L_{5100} observational space for 117 AGNs that underwent reverberation mapping. Color-coded according to their Eddington ratios (L_{bol}/L_{Edd}), this plot reveals the best-fit relation for the aforementioned dataset as $\log R_{H\beta} = 0.387 \times (\log L_{5100}) - 15.702$. It exhibits a Spearman's correlation coefficient (ρ) of 0.733 and a p-value of 2.733×10^{-21}, leading to a relation with a significantly shallower slope than the previous findings of Bentz et al. [23]. This prompts questioning the validity of the empirical R_{BLR}-L_{5100} relation.

Interestingly, the sources that contribute to the increased dispersion in the relation exhibit a discernible relation with the Eddington ratio. Namely, sources deviating more from the empirical R_{BLR}-L_{5100} relation tend to possess higher Eddington ratios. In Martínez-Aldama et al. [115], it was demonstrated that this dispersion can be reconciled within the standard R_{BLR}-L_{5100} relation by introducing a dependency on the Eddington ratio (L_{bol}/L_{Edd}). This phenomenon is highlighted in the right panel of Figure 2 in Panda and Marziani [114], where the offset (ΔR_{BLR}) between the observed time delays and those calculated using the empirical R_{BLR}-L_{5100} relation (ΔR_{BLR}, [23]), is plotted against their respective L_{5100} values for the sources presented in the left panel. Notably, there is a distinct reduction in ΔR_{BLR} values for higher Eddington ratio sources, suggesting that objects with smaller observed time lags, particularly at the high-luminosity end of the relation, are accreting at or about the Eddington limit. Consequently, these objects are anticipated to host SMBHs with lower M_{BH} [25,28]. While certain high-Eddington sources exhibit reduced R_{BLR} sizes, some sources with comparable R_{BLR} sizes to those predicted by Bentz et al. [23] remain despite their high Eddington nature. Building upon this, Du and Wang [31] further demonstrated that the relation can be restored to its original slope of $\sim$0.5 by introducing an additional correction term. This supplementary term is based on the optical Fe II strength measured to the Hβ emission (R_{FeII}), that has been validated as a reliable observational proxy for the Eddington ratio in previous studies [17,31,84,87,116,117], as previously discussed. The analytical relation thus becomes [31]:

$$\log\left(\frac{R_{BLR}}{1 \text{ light-day}}\right) = \kappa + \alpha \log\left(\frac{L_{5100}}{10^{44} \text{ erg s}^{-1}}\right) + \gamma R_{FeII} \quad (1)$$

By incorporating this correction in terms of a measurable quantity, R_{FeII}, we circumvent the circularity issue that arose when the correction was explicitly based on the Eddington ratio [115]. This approach provides an independent estimation of the luminosity distance (D_L). Consequently, we can build a Hubble diagram using quasars for which we know both their luminosity distances and redshifts. This setup allows us to test the validity of the standard cosmological model as well as alternative models (see, e.g., [131,134,135]). Notably, super-Eddington accreting sources are anticipated to be favorably selected as redshift increases, particularly in a flux-limited sample [136]. This selection bias, combined with intrinsic evolution in the Eddington ratio, e.g., [137,138], often results in high-redshift quasar spectra resembling extreme Population A spectra found at lower redshifts [17]. Thus, the inclusion of such sources is pivotal to extend the R_{BLR}-L_{5100} relation to the higher luminosity regime, as illustrated in the left panel of Figure 3, where the parameters $\kappa = 1.65 \pm 0.06$, $\alpha = 0.45 \pm 0.03$, and $\gamma = -0.35 \pm 0.08$ are employed.

The incorporation of R_{FeII}, particularly for objects exhibiting intense Fe II emission, effectively compensates for the shorter time delays, thereby accounting for the smaller R_{BLR} sizes observed. Furthermore, this approach enables the recovery of a slope (α) that aligns more closely with the theoretical predictions, specifically 0.5 [127,139]. Consequently, we can confidently employ the modified R_{BLR}-L_{5100} relation (inclusive of the correction term involving R_{FeII}) to determine the source's monochromatic luminosity, independently. By combining the knowledge from directly observed flux with this approach, we can eventually derive accurate estimates for the luminosity distance to the target source and avoid the circularity problem:

$$D_L = \sqrt{L_{5100}/4\pi F} \propto \frac{c\tau}{\sqrt{4\pi F}} \quad (2)$$

4. Coronal Lines as BH Mass Tracers Using Optical and NIR Spectroscopy

The Narrow-Line Region (NLR) is another important component of AGNs, distinct from the BLR and the accretion disk. It is located farther out from the black hole than the BLR and is characterized by narrower emission lines in its spectrum indicating that the gas in the NLR moves at lower velocities in case of a virial velocity field. Due to the characteristic low particle densities ($\lesssim 10^7$ cm^{-3}), the NLR also emits forbidden lines (e.g., the coronal lines) of ionized elements like oxygen, nitrogen, and sulfur. The emission lines from the NLR provide valuable diagnostic tools for studying the physical conditions of the gas, e.g., its density, temperature, and metallicity. The ratios of different emission lines can be used to deduce the properties of the ionizing radiation field and the gas dynamics.

Coronal Lines (CLs) are a special class of emission lines with a majority of them originating in the NLR and are characterized by their elevated ionization potentials (IPs), spanning from approximately 50 eV up to a few hundred eV. This unique property renders them exceptional indicators of the ionizing continuum [140–145]. While they may appear fainter compared to the conventional medium-ionization lines frequently employed for photoionization analysis, advancements in high angular resolution techniques for nearby AGNs have unveiled that CLs, notably those observable in the NIR spectrum, stand out as some of the most prominent features, e.g., [146–149]. The selection of objects in this study is based on two main criteria: M_{BH} estimated through reverberation mapping and the availability of single-epoch optical and/or NIR spectra with high-quality measurements of the CLs. The initial basis confines that our sources belong to the Type-1 AGNs exclusively, while the latter criterion aims to mitigate potential issues stemming from variability. While a preference is given to sources offering both optical and NIR spectra, this specific requirement could not always be met. The utilized coronal lines include [Fe VII] λ6087 Å in the optical spectrum and [S VIII] 0.991 µm, [Si X] 1.432 µm, and [Si VI] 1.964 µm in the NIR spectrum. These lines are recognized as among the most potent CLs in AGNs [150,151] and span a substantial ionization potential (IP) range of 100 to 350 eV. Additionally, H II lines, specifically Hβ, Paβ, and Brγ, are incorporated. Collectively, this set of lines captures the ionizing continuum over a range spanning from 13.6 to 351 eV. The preference for NIR CLs is attributed to their reduced susceptibility to extinction, while the choice of optical CL [Fe VII] is attributed to its robust strength, modest extinction, and an IP in the vicinity of [Si VI]. The final selection comprises 31 AGNs (Table 1 in Prieto et al. [152]). This dataset encompasses BH masses, primarily sourced from the compilation by Bentz and Katz [153], along with CL ratios extracted from spectra studied in Prieto et al. [152] or derived from previously documented spectroscopic data, along with the corresponding data sources. The majority of the line flux ratios in the near-infrared (NIR) utilized in this study are derived from Riffel et al. [154]. In cases where the mentioned publication did not provide data for specific targets, we acquired observations through the use of the Gemini Near-Infrared Spectrograph (GNIRS) on the Gemini North Observatory, or the ARCOiRIS/TripleSpec4 near-infrared spectrograph, which was installed on either the 4 m Blanco or 4 m SOAR telescopes.

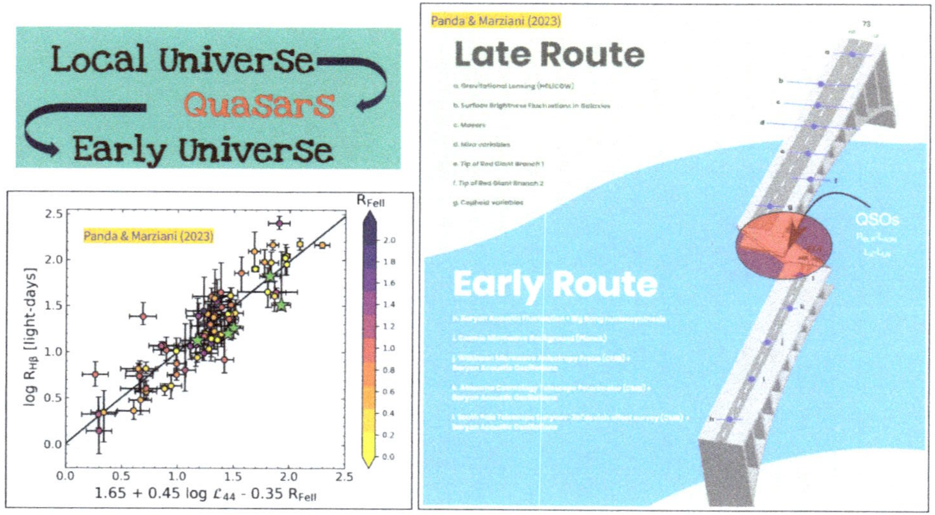

Figure 3. Collage demonstrating the use of quasars as standardizable distance indicators (**Left panel**): A summarized rendition of Figure 5 from Du and Wang [31] displays the updated relationship between R_{BLR} versus L_{5100} and incorporating the parameter R_{FeII} for 75 sources. The black solid line represents the best-fit relation, characterized by a ρ = 0.894 and a p-value = 3.466×10^{-27}. The region marked in blue shows the 99% confidence interval around the best-fitting line, encompassing a scatter of approximately 0.196 dex. In this context, $\mathcal{L}_{44}$ denotes the L_{5100} normalized by 10^{44} erg s^{-1}. Sources exceeding $L_{bol}/L_{Edd} > 3$ are highlighted using green star symbols. Abridged figure based on Panda and Marziani [114]. (**Right panel**): This diagram illustrates the diverse range of methodologies utilized for determining the Hubble constant (H_0), one that characterizes the expansion rate of the Universe. The observations can be divided into two categories: one involving investigations of the very early Universe (depicted in the lower half of the diagram), and the other focusing on the Universe's expansion estimated from the local distance indicators (shown in the upper half). On the right-hand side of the graphic, the letters associated with each technique (enumerated on the left) are placed along the bridge. Collating data from the various methods within the low-z Universe exploration gives an average H_0 estimate of 73 km s^{-1} Mpc^{-1}. For the early Universe techniques, the combined H_0 value is determined to be 67.4 km s^{-1} Mpc^{-1}. Selected quasar-based techniques, the radius-luminosity relation, and another - X-ray to UV luminosity relation are highlighted in the red circle. Abridged figure based on Panda and Marziani [114].

We demonstrate the relationship of the $M_{\rm BH}$ with a prominent CL, [Si VI] 1.963 μm which has an IP [Si VI] = 167 eV[3] after normalizing it to the nearest H I broad line emission, i.e., Brγ, where the dependence to the $M_{\rm BH}$ was found to be the strongest. We observe a tight correlation between $M_{\rm BH}$ and the CL ratio [Si VI]/Br$\gamma_{\rm broad}$. Figure 4 introduces a novel diagnostic diagram, showcasing the correlation between the black hole (BH) mass of the objects in our sample and the ratio of [Si VI] 1.963 μm to Br$\gamma_{\rm broad}$[4] . This ratio exhibits a distinct trend with BH mass spanning three orders of magnitude. A linear regression analysis applied to the data yields the following result:

$$\log M_{\rm BH} = (6.40 \pm 0.17) - (1.99 \pm 0.37) \times \log\left(\frac{[\rm Si\ VI]}{\rm Br\gamma_{broad}}\right), \tag{3}$$

which has a 1σ scatter of 0.235 dex. We used the LTSFIT package[5] [156] for the regression analysis accounting also for the errors in the two parameters. The Pearson correlation coefficient for the correlation is $r = -0.76$, with a p-value = 3.8×10^{-5}. In contrast, when examining 49 galactic bulges with directly measured dynamical black hole (BH) masses, a dispersion of 0.22 dex is revealed in the $M - \sigma$ relation [157]. It is worth noting that the intrinsic scatter in mass-luminosity relationships is approximately 40%, as reported by Kaspi et al. [158]. This variability is primarily attributed to disparities in the shape of the optical-UV continuum.

The current scaling relation for BH masses is constrained to Type-1 AGNs, encompassing narrow-line Seyfert galaxies as well. This limitation arises from the necessity to include only reliably determined BH masses and to normalize them against broad H I gas measurements. Nevertheless, efforts are underway to explore the feasibility of extending this relation to Type-2, obscured AGNs. The newly proposed scaling relation offers an efficient and theoretically-inspired alternative for estimating $M_{\rm BH}$ using single epoch spectroscopic data. This approach eliminates the need for extensive telescope time, as required in techniques like reverberation mapping, or the necessity for highly precise absolute flux calibration, which is essential in methods such as the continuum luminosity approach. By making use of the capabilities of the JWST and extensive infrared surveys, this method has the promise of efficiently determining the masses of large samples of AGNs. For a detailed examination of the observed spectra of the sources and comprehensive photoionization modeling to validate the underlying principles of the observed scaling relationship, we recommend referring to the work conducted by Prieto et al. [152].

4.1. Standard Disk Predictions and Radiative Transfer Modeling

The correlation observed between coronal line (CL) emission and BH mass prompted an investigation into the potential connection between the aforementioned CL emission and predictions derived from standard accretion disk theory. This exploration aimed to establish a link between CL emission and a key attribute of the accretion disk—specifically, the disk's temperature peak. To test this hypothesis, the photoionization code CLOUDY v17.02 [159] was employed to generate a range of synthetic quasar spectra encompassing varied physical conditions. The primary objective was to assess whether the observed CL line ratios exhibit a discernible dependence on $T_{\rm disk}$, which is introduced to CLOUDY through the ionizing continua depicted in Figure 4 (upper left panel). In this figure, the CLs are attuned to distinct energy ranges within the AGN ionizing continuum. This continuum arises from a composite of two components: a standard Shakura-Sunyaev (SS) accretion disk [89], describing the spectral transition from ultraviolet (UV) to soft X-rays involves capturing a power-law segment featuring both lower and upper energy limits, addressing the spectrum's rise at higher energies. It follows the equation

$$F_\nu = \nu^{\alpha_{uv}} \exp\left(\frac{-h\nu}{kT_{\rm disk}}\right) \exp\left(\frac{-kT_{\rm IR}}{h\nu}\right) + a\nu^{\alpha_x}. \tag{4}$$

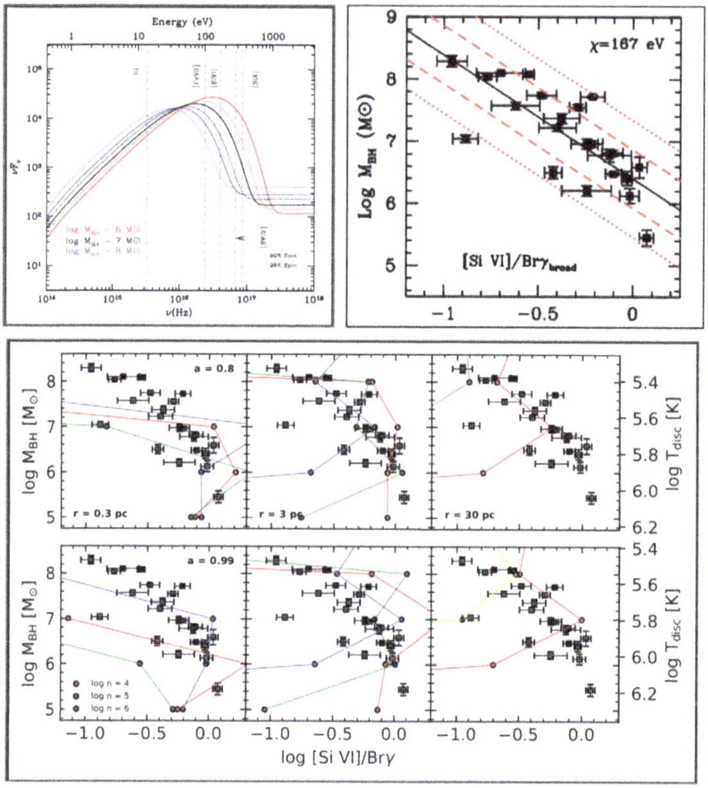

Figure 4. Coronal line (CL) emission driven by SMBH-dependent properties in an AGN SED (**Top left**): In this study, the generic ionizing continuum for AGNs is formulated according to Equation (4), accommodating the range of BH masses taken from the AGN sample [153] where the masses are derived from the Hβ-based reverberation mapping (see **top right panel**). For each BH mass, an associated temperature of the accretion disk ($T_{\rm disk}$) is considered, adhering to the thin disk approximation as delineated in Equation (5). This representation results in a series of curves, where each curve corresponds to a distinct BH mass. We also include the dependence of the BH spin in these SEDs, i.e., two values of the BH spin: a = 0.8 (dashed) and a = 0.99 are incorporated. The analysis encompasses three representative BH masses: 10^6 (in red), 10^7 (in black), and 10^8 (in blue). The BH masses are reported in solar masses. To provide context, vertical dashed lines are employed to signify the ionization potentials (IPs) of the lines incorporated in the analysis; (**Top right**): Observed [Si VI] 1.963μm/Br$\gamma_{\rm broad}$ ratio (denoted by R) versus BH mass for the sources analyzed in Prieto et al. [152]. The linear best-fit is shown using the black line and the 1σ and 2σ deviation are shown using the red-dashed and -dotted lines, respectively; (**Bottom panel**): CLOUDY predictions for [Si VI]/Br$\gamma_{\rm broad}$ vs BH mass using the ionizing continua shown in the top-right panel. Models are simulated for local cloud densities, $n_H = 10^4$ cm^{-3} (red), 10^5 cm^{-3} (green), and 10^6 cm^{-3} (blue). Each **sub-panel** shows the results from the CLOUDY modeling for varying distance from the ionizing source, (from left to right) r = 0.3, 3, and 30 pc, respectively. We consider two BH spins: a = 0.8 (**upper panel**) and a = 0.99 (**lower panel**). We depict the corresponding $T_{\rm disk}$ per each BH mass, on the right side. Observed data are marked in black squares. Abridged figure based on Prieto et al. [152].

The first component is an expression of the SS disk, formulated as an exponential function featuring a cutoff at the effective temperature of the disk, $T_{\rm disk}$, coupled with a power-law segment characterized by $\alpha_{\rm uv}$ = 0.33 to account for the disk's low-energy tail. The lower boundary of the disk's energy range is defined by an infrared-exponential component with a cutoff at 0.01 Ryd. In the high-energy regime, the distribution is described by a broken power law with a spectral index denoted as $\alpha_x = -1$, which exhibits a cutoff at 100 keV. The relationship between the SS disk and the high-energy power law is governed by the parameter represented as a in Equation (4), indicating the ratio of luminosities at 2 keV and 2500 Å. This ratio is expressed through a power-law function with a spectral index α_{ox} set at −1.4. Except for α_{uv} = 0.33, all other parameters mentioned here adhere to the general AGN continuum utilized in CLOUDY [160]. In the ionizing continuum presented in Figure 4 (upper left panel), the peak emission effectively corresponds to $T_{\rm disk}$, with the disk's temperature increasing as the BH mass decreases, aligning with predictions from standard accretion disk theory. Importantly, it is worth noting that the ionization potentials of the coronal lines align closely with the peak temperatures, or slightly beyond, depending on the BH mass. If the accretion disk primarily acts as the source of photons for ionizing the coronal gas, we can anticipate a relationship between the strength of coronal lines and the temperature of the accretion disk, denoted as $T_{\rm disk}$. Given the relationship between $T_{\rm disk}$ and BH mass, this connection might consequently result in a correlation between the strength of coronal lines and BH mass. We follow the standard accretion disk prescription, and employing it for a BH in Kerr metric, the disk temperature, $T_{\rm disk}$, can be written as follows (Frank et al. [161], Lasota [162] prescription is used):

$$T_{\rm disk} = 3.4\times10^5 K\left(\frac{M_{\rm BH}}{10^8 M_\odot}\right)^{1/4}\times\left(\frac{\left(\frac{dM}{dt}\right)}{0.1}\right)^{1/4} \times\left(\frac{\eta}{0.26}\right)^{-1/4}\times\left(\frac{R_{G,in}}{1.4}\right)^{-3/4} \quad (5)$$

Here, $M_{\rm BH}$ represents the black hole mass, (dM/dt) signifies the accretion rate in Eddington units, η stands for the black hole accretion efficiency, and $R_{G,in}$ represents the innermost stable circular orbit (ISCO), measured in units of gravitational radii, $R_G = GM_{\rm BH}/c^2$ units (where G represents the gravitational constant and c is the speed of light). This equation is normalized for a BH mass of $M_{\rm BH} = 10^8$ $M_\odot$, accreting at 10% of the Eddington limit (i.e., $dM/dt \sim 0.1\ (dM/dt)_{Edd}$), and η = 0.26, that corresponds to a BH that is maximally spinning, i.e., $a = 0.99cJ/GM_{\rm BH}^2$ (hereafter referred to as a = 0.99, where J is the spin angular momentum), where the rotation is in clockwise direction. Recent estimations of BH spin from various methods tend to converge towards values close to 1 (see [163] for a comprehensive compilation). As the spin of the object increases, the ISCO contracts, resulting in higher values for the accretion disk temperature, denoted as $T_{\rm disk}$. When the condition of co-rotation is met, the disk temperature reaches its maximum value.

The ionizing continua follow the correlation between the accretion disk temperature ($T_{\rm disk}$) and BH mass, as described in Equation (5). Curves are presented for three specific BH masses and two spin values, namely 0.8 and 0.99. Spin values below 0.8 result in minimal deviations in T_{disk} when compared to the a = 0.8 scenario, and are consequently omitted from presentation (see, for instance, a comprehensive study on spin parameter space by [164]).

With such an incident continuum as the predominant contributor to ionize the emitting regions, and considering an appropriate range of local cloud densities ($n_e \sim [10^4, 10^6]$ cm^{-3}) as well as cloud distances (ranging from 0.3 pc to approximately 30 pc) to ensure the continuance of CLs, photoionization models conducted using CLOUDY [159] yield a reasonably accurate depiction of the correlation between $M_{\rm BH}$ and the [Si VI]/Br$\gamma_{\rm broad}$ (see bottom panel of Figure 4).

It is discernible that the peak emission from the disk progressively approaches the ionization potentials of Si^{+5} and Fe^{+6} as the BH mass decreases and the spin increases. However, as we transition to higher BH masses (10^7–10^8 $M_\odot$) or beyond, the peak energies obtained from the accretion disk alone is insufficient for the ions with higher IPs, Si^{+9} and S^{+7}, with T_{disk} becoming cooler as the BH mass rises. This is where we bring in the SEDs that include the warm comptonizing component reflected in the soft X-ray regime. We make a re-run of the CLOUDY models by replacing the incident continuum shape with these modified accretion disk SEDs. We notice a clear increase in the flux available at the locations for the higher ionization CLs, i.e., Si^{+9} and S^{+7} ions. These new models provide better agreement to the observed correlations between the CL ratios and the BH mass for these other ionic species. The results of the Si^{+5} and Fe^{+6} ions are relatively unaffected.

5. Getting Ready for LSST: Predictions for BLR and AD Time-Lag Recovery

5.1. BLR Time-Lag Recovery Expectations from LSST

The Vera C. Rubin Observatory's Legacy Survey of Space and Time (LSST; [165–171]) is poised to revolutionize our understanding of the universe by delivering an unparalleled volume of data across various fields. Observations are set to commence relatively soon, likely around mid-2024, underscoring the significance of optimizing data utilization, especially during the initial phases of operation. Amidst its myriad outcomes, LSST is anticipated to unveil up to ten million quasars, paving the way for extensive reverberation monitoring of AGNs over a redshift range spanning from 0 to 7 [172,173].

Broad emission lines are a hallmark of nearly all luminous AGN (see, e.g., [1] for a comprehensive review). Nonetheless, the exact location and configuration of the corresponding region, known as the Broad Line Region (BLR), remain largely unresolved, except for recent infrared observations of three active galactic nuclei (AGN) using the GRAVITY Very Large Telescope Interferometer (VLTI) instrument. These AGN include 3C 273 (a well-known radio-loud source, as seen in Gravity Collaboration et al. [174]), and two radio-quiet sources: NGC 3783 (as observed in GRAVITY Collaboration et al. [175]), and IRAS 09149-6206 (as reported in GRAVITY Collaboration et al. [149]). For most other objects, our understanding of BLR structure and dynamics heavily relies on reverberation monitoring. While the LSST will provide photometric light curves for numerous AGNs, the translation from photometry to line time delays is intricate and demands precision. It is essential to assess the reliability of measurements, taking into account source luminosity, redshift, and observing cadence, particularly when using LSST's photometric bands. As a response, we have developed a simulation tool aimed at optimizing measurement success across the entire survey, the first two years, and the inaugural year. This optimization effort encompasses the Main Survey (MS) and the planned Deep Drilling Fields (DDFs). The LSST's main survey (MS) will encompass roughly 18,000 square degrees of the southern sky, employing six filters. Each location within this survey area is anticipated to receive approximately 825 observations over the entire campaign duration. The expected magnitude limits, with a signal-to-noise ratio (SNR) greater than 5σ, are r $<$ 24.5 in individual images and r $<$ 27.8 when considering the fully stacked data. Apart from the Rubin Observatory's extensive main survey, conducted using a "universal cadence" a notable portion of Rubin's observing time, approximately 10%, will be allocated to various additional programs. This allocation includes the intensive observation of specific DDFs, namely COSMOS, Extended Chandra Deep Field-South (ECDFS), XMM-LSS, ELAIS S1, and Euclid Deep Field Survey (EDFSa and EDFSb). These DDFs will receive more comprehensive coverage and more frequent temporal sampling, particularly in some of the LSST's ugrizy filters, in comparison to the standard observations of the broader sky (see, e.g., [173,176]). The LSST DDFs will have the deepest coadded depths of the survey [176], and are expected to exceed those achieved by 8-meter class observatories in the same areas (see [177]). Roughly 10% of the LSST observing time will be used in mini-surveys and the six DDFs. Through a combination of depth and cadence, the DDFs are expected to provide the most complete catalogs of AGNs and their host galaxies [173,176,178].

Our modeling approach relies on stochastic light curves, which provide a robust representation of active galactic nuclei (AGN) variability. We utilize the Timmer-König algorithm to generate synthetic light curves, which are subsequently sampled according to specific cadences. We explore various cadence simulations provided by OpSims from LSST [165,173,179–181] for the MS and the DDFs as listed in Figure 5 (top right panel).

For more details on the properties of each of these simulations we refer the readers to Czerny et al. [182]. These simulated light curves are designed to include the expected contamination from prominent emission lines (Hβ, Mg II, CIV), the Fe II pseudo-continuum, and stellar emissions. By selecting appropriate photometric bands suitable for the source's redshift, we compare the predicted line time delay with the retrieved delay across 100 statistical realizations of the light curves.

The outcomes representing the entirety of the 10-year monitoring within one of the Deep Drilling Fields (DDFs), specifically ELAIS S1, are displayed in the left panels of Figure 5[6]. It is evident that for bright quasars, the results obtained from the DDF exhibit improvements in the smallest redshift range (below 1.0) compared to the MS (refer to Figure 4 in Czerny et al. [182]), wherein more frequent coverage facilitates accurate determination of the time delay, especially in cases where it is relatively short. This difference becomes highly significant for faint quasars. While the MS lacks reliable sampling for them, those situated within the DDFs can be robustly measured for redshifts surpassing 1.8. This becomes intriguing and pivotal, considering that fainter quasars are anticipated to dominate the quasar population, thus the DDFs have the potential to detect numerous faint quasars. Notably, about 80% of quasars in recent SDSS data releases are brighter than $\log L_{3000} = 44.7$, rendering them well-measurable within DDFs. In contrast, only approximately 15% of quasars in the MS possess sufficient luminosity to yield adequately measured delay timescales. For additional insights, readers are directed to Figure 3 and Table 3 in Czerny et al. [182]. The decreased accuracy at lower redshifts can be partly attributed to the significant intervals between observing seasons. The noticeable underestimation of the time delay occurs for redshifts between 1.0 and 1.5, primarily because the simulations for the chosen luminosity level yield an expected time delay of roughly 180 days. For redshifts below 0.5, segmenting data into separate seasons could offer improvement.

We also scrutinize the feasibility of yielding informative outcomes from the first two years of LSST data. For the MS, reliable results are not anticipated due to the prevailing sampling. In contrast, the denser cadence in the DDFs renders it possible to measure time delays of the order of a year. Especially when dealing with faint AGN, where anticipated time delays are below 400 days, measurements are viable (see Figure 5, top panel in the middle). The actual measurements and the expected values agree within a 1σ margin of error at all redshifts. However, a systematic mean offset from the expected values across all redshifts is evident. This offset is attributed to the relatively short data sets. Nevertheless, these measurements offer value for specific statistical investigations, even though the presence of a systematic offset of approximately $\sim$40% should be taken into account. This offset depends on the precise source luminosity, indicating the need for additional simulations to refine the results. In the case of bright AGN, the expected time delays are so extended that meaningful results are only attainable for objects with redshifts below 0.3 when there are only two years of data available. Otherwise, the recovered delay reaches saturation at the code's maximum limit, which is determined as the lesser of half the duration of the data set and twice the expected time delay (see Figure 5, bottom panel in the middle).

Considering a single year of data usage renders the situation even more challenging. Reliable measurements are not viable for bright sources across any redshift range. For faint sources, only promising results from low redshifts emerge. In both scenarios, AGN with redshifts surpassing 0.7 is inaccessible, which encourages cautious searches that may consider setting a lower redshift limit, approximately around $\sim$0.4. Nonetheless, it is promising that specific AGN emission line time delays can still be measured, even within shortened monitoring periods characterized by highly irregular cadences.

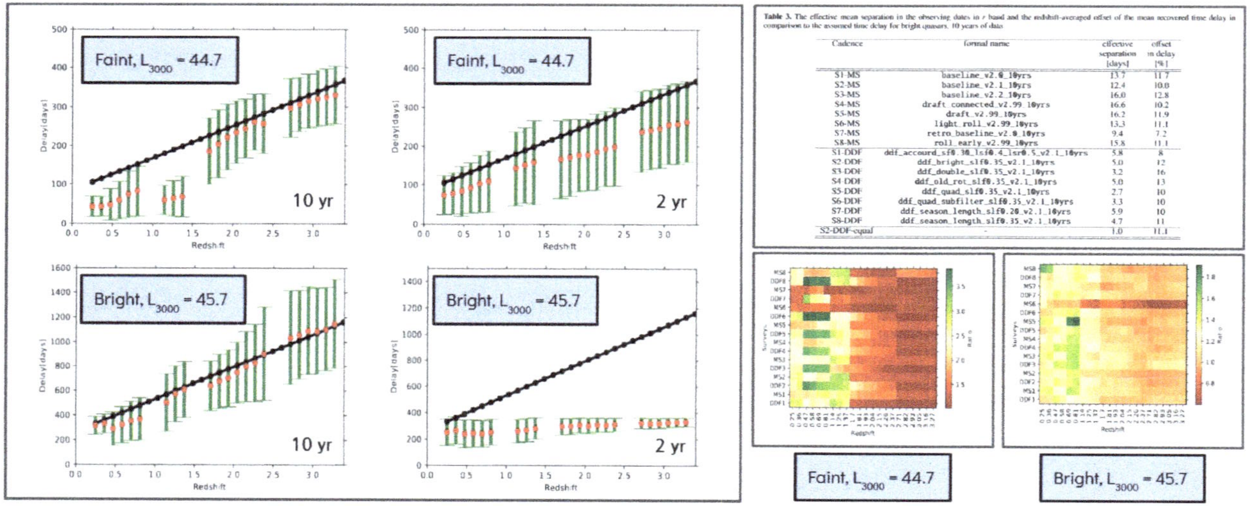

Table 3. The effective mean separation in the observing dates in *r* band and the redshift-averaged offset of the mean recovered time delay in comparison to the assumed time delay for bright quasars. 10 years of data.

Cadence	formal name	effective separation [days]	offset in delay [%]
S1-MS	baseline_v2.0_10yrs	13.7	11.7
S2-MS	baseline_v2.1_10yrs	12.4	10.0
S3-MS	baseline_v2.2_10yrs	16.0	12.8
S4-MS	draft_connected_v2.99_10yrs	16.6	10.2
S5-MS	draft_v2.99_10yrs	16.2	11.9
S6-MS	light_roll_v2.99_10yrs	13.3	11.1
S7-MS	retro_baseline_v2.0_10yrs	9.4	7.2
S8-MS	roll_early_v2.99_10yrs	15.8	11.1
S1-DDF	ddf_accourd_sf0.30_lsf0.4_lsr0.5_v2.1_10yrs	5.8	8
S2-DDF	ddf_bright_slf0.35_v2.1_10yrs	5.0	12
S3-DDF	ddf_double_slf0.35_v2.1_10yrs	3.2	16
S4-DDF	ddf_old_rot_slf0.35_v2.1_10yrs	5.0	13
S5-DDF	ddf_quad_slf0.35_v2.1_10yrs	2.7	10
S6-DDF	ddf_quad_subfilter_slf0.35_v2.1_10yrs	3.3	10
S7-DDF	ddf_season_length_slf0.20_v2.1_10yrs	5.9	10
S8-DDF	ddf_season_length_slf0.35_v2.1_10yrs	4.7	11
S2-DDF-equal	-	1.0	11.1

Figure 5. Expectations from the time-lag recovery for AGNs to be observed with LSST: rest-frame BLR time-lag versus AGN luminosity (Clockwise from top left:) In the upper panel, we compare the adopted and recovered time delays as a function of redshift for a faint AGN (log L_{3000} = 44.7), while in the lower panel, we present the same comparison for a bright AGN (log L_{3000} = 45.7). These data points are obtained from 10 years of simulated observations in the ELAIS-S1 field, one of the Deep Drilling Fields (DDFs). The green error bars represent the standard deviation expected in a single-source measurement. This is determined by utilizing 100 statistically equivalent simulations; (**right panels**) Same as the left panel but from the 2 years of simulated observations in the DDFs. The table situated in the top right corner provides information about two key aspects: It showcases the effective mean separation in observing dates in the *r* band and the redshift-averaged offset of the mean recovered time delay when compared to the assumed time delay for bright quasars. This data is derived from a 10-year simulation, considering various survey strategies proposed by the scientific collaborations within LSST, and in the bottom left and bottom right panels, one can observe the color-coded relative systematic error in determining the time delay. This information pertains to both faint (**left panel**) and bright (**right panel**) AGN, and it's based on a 10-year simulation across 8 Main Survey (MS) and 8 DDF cadences. For these simulations, an arbitrary location in the sky, specifically (0, −30), was chosen for the MS, while for DDF, the ELAIS-S1 field centered at (9.45, −44.025) was used as the reference location. Abridged figure based on Czerny et al. [182].

To evaluate the potential cadences suggested for LSST's various survey strategies, we tested eight different scenarios for both the MS and DDFs (see top right panel of Figure 5). For succinct representation, error bars were omitted as their impact remains relatively stable. Instead, we employed color-coded plots that depict the relative change between the mean derived time delay, τ_{derived}, and the time delay assumed in simulations, τ_{adopted}, expressed as $\delta = \frac{\tau_{\text{adopted}}}{\tau_{\text{derived}}}$. The dimensionless value exhibits a strong dependence on source luminosity and is depicted as a function of redshift separately for faint and bright AGN (refer to the bottom right panels of Figure 5). These assessments were performed using a 10-year simulated cadence. Across most cadences, results generally align with those previously obtained. The results for bright AGN are quite gratifying, particularly for moderate to high redshifts. However, there is a recurring issue of underestimating the time delay for AGN located near redshifts of approximately z $\sim$ 0.7 wherein its derived value is too short.As discussed by Czerny et al. [135], obtaining meaningful results for bright quasars requires five measurements per year with uniform spectroscopic coverage or twice that number for photometric measurements with non-uniform sampling. Lightcurves providing fewer data points cannot be effectively utilized. For fainter quasars, the majority of cadences anticipate time delays that are too short to be sufficiently covered in the MS mode. In the case of DDFs, the outcomes exhibited a similar qualitative trend, permitting a quantitative comparison through two global parameters computed for each cadence. One parameter indicates the mean time interval between visits, considering observations on the same day as a single exposure. The second parameter signifies the redshift-averaged value of δ. These values for bright quasars are presented in the top right panel of Figure 5.

We illustrate that it is feasible to obtain emission line time delays from the photometric data provided by LSST for a significant proportion of quasars. Our findings reveal the following insights: (i) for quasars surpassing $\log L_{3000} = 45.7$, the cadence within the MS generally proves sufficient for measuring line time delays relative to the continuum. Although individual measurement errors in time delays are relatively high, around 40%, most cadences exhibit a systematic offset of 10–15%. Hence, aggregating measurements from numerous quasars enables statistical studying trends like the radius-luminosity relation; (ii) for quasars with luminosities dimmer than $\log L_{3000} = 44.7$, it is not advisable to use the MS. Determining a practical redshift limit for usability in the case of intermediate luminosity quasars will require simulations; (iii) successfully measuring line time delays in quasars with luminosities below $\log L_{3000} = 44.7$ is achievable within the DDFs. In such cases, even the initial two years of data are sufficient, and extending the dataset does not significantly improve delay measurements at lower redshifts unless the data is sampled over shorter intervals or subjected to detrending; (iv) dense sampling can also aid in studying bright quasars situated within DDF fields, marginally reducing individual errors to below $\sim$30%; (v) each of the cadences investigated introduces a systematic offset of approximately 10% between the assumed setup delay and the recovered time delay. The offset persists even with numerous measured quasars, necessitating correction via numerical simulations. The offset's magnitude hinges on quasar properties, cadence, photometric errors, and the chosen time delay measurement method; (vi) while certain cadences fare better than others, the cadence does not appear to be a critical determinant for line delay measurements.

Hence, by carefully selecting source luminosities and their associated redshift ranges, we can anticipate reliable time delay measurements for a substantial portion of quasars observed within the Main Survey (MS) upon its completion. The exact fraction can be determined by configuring the observational cadence. Significantly, results for intrinsically fainter sources in Deep Drilling Field (DDF) locations with lower redshifts can be derived from data acquired even during the project's inaugural year. For more comprehensive information on these results and how the choice of survey strategies influences the recovery of time delays in the Broad Line Region (BLR) for the observed AGNs throughout the campaign's duration, we direct readers to the study by Czerny et al. [182].

5.2. Accretion Disk Continuum Modeling and Contamination from the BLR

The reverberation technique has been employed to measure the size of the accretion disk in AGNs using continuum time delays (see, e.g., [40,183–193]). Initially proposed by Collier et al. [194] (with a torus-based variation by Oknyanskij [195]), this approach involves evaluating the size of the accretion disk across different wavelengths and then comparing it to the classical accretion disk model proposed by Shakura and Sunyaev [89]. This comparison is made possible through a particular scaling of monochromatic flux and effective temperature relative to the product of the black hole mass and accretion rate (as discussed in [196]), allows for directly determining the source's distance using the measured time delay (τ) between two wavelengths and the observed monochromatic flux (f_ν) at one of these wavelengths. This distance estimation remains independent of the black hole mass and accretion rate. Theoretically, the time delay is expected to scale with wavelength λ as $\tau \propto \lambda^{4/3}$, and the coefficient of proportionality encompasses observed flux and distance, enabling redshift-independent distance determination.

Empirical monitoring of several sources corroborated the anticipated delay pattern, e.g., [194,197–199], specifically the proportionality $\tau \propto \lambda^{4/3}$. Yet, the determined proportionality constant from observations often significantly exceeded the theoretical prediction, varying by roughly 40% up to several factors, when employing the standard cosmology, e.g., [189,194,197,199,200]. Certain sources/papers achieved alignment with expectations. For instance, a correction for internal AGN reddening is necessary to be consistent with $\nu^{1/3}$ (see, e.g., Weaver and Horne [201]). Neglecting to incorporate this correction can lead to a notable underestimation of AGN luminosity, such as a tenfold underestimation in the ultraviolet and approximately a fourfold underestimation in the optical (Gaskell [202]). This bias in luminosity can consequently lead to underestimated AD sizes as recently shown by Gaskell et al. [203] for NGC5548. Conversely, as proposed by Pozo Nuñez et al. [198], an underestimation of the black hole mass owing to the unknown geometry of the BLR is suggested as the root of a biased time-delay spectrum (as depicted in Figures 4 and 6 of their study). Models considering the X-ray illumination of the accretion disk (AD) could also explain the observed longer delays in specific cases, taking into consideration factors such as the height of the corona or the extended nature of the reprocessor [204,205], while others reported disparities at specific wavelength bands near the Balmer edge, e.g., [206–211]. Indications arose of substantial contamination from the more distant reprocessing region—the BLR—especially around the Balmer (∼3647 Å) and Paschen (∼8206 Å) edges (e.g., see Figure 1 left panel in Chelouche et al. [212]), collectively referred to as the diffuse continuum emission (DCE, [213,214]). In his recent research, Netzer [215] concluded that in the case of most examined sources, reprocessing primarily takes place in the BLR, thereby extending the continuum time delay beyond the anticipated disk value. A similar hypothesis had previously arisen for NGC 5548, as presented by Lawther et al. [216]. Contributions from non-disk components, such as the Fe II pseudo-continuum and the Balmer continuum, were also considered [126], also contribute to the contamination, which becomes significant when employing photometric broadband filters, as in the case of LSST. For instance, Netzer [215] recently revealed a new delay-luminosity relationship for specific sources driven solely by the DCE of the BLR, unrelated to the accretion disk itself. This intricate relationship's origin remains subject to debate, making it challenging to discern via observations. Thus, comprehending these effects is vital to evaluate their impact and develop methods to disentangle genuine continuum information from the accretion disk. The time delay of the accretion disk, studied across numerous objects, will aid in untangling these effects, underscoring LSST's pivotal role.

In our recent study (Jaiswal et al. [217]), we employ numerical simulations to determine if the assumed disk and BLR region geometries can be extracted from time delay measurements. We generate synthetic light curves to replicate incident radiation, taking into account parameters that describe both the accretion disk and the BLR reprocessing. In our modeling, we estimate the expected time delays, incorporating contributions from both the reprocessing of incident X-ray flux by the disk and subsequent reprocessing by the

BLR. Our goal is to ascertain whether these two effects can be differentiated. Our findings indicate that the impact of BLR scattering on predicted time delays mirrors the effect of increasing X-ray source elevation without BLR contribution. Consequently, continuum time delays rise linearly with larger BLR contribution, introducing added degeneracy when attempting to deduce system parameters from observed time delays. Nonetheless, both effects alter the slope of the delay-wavelength curve when viewed in logarithmic space. This alteration provides a means to extract the pure disk time delay, which is essential for cosmological applications.

In another recent study (Pozo Nuñez et al. [214]), we explore determining the AD sizes of quasars and assess the LSST survey's proposed cadence using photometric reverberation mapping (PRM). Our extensive simulations factor in LSST survey characteristics and quasars' intrinsic properties. These simulations characterize light curves, enabling the determination of AD sizes using various algorithms—we account for contamination by a host galaxy in the photometric aperture using the flux variation gradient method [132,218–222]. We also refer the readers to Figure 2 in Pozo Nuñez et al. [214]. We include the nuclear reddening contributions caused by the internal reddening of the AGN [221,222], as well as the contribution from the variable DCE [213,223] and BLR emission lines [224,225]. We determine that: (i) reaching a signal-to-noise ratio (S/N) of a minimum of 100, with BLR emission line contribution staying below 10% within the bandpasses, facilitates the recovery of time delays with 5% and 10% accuracy for 2-day and 5-day time sampling, respectively, for quasars in the range $1.5 < z < 2.0$. Accuracy of 10–20% is attainable for quasars at $z < 1.5$ only if BLR emission line contribution remains under 5%. Augmenting S/N does not notably enhance results; (ii).assuming an optically thick, geometrically thin accretion disk model, the obtained time delay spectrum aligns well with estimated black hole masses, with an accuracy of approximately 30%. Under the prevailing observational circumstances of LSST, this accuracy range encompasses intermediate black hole masses in the vicinity of $\sim 10^8$–10^9 $M_\odot$. For the DDFs, with possible 12 day time sampling, the range could extend to $\sim 5 \times 10^7$ $M_\odot$. We recommend strategies for LSST's observing approach to effectively recover continuum time lags for a substantial number of AGNs, including correcting fluxes in each photometric band for host galaxy contamination and internal reddening, as well as scaling the BLR's variable DCE lag spectrum to account for luminosity differences. It is important to emphasize that our analysis extends beyond LSST and can be applied to any future photometric reverberation mapping survey of accretion disks. Considering cosmological time dilation, studies involving the reverberation mapping of continuum-emission lines in luminous quasars at around $z \sim 2$ require approximately 15 years of observations (for example, using CIV lines). Accretion disk time delays are roughly 10 times shorter than those of the BLR (as discussed in Wang et al. [226]), which holds the potential to establish a more precise AD size-luminosity relationship. This, in turn, enables quasars to function as effective standard cosmological candles up to high redshifts. Nevertheless, achieving sufficient time sampling and conducting specialized surveys are crucial for the exploration of specific quasar populations. In this context, we are planning a photometric reverberation mapping campaign for accretion disks (AD PRM) targeting quasars up to $z \leq 1.5$ utilizing optical and near-infrared telescopes at the Cerro Armazones Observatory in Chile. Further information about this upcoming survey will be presented in a forthcoming paper.

6. Concluding Remarks

Through this review, we have highlighted some recent progress in the studies of accretion disk structure and the development of a database of AGN SEDs, especially for the sources accreting at or above the Eddington limit (see Section 2). We showcase how the increase in the accretion rate needs to be accounted for changes in the accretion disk structure which then gives rise to the additional anisotropic emission. At close to the Eddington accretion rates, the accretion disk transitions towards a "slim" disk scenario wherein we see the formation of a *funnel-like* feature at the innermost region closest to the SMBH. Such a change in the disk structure could explain the dichotomy of the incident flux received by the high- and low-ionization line-emitting regions. The low-ionization lines located closer to the disk's surface end up receiving a much weaker, filtered continuum which in turn allows for these clouds to move in closer to the central source (see Figure 1)—this would clarify why the high-accreting sources tend to show shorter BLR lags in the R_{BLR}-L_{5100} relation. We additionally derive the modeled SEDs which are dependent on the accretion rates (going from sub-Eddington to super-Eddington rates) and also highlight how the continuum emission varies as a function of the viewing angle (see Figure 2).

We then discuss the effectiveness of the BLR radius vs. AGN luminosity relation(s) in providing us with an independent way to estimate the luminosity distances to these cosmic sources, help build the Hubble diagram, and eventually test and retrieve cosmological parameters of our universe (see Section 3). Stepping back to consider the broader perspective, we can build the Hubble diagram utilizing the corrected luminosity distances from Equation (2), adjusted following Equation (1), along with the corresponding redshift values for each source. The crux of this approach lies in possessing measurements of time delays (e.g., derived from Equation (1)) and the monochromatic flux of the AGN derived from spectroscopy. This enables the estimation of luminosity distances independent of accretion properties. Consequently, extensive sets of AGNs subject to reverberation mapping can be harnessed as effective cosmological distance indicators [130,131,194,227,228]. This inclusive approach opens the door to exploring the redshift-dependent cosmological parameters' evolution, thereby aiding in reconciling the Hubble tension—the discrepancy between the estimated Hubble constant values in the late and early universe (we refer the readers to the right panel in Figure 3). The ideas and findings presented in this study hold immense potential, setting the foundational platforms for the extensive array of AGNs which will be investigated using ongoing and future ground-based telescopes with 10-meter-class capabilities (e.g., Maunakea Spectroscopic Explorer, [229]) and 40 meter-class capabilities (e.g., The Extremely Large Telescope, [230]); in addition to space-based missions such as the JWST [231,232] and the Roman Space Telescope [233]. It is crucial to have access to a growing database of high-quality data, including photometric, spectroscopic, and interferometric information spanning higher redshifts which are essential for advancing our ever-expanding theoretical understanding of the mechanisms and evolution of AGNs and for utilizing them as tools for estimating cosmic distances.

We then demonstrate a novel BH mass scaling relation traced by the intensities of high-ionization, forbidden emission lines (known as coronal lines) that we have discovered (see Section 4, also Figure 4). These coronal lines, a bulk of them produced in the NLR, are shown to evolve with the change in the underlying accretion disk structure, which is the primary source of the photons that produce emission of such lines (see Section 4.1). A recent intriguing avenue that has gained prominence is the potential of utilizing high-ionization emission lines within the optical and near-infrared spectral ranges as effective indicators for accurately characterizing the dynamics and physical properties of gas ionized by the AGN, and further associated with X-ray emission lines [234]. These emission lines, often referred to as "footprint lines", emanate from gas within a comparable range of ionization states as those of oxygen and neon ions configured in H- and He-like states—these same ions are responsible for generating X-ray emission lines. Instruments such as the Hubble Space Telescope/STIS and JWST/NIRSpec, which operate in the optical and infrared regimes, possess the capacity to detect these footprint lines. Consequently, these lines present a promising opportunity to glean insights into the kinematics of the extended gas responsible for emitting X-rays. Building upon prior foundational research [152,234], a natural progression is to establish an extensive database encompassing properties of coronal lines across a diverse spectrum of active galaxies. Such a compilation would prove invaluable in identifying new AGNs, particularly in higher redshift domains, a potential magnified by the capabilities of the forthcoming JWST data. An especially noteworthy application of these coronal lines' diagnostics is their potential to detect and unveil accreting black holes within the intermediate mass range, known as Intermediate Mass Black Holes (IMBHs). As the JWST builds on its recent success [235–251], we find ourselves in an opportune position to validate the theoretical predictions of coronal line emissions and their fundamental linkage with central accreting black holes.

We showcased our efforts in preparation for the upcoming Vera Rubin Observatory's Legacy Survey for Space and Time (LSST) in the context of the accretion disk and BLR time delay measurements using photometric monitoring and scrutinizing the survey strategies that are under discussion (see Sections 5.1 and 5.2). We critically assess the various survey strategies for the LSST by accounting for them in our time-lag recovery pipeline as part of our in-kind contribution to LSST. We test the reliability of the time-lag measurements for multiple broad emission lines across a wide redshift and luminosity range and include the contribution from the host galaxy and other contaminants, such as the Fe II pseudocontinuum (see Figure 5). Through our analyses, we find that the success rate of lag-recovery is significantly higher in the Deep Drilling Fields (DDFs) where the dense sampling is beneficial to reduce the uncertainties in the lag measurements [182]. In a corollary work, we explore determining the accretion disk (AD) sizes of quasars and assess the LSST survey's proposed cadence using photometric reverberation mapping. Our extensive simulations factor in LSST survey characteristics and quasars' intrinsic properties. Similar to the previous work, we account for the contamination from the host galaxy, including nuclear reddening contributions caused by the internal reddening of the AGN, as well as the contribution from the variable diffuse continuum emission (DCE) and BLR emission lines. We find from our numerical simulations, that reliable time-lags for the accretion disk can be obtained from the LSST for a few thousands of quasars, specifically in the DDFs with a possible 1–2 day time sampling. We test also how the contribution from the BLR lines can constrain the redshift windows which should be prioritized to recover accurate AD time lags [214]. Reverberation mapping of high-z quasars ($z \sim 2$) requires approximately 10–15 years of observations. Accretion disk time delays are roughly 10 times shorter than those of the BLR, which holds the potential to establish a more precise AD size-luminosity relationship. This, in turn, enables quasars to function as effective standard cosmological candles up to high redshifts.

Author Contributions: The idea for the article and manuscript preparation was made by S.P., P.M., B.C., A.R.-A. and F.P.N. assisted with the writing in different sections presented in the manuscript. Each of the co-authors helped with the review and editing of the presented version. All authors have read and agreed to the published version of the manuscript.

Funding: This research was funded by the Conselho Nacional de Desenvolvimento Científico e Tecnológico (CNPq) Fellowships, grant number 164753/2020-6 and 313497/2022-2, the European Research Council (ERC) under the European Union's Horizon 2020 research and innovation program, grant number 951549.

Data Availability Statement: Data are contained within the article.

Acknowledgments: The authors thank the anonymous reviewers whose suggestions helped us improve the overall presentation of the results and findings in this review. SP is grateful to the organizers of the 14th Serbian Conference on Spectral Line Shapes in Astrophysics, Bajina Bašta, Serbia, 19–23 June 2023, where the work was presented in the form of an invited talk. FPN gratefully acknowledges the generous and invaluable support of the Klaus Tschira Foundation.

Conflicts of Interest: The authors declare no conflict of interest.

Abbreviations

The following abbreviations are used in this manuscript:

AD	Accretion disk
AGN	Active galactic nuclei
BH	Black hole

BLR	Broad-line region
CL	Coronal lines
DCE	Diffuse Continuum Emission
DDF	Deep Drilling Field
DL	Luminosity distance
ECDFS	Extended Chandra Deep Field-South
EDFS	Euclid Deep Field Survey
FWHM	Full width at half maximum
GNIRS	Gemini Near-Infrared Spectrograph
IMBH	Intermediary mass black hole
IP	Ionization potential
ISCO	Innermost stable circular orbit
LSST	Legacy Survey of Space and Time
MS	Main Survey
NLR	Narrow line region
NLS1	Narrow-Line Seyfert 1
NIR	Near infrared
PRM	Photometric Reverberation Mapping
RM	Reverberation Mapping
SDSS	Sloan Digital Sky Survey
SED	Spectral energy distribution
SEAMBH	Super-Eddington Accreting Massive Black Holes
SMBH	Supermassive black hole
SNe	Supernovae
SS	Shakura and Sunyaev
VLTI	Very Large Telescope Interferometer
WC	Warm Corona

Notes

1 The left panel of Figure 1 corresponds to the optical domain of Eigenvector 1, often referred to as the quasar main sequence. It defines the parameter space based on FWHM(Hβ) and the strength of optical Fe II emission (denoted as R_{FeII}). The latter is determined by the ratio of Fe II flux within 4434–4684 Å (characterizing the blueward bump of the Hβ profile) to the broad Hβ flux.

2 This relationship supposes the radius of the BLR for Hβ and the adjacent continuum luminosity at 5100 Å

3 This CL ranks as one of the most prevalent and luminous features observed in the spectra of AGNs, as discussed in previous studies [151]. For a comprehensive summary of our findings, we direct readers to the extensive analysis provided by Prieto et al. [152].

4 We note in passing that there is a good agreement between the FWHMs obtained for Brγ_{broad} and Hβ where the Brγ_{broad}-based FWHMs are narrower by $\sim$300 km s^{-1} (see, e.g., [155])

5 http://www-astro.physics.ox.ac.uk/mxc/software/lts (accessed on 24 September 2022)

6 For the Main Survey (MS), we adopted the sky location (0, $-$30), while for the DDF, the central coordinates of ELAIS-S1 at (9.45, $-$44.025) were utilized.

References

1. Krolik, J.H. *Active Galactic Nuclei: From the Central Black Hole to the Galactic Environment*; Princeton University Press: Princeton, NJ, USA, 1999.
2. Seyfert, C.K. Nuclear Emission in Spiral Nebulae. *Astrophys. J.* **1943**, *97*, 28. [CrossRef]
3. Schmidt, M. 3C 273: A Star-Like Object with Large Red-Shift. *Nature* **1963**, *197*, 1040. [CrossRef]
4. Greenstein, J.L. Red-Shift of the Unusual Radio Source: 3C 48. *Nature* **1963**, *197*, 1041–1042. [CrossRef]
5. Greenstein, J.L.; Schmidt, M. The Quasi-Stellar Radio Sources 3C 48 and 3C 273. *Astrophys. J.* **1964**, *140*, 1. [CrossRef]
6. Souffrin, S. Study of Seyfert Galaxies. Ill. Ionization Mechanism. *Astron. Astrophys.* **1969**, *1*, 414.
7. Lynden-Bell, D. Galactic Nuclei as Collapsed Old Quasars. *Nature* **1969**, *223*, 690–694. [CrossRef]
8. Weedman, D.W. High-Velocity Gas Motions in Galactic Nuclei. *Astrophys. J.* **1970**, *159*, 405. [CrossRef]
9. Khachikian, E.E.; Weedman, D.W. A spectroscopic study of luminous galactic nuclei. *Astrofizika* **1971**, *7*, 389–406. [CrossRef]
10. Weedman, D.W. Luminosities of Seyfert galaxies and QSOs. *Astrophys. J.* **1976**, *208*, 30–36. [CrossRef]
11. Weedman, D.W. Seyfert galaxies. *Annu. Rev. Astron. Astrophys.* **1977**, *15*, 69–95. [CrossRef]
12. Blandford, R.D.; McKee, C.F. Reverberation mapping of the emission line regions of Seyfert galaxies and quasars. *Astrophys. J.* **1982**, *255*, 419–439. [CrossRef]

13. Schmidt, M.; Green, R.F. Quasar evolution derived from the Palomar bright quasar survey and other complete quasar surveys. *Astrophys. J.* **1983**, *269*, 352–374. [CrossRef]
14. Peterson, B.M. Emission-Line Variability in Seyfert Galaxies. *Publ. Astron. Soc. Pac.* **1988**, *100*, 18. [CrossRef]
15. Peterson, B.M. Reverberation Mapping of Active Galactic Nuclei. *Publ. Astron. Soc. Pac.* **1993**, *105*, 247. [CrossRef]
16. Kaspi, S.; Smith, P.S.; Netzer, H.; Maoz, D.; Jannuzi, B.T.; Giveon, U. Reverberation Measurements for 17 Quasars and the Size-Mass-Luminosity Relations in Active Galactic Nuclei. *Astrophys. J.* **2000**, *533*, 631–649. [CrossRef]
17. Sulentic, J.W.; Marziani, P.; Dultzin-Hacyan, D. Phenomenology of Broad Emission Lines in Active Galactic Nuclei. *Annu. Rev. Astron. Astrophys.* **2000**, *38*, 521–571. [CrossRef]
18. Peterson, B.M.; Ferrarese, L.; Gilbert, K.M.; Kaspi, S.; Malkan, M.A.; Maoz, D.; Merritt, D.; Netzer, H.; Onken, C.A.; Pogge, R.W.; et al. Central Masses and Broad-Line Region Sizes of Active Galactic Nuclei. II. A Homogeneous Analysis of a Large Reverberation-Mapping Database. *Astrophys. J.* **2004**, *613*, 682–699. [CrossRef]
19. Vestergaard, M.; Peterson, B.M. Determining Central Black Hole Masses in Distant Active Galaxies and Quasars. II. Improved Optical and UV Scaling Relationships. *Astrophys. J.* **2006**, *641*, 689–709. [CrossRef]
20. Tadhunter, C. An introduction to active galactic nuclei: Classification and unification. *New Astron.* **2008**, *52*, 227–239. [CrossRef]
21. Bentz, M.C.; Peterson, B.M.; Netzer, H.; Pogge, R.W.; Vestergaard, M. The Radius-Luminosity Relationship for Active Galactic Nuclei: The Effect of Host-Galaxy Starlight on Luminosity Measurements. II. The Full Sample of Reverberation-Mapped AGNs. *Astrophys. J.* **2009**, *697*, 160–181. [CrossRef]
22. Antonucci, R. A panchromatic review of thermal and nonthermal active galactic nuclei. *Astron. Astrophys. Trans.* **2012**, *27*, 557–602. [CrossRef]
23. Bentz, M.C.; Denney, K.D.; Grier, C.J.; Barth, A.J.; Peterson, B.M.; Vestergaard, M.; Bennert, V.N.; Canalizo, G.; De Rosa, G.; Filippenko, A.V.; et al. The Low-luminosity End of the Radius-Luminosity Relationship for Active Galactic Nuclei. *APJ* **2013**, *767*, 149. [CrossRef]
24. Kormendy, J.; Ho, L.C. Coevolution (Or Not) of Supermassive Black Holes and Host Galaxies. *Annu. Rev. Astron. Astrophys.* **2013**, *51*, 511–653. [CrossRef]
25. Du, P.; Hu, C.; Lu, K.X.; Wang, F.; Qiu, J.; Li, Y.R.; Bai, J.M.; Kaspi, S.; Netzer, H.; Wang, J.M.; et al. Supermassive Black Holes with High Accretion Rates in Active Galactic Nuclei. I. First Results from a New Reverberation Mapping Campaign. *Astrophys. J.* **2014**, *782*, 45. [CrossRef]
26. Du, P.; Hu, C.; Lu, K.X.; Huang, Y.K.; Cheng, C.; Qiu, J.; Li, Y.R.; Zhang, Y.W.; Fan, X.L.; Bai, J.M.; et al. Supermassive Black Holes with High Accretion Rates in Active Galactic Nuclei. IV. Hβ Time Lags and Implications for Super-Eddington Accretion. *Astrophys. J.* **2015**, *806*, 22. [CrossRef]
27. Netzer, H. Revisiting the Unified Model of Active Galactic Nuclei. *Annu. Rev. Astron. Astrophys.* **2015**, *53*, 365–408. [CrossRef]
28. Du, P.; Lu, K.X.; Zhang, Z.X.; Huang, Y.K.; Wang, K.; Hu, C.; Qiu, J.; Li, Y.R.; Fan, X.L.; Fang, X.E.; et al. Supermassive Black Holes with High Accretion Rates in Active Galactic Nuclei. V. A New Size-Luminosity Scaling Relation for the Broad-line Region. *Astrophys. J.* **2016**, *825*, 126. [CrossRef]
29. Padovani, P.; Alexander, D.M.; Assef, R.J.; De Marco, B.; Giommi, P.; Hickox, R.C.; Richards, G.T.; Smolčić, V.; Hatziminaoglou, E.; Mainieri, V.; et al. Active galactic nuclei: What's in a name? *Astron. Astrophys. Rev.* **2017**, *25*, 2. [CrossRef]
30. Grier, C.J.; Trump, J.R.; Shen, Y.; Horne, K.; Kinemuchi, K.; McGreer, I.D.; Starkey, D.A.; Brandt, W.N.; Hall, P.B.; Kochanek, C.S.; et al. The Sloan Digital Sky Survey Reverberation Mapping Project: Hα and Hβ Reverberation Measurements from First-year Spectroscopy and Photometry. *Astrophys. J.* **2017**, *851*, 21. [CrossRef]
31. Du, P.; Wang, J.M. The Radius-Luminosity Relationship Depends on Optical Spectra in Active Galactic Nuclei. *Astrophys. J.* **2019**, *886*, 42. [CrossRef]
32. Popović, L.Č. Broad spectral lines in AGNs and supermassive black hole mass measurements. *Open Astron.* **2020**, *29*, 1–14. [CrossRef]
33. Zou, S.; Jiang, L.; Shen, Y.; Wu, J.; Bañados, E.; Fan, X.; Ho, L.C.; Riechers, D.A.; Venemans, B.; Vestergaard, M.; et al. Strong Mg II and Fe II Absorbers at $2.2 < z < 6.0$. *Astrophys. J.* **2021**, *906*, 32. [CrossRef]
34. Matthews, B.M.; Shemmer, O.; Dix, C.; Brotherton, M.S.; Myers, A.D.; Andruchow, I.; Brandt, W.N.; Ferrero, G.A.; Gallagher, S.C.; Green, R.; et al. Placing High-redshift Quasars in Perspective: A Catalog of Spectroscopic Properties from the Gemini Near Infrared Spectrograph-Distant Quasar Survey. *Astrophys. J. Suppl.* **2021**, *252*, 15. [CrossRef]
35. Wang, S.; Jiang, L.; Shen, Y.; Ho, L.C.; Vestergaard, M.; Bañados, E.; Willott, C.J.; Wu, J.; Zou, S.; Yang, J.; et al. Metallicity in Quasar Broad-line Regions at Redshift 6. *Astrophys. J.* **2022**, *925*, 121. [CrossRef]
36. Ramos Padilla, A.F.; Wang, L.; Małek, K.; Efstathiou, A.; Yang, G. The viewing angle in AGN SED models: A data-driven analysis. *Mon. Not. R. Astron. Soc.* **2022**, *510*, 687–707. [CrossRef]
37. Lai, S.; Bian, F.; Onken, C.A.; Wolf, C.; Mazzucchelli, C.; Bañados, E.; Bischetti, M.; Bosman, S.E.I.; Becker, G.; Cupani, G.; et al. Chemical abundance of z 6 quasar broad-line regions in the XQR-30 sample. *Mon. Not. R. Astron. Soc.* **2022**, *513*, 1801–1819. [CrossRef]
38. Bianchi, S.; Mainieri, V.; Padovani, P. Active Galactic Nuclei and Their Demography Through Cosmic Time. In *Handbook of X-ray and Gamma-ray Astrophysics*; Springer: Singapore, 2022; p. 4. [CrossRef]
39. Sacchi, A.; Risaliti, G.; Signorini, M.; Lusso, E.; Nardini, E.; Bargiacchi, G.; Bisogni, S.; Civano, F.; Elvis, M.; Fabbiano, G.; et al. Quasars as high-redshift standard candles. *Astron. Astrophys.* **2022**, *663*, L7. [CrossRef]

40. Guo, H.; Barth, A.J.; Wang, S. Active Galactic Nuclei Continuum Reverberation Mapping Based on Zwicky Transient Facility Light Curves. *Astrophys. J.* **2022**, *940*, 20. [CrossRef]
41. Schindler, J.T.; Bañados, E.; Connor, T.; Decarli, R.; Fan, X.; Farina, E.P.; Mazzucchelli, C.; Nanni, R.; Rix, H.W.; Stern, D.; et al. The Pan-STARRS1 $z > 5.6$ Quasar Survey. III. The $z \approx 6$ Quasar Luminosity Function. *Astrophys. J.* **2023**, *943*, 67. [CrossRef]
42. Wu, Q.; Shen, Y. A Catalog of Quasar Properties from Sloan Digital Sky Survey Data Release 16. *Astrophys. J. Suppl.* **2022**, *263*, 42, [CrossRef]
43. Onken, C.A.; Wolf, C.; Hon, W.J.; Lai, S.; Tisserand, P.; Webster, R. AllBRICQS: The All-sky BRIght, Complete Quasar Survey. *Publ. Astron. Soc. Aust.* **2023**, *40*, e010. [CrossRef]
44. Kynoch, D.; Mitchell, J.A.J.; Ward, M.J.; Done, C.; Lusso, E.; Landt, H. The SOUX AGN sample: SDSS-XMM-Newton optical, ultraviolet, and X-ray selected active galactic nuclei spanning a wide range of parameter space - sample definition. *Mon. Not. R. Astron. Soc.* **2023**, *520*, 2781–2805. [CrossRef]
45. Mazzucchelli, C.; Bischetti, M.; D'Odorico, V.; Feruglio, C.; Schindler, J.T.; Onoue, M.; Bañados, E.; Becker, G.D.; Bian, F.; Carniani, S.; et al. XQR-30: Black hole masses and accretion rates of 42 $z \gtrsim 6$ quasars. *Astron. Astrophys.* **2023**, *676*, A71. [CrossRef]
46. Yang, J.; Wang, F.; Fan, X.; Hennawi, J.F.; Barth, A.J.; Bañados, E.; Sun, F.; Liu, W.; Cai, Z.; Jiang, L.; et al. A SPectroscopic Survey of Biased Halos in the Reionization Era (ASPIRE): A First Look at the Rest-frame Optical Spectra of $z > 6.5$ Quasars Using JWST. *Astrophys. J. Lett.* **2023**, *951*, L5. [CrossRef]
47. Bischetti, M.; Fiore, F.; Feruglio, C.; D'Odorico, V.; Arav, N.; Costa, T.; Zubovas, K.; Becker, G.; Bosman, S.E.I.; Cupani, G.; et al. The Fraction and Kinematics of Broad Absorption Line Quasars across Cosmic Time. *Astrophys. J.* **2023**, *952*, 44. [CrossRef]
48. López, I.E.; Brusa, M.; Bonoli, S.; Shankar, F.; Acharya, N.; Laloux, B.; Dolag, K.; Georgakakis, A.; Lapi, A.; Ramos Almeida, C.; et al. The miniJPAS survey: AGN and host galaxy coevolution of X-ray-selected sources. *Astron. Astrophys.* **2023**, *672*, A137. [CrossRef]
49. Sun, M. The mass distribution of quasars in optical time-domain surveys. *Mon. Not. R. Astron. Soc.* **2023**, *521*, 2954–2961. [CrossRef]
50. D'Odorico, V.; Bañados, E.; Becker, G.D.; Bischetti, M.; Bosman, S.E.I.; Cupani, G.; Davies, R.; Farina, E.P.; Ferrara, A.; Feruglio, C.; et al. XQR-30: The ultimate XSHOOTER quasar sample at the reionization epoch. *Mon. Not. R. Astron. Soc.* **2023**, *523*, 1399–1420. [CrossRef]
51. Secunda, A.; Greene, J.E.; Jiang, Y.F.; Yao, P.Z.; Zoghbi, A. Negative Lags on the Viscous Timescale in Quasar Photometry and Prospects for Detecting More with LSST. *Astrophys. J.* **2023**, *956*, 81. [CrossRef]
52. Fan, X.; Bañados, E.; Simcoe, R.A. Quasars and the Intergalactic Medium at Cosmic Dawn. *Annu. Rev. Astron. Astrophys.* **2023**, *61*, 373–426. [CrossRef]
53. Kokorev, V.; Fujimoto, S.; Labbe, I.; Greene, J.E.; Bezanson, R.; Dayal, P.; Nelson, E.J.; Atek, H.; Brammer, G.; Caputi, K.I.; et al. UNCOVER: A NIRSpec Identification of a Broad-line AGN at $z = 8.50$. *Astrophys. J. Lett.* **2023**, *957*, L7. [CrossRef]
54. Jin, C.; Lusso, E.; Ward, M.; Done, C.; Middei, R. Wavelength Dependences of the Optical/UV and X-ray Luminosity Correlations of Quasars. *Mon. Not. R. Astron. Soc.* **2023**, *527*, 356–373. [CrossRef]
55. Arevalo, P.; Churazov, E.; Lira, P.; Sanchez-Saez, P.; Bernal, S.; Hernandez-Garcia, L.; Lopez-Navas, E.; Patel, P. The universal power spectrum of Quasars in optical wavelengths: Break timescale scales directly with both black hole mass and accretion rate. *arXiv* **2023**, arXiv:2306.11099. [CrossRef]
56. Ward, C.; Gezari, S.; Nugent, P.; Kerr, M.; Eracleous, M.; Frederick, S.; Hammerstein, E.; Graham, M.J.; van Velzen, S.; Kasliwal, M.M.; et al. Panic at the ISCO: The visible accretion disks powering optical variability in ZTF AGN. *arXiv* **2023**, arXiv:2309.02516. [CrossRef]
57. Juráňová, A.; Costantini, E.; Di Gesu, L.; Ebrero, J.; Kaastra, J.; Korista, K.; Kriss, G.A.; Mehdipour, M.; Piconcelli, E.; Rogantini, D. Study of the optical to X-ray broad emission lines of Mrk 110. *arXiv* **2023**, arXiv:2311.03459. [CrossRef]
58. Rosborough, S.; Robinson, A.; Almeyda, T.; Noll, M. Modeling the Reverberation Response of the Broad Line Region in Active Galactic Nuclei. *arXiv* **2023**, arXiv:2311.03590. [CrossRef]
59. Strittmatter, P.A.; Hill, P.; Pauliny-Toth, I.I.K.; Steppe, H.; Witzel, A. Radio observations of optically selected quasars. *Astron. Astrophys.* **1980**, *88*, L12–L15.
60. Peacock, J.A.; Miller, L.; Longair, M.S. The statistics of radio emission from quasars. *Mon. Not. R. Astron. Soc.* **1986**, *218*, 265–278. [CrossRef]
61. Kellermann, K.I.; Sramek, R.; Schmidt, M.; Shaffer, D.B.; Green, R. VLA Observations of Objects in the Palomar Bright Quasar Survey. *Astron. J.* **1989**, *98*, 1195. [CrossRef]
62. Miller, L.; Peacock, J.A.; Mead, A.R.G. The bimodal radio luminosity function of quasars. *Mon. Not. R. Astron. Soc.* **1990**, *244*, 207–213.
63. Antonucci, R. Unified models for active galactic nuclei and quasars. *Annu. Rev. Astron. Astrophys.* **1993**, *31*, 473–521. [CrossRef]
64. Kellermann, K.I.; Sramek, R.A.; Schmidt, M.; Green, R.F.; Shaffer, D.B. The Radio Structure of Radio Loud and Quiet Quasars in the Palomar Bright Quasar Survey. *Astron. J.* **1994**, *108*, 1163. [CrossRef]
65. Wilson, A.S.; Colbert, E.J.M. The Difference between Radio-loud and Radio-quiet Active Galaxies. *Astrophys. J.* **1995**, *438*, 62. [CrossRef]
66. Xu, C.; Livio, M.; Baum, S. Radio-loud and Radio-quiet Active Galactic Nuclei. *Astron. J.* **1999**, *118*, 1169–1176. [CrossRef]
67. Laor, A. On Black Hole Masses and Radio Loudness in Active Galactic Nuclei. *Astrophys. J. Lett.* **2000**, *543*, L111–L114. [CrossRef]

68. Véron-Cetty, M.P.; Véron, P. The emission line spectrum of active galactic nuclei and the unifying scheme. *Astron. Astrophys. Rev.* **2000**, *10*, 81–133. [CrossRef]
69. Boroson, T.A. Black Hole Mass and Eddington Ratio as Drivers for the Observable Properties of Radio-loud and Radio-quiet QSOs. *Astrophys. J.* **2002**, *565*, 78–85. [CrossRef]
70. Sikora, M.; Stawarz, Ł.; Lasota, J.P. Radio Loudness of Active Galactic Nuclei: Observational Facts and Theoretical Implications. *Astrophys. J.* **2007**, *658*, 815–828. [CrossRef]
71. Tchekhovskoy, A.; Narayan, R.; McKinney, J.C. Black Hole Spin and The Radio Loud/Quiet Dichotomy of Active Galactic Nuclei. *Astrophys. J.* **2010**, *711*, 50–63. [CrossRef]
72. Kellermann, K.I.; Condon, J.J.; Kimball, A.E.; Perley, R.A.; Ivezić, Ž. Radio-loud and Radio-quiet QSOs. *Astrophys. J.* **2016**, *831*, 168. [CrossRef]
73. Padovani, P. On the two main classes of active galactic nuclei. *Nat. Astron.* **2017**, *1*, 0194. [CrossRef]
74. Panessa, F.; Baldi, R.D.; Laor, A.; Padovani, P.; Behar, E.; McHardy, I. The origin of radio emission from radio-quiet active galactic nuclei. *Nat. Astron.* **2019**, *3*, 387–396. [CrossRef]
75. Schmidt, M. Space Distribution and Luminosity Functions of Quasars. *Astrophys. J.* **1970**, *162*, 371. [CrossRef]
76. Padovani, P.; Miller, N.; Kellermann, K.I.; Mainieri, V.; Rosati, P.; Tozzi, P. The VLA Survey of Chandra Deep Field South. V. Evolution and Luminosity Functions of Sub-millijansky Radio Sources and the Issue of Radio Emission in Radio-quiet Active Galactic Nuclei. *Astrophys. J.* **2011**, *740*, 20. [CrossRef]
77. Bonzini, M.; Padovani, P.; Mainieri, V.; Kellermann, K.I.; Miller, N.; Rosati, P.; Tozzi, P.; Vattakunnel, S. The sub-mJy radio sky in the Extended Chandra Deep Field-South: Source population. *Mon. Not. R. Astron. Soc.* **2013**, *436*, 3759–3771. [CrossRef]
78. Padovani, P. The faint radio sky: Radio astronomy becomes mainstream. *Astron. Astrophys. Rev.* **2016**, *24*, 13. [CrossRef]
79. Boroson, T.A.; Green, R.F. The Emission-Line Properties of Low-Redshift Quasi-stellar Objects. *Astrophys. J. Suppl.* **1992**, *80*, 109. [CrossRef]
80. Marziani, P.; Sulentic, J.W. Highly accreting quasars: Sample definition and possible cosmological implications. *Mon. Not. R. Astron. Soc.* **2014**, *442*, 1211–1229. [CrossRef]
81. Marziani, P.; Panda, S.; Deconto Machado, A.; Del Olmo, A. Metal Content in Relativistically Jetted and Radio-Quiet Quasars in the Main Sequence Context. *Galaxies* **2023**, *11*, 52. [CrossRef]
82. Sulentic, J.W.; Marziani, P.; Zamanov, R.; Bachev, R.; Calvani, M.; Dultzin-Hacyan, D. Average Quasar Spectra in the Context of Eigenvector 1. *Astrophys. J. Lett.* **2002**, *566*, L71–L75. [CrossRef]
83. Zamfir, S.; Sulentic, J.W.; Marziani, P.; Dultzin, D. Detailed characterization of Hβ emission line profile in low-z SDSS quasars. *Mon. Not. R. Astron. Soc.* **2010**, *403*, 1759. [CrossRef]
84. Marziani, P.; Dultzin, D.; Sulentic, J.W.; Del Olmo, A.; Negrete, C.A.; Martínez-Aldama, M.L.; D'Onofrio, M.; Bon, E.; Bon, N.; Stirpe, G.M. A main sequence for quasars. *Front. Astron. Space Sci.* **2018**, *5*, 6. [CrossRef]
85. Marziani, P.; Berton, M.; Panda, S.; Bon, E. Optical Singly-Ionized Iron Emission in Radio-Quiet and Relativistically Jetted Active Galactic Nuclei. *Universe* **2021**, *7*, 484. [CrossRef]
86. Dias dos Santos, D.; Rodríguez-Ardila, A.; Panda, S.; Marinello, M. First Observation of a Double-peaked O I Emission in the Near-infrared Spectrum of an Active Galaxy. *Astrophys. J. Lett.* **2023**, *953*, L3. [CrossRef]
87. Panda, S.; Marziani, P.; Czerny, B. The Quasar Main Sequence Explained by the Combination of Eddington Ratio, Metallicity, and Orientation. *Astrophys. J.* **2019**, *882*, 79. [CrossRef]
88. Sulentic, J.; Marziani, P. Quasars in the 4D Eigenvector 1 Context: A stroll down memory lane. *Front. Astron. Space Sci.* **2015**, *2*, 6. [CrossRef]
89. Shakura, N.I.; Sunyaev, R.A. Black holes in binary systems. Observational appearance. *Astron. Astrophys.* **1973**, *24*, 337–355.
90. Gaskell, C.M. What broad emission lines tell us about how active galactic nuclei work. *New Astron.* **2009**, *53*, 140–148. [CrossRef]
91. Panda, S. Physical Conditions in the Broad-line Regions of Active Galaxies. Ph.D. Thesis, Polish Academy of Sciences, Center for Theoretical Physics, Warsaw, Poland, 2021.
92. Netzer, H.; Peterson, B.M. Reverberation Mapping and the Physics of Active Galactic Nuclei. In *Proceedings of the Astronomical Time Series*; Maoz, D., Sternberg, A., Leibowitz, E.M., Eds.; Astrophysics and Space Science Library; Spriger: Berlin/Heidelberg, Germany, 1997; Volume 218, p. 85. [CrossRef]
93. Wandel, A.; Peterson, B.M.; Malkan, M.A. Central Masses and Broad-Line Region Sizes of Active Galactic Nuclei. I. Comparing the Photoionization and Reverberation Techniques. *Astrophys. J.* **1999**, *526*, 579–591. [CrossRef]
94. Horne, K.; Peterson, B.M.; Collier, S.J.; Netzer, H. Observational Requirements for High-Fidelity Reverberation Mapping. *Publ. Astron. Soc. Pac.* **2004**, *116*, 465–476. [CrossRef]
95. Denney, K.D.; Watson, L.C.; Peterson, B.M.; Pogge, R.W.; Atlee, D.W.; Bentz, M.C.; Bird, J.C.; Brokofsky, D.J.; Comins, M.L.; Dietrich, M.; et al. A Revised Broad-line Region Radius and Black Hole Mass for the Narrow-line Seyfert 1 NGC 4051. *Astrophys. J.* **2009**, *702*, 1353–1366. [CrossRef]
96. Shapovalova, A.I.; Popović, L.Č.; Burenkov, A.N.; Chavushyan, V.H.; Ilić, D.; Kovačević, A.; Bochkarev, N.G.; León-Tavares, J. Long-term variability of the optical spectra of NGC 4151. II. Evolution of the broad Hα and Hβ emission-line profiles. *Astron. Astrophys.* **2010**, *509*, A106. [CrossRef]

97. Woo, J.H.; Treu, T.; Barth, A.J.; Wright, S.A.; Walsh, J.L.; Bentz, M.C.; Martini, P.; Bennert, V.N.; Canalizo, G.; Filippenko, A.V.; et al. The Lick AGN Monitoring Project: The M $_{BH}$-σ_* Relation for Reverberation-mapped Active Galaxies. *Astrophys. J.* **2010**, *716*, 269–280. [CrossRef]
98. Denney, K.D.; Peterson, B.M.; Pogge, R.W.; Adair, A.; Atlee, D.W.; Au-Yong, K.; Bentz, M.C.; Bird, J.C.; Brokofsky, D.J.; Chisholm, E.; et al. Reverberation Mapping Measurements of Black Hole Masses in Six Local Seyfert Galaxies. *Astrophys. J.* **2010**, *721*, 715–737. [CrossRef]
99. Grier, C.J.; Peterson, B.M.; Pogge, R.W.; Denney, K.D.; Bentz, M.C.; Martini, P.; Sergeev, S.G.; Kaspi, S.; Minezaki, T.; Zu, Y.; et al. Reverberation Mapping Results for Five Seyfert 1 Galaxies. *Astrophys. J.* **2012**, *755*, 60. [CrossRef]
100. Wang, J.M.; Du, P.; Hu, C.; Netzer, H.; Bai, J.M.; Lu, K.X.; Kaspi, S.; Qiu, J.; Li, Y.R.; Wang, F.; et al. Supermassive Black Holes with High Accretion Rates in Active Galactic Nuclei. II. The Most Luminous Standard Candles in the Universe. *Astrophys. J.* **2014**, *793*, 108. [CrossRef]
101. Pancoast, A.; Brewer, B.J.; Treu, T.; Park, D.; Barth, A.J.; Bentz, M.C.; Woo, J.H. Modelling reverberation mapping data—II. Dynamical modelling of the Lick AGN Monitoring Project 2008 data set. *Mon. Not. R. Astron. Soc.* **2014**, *445*, 3073–3091. [CrossRef]
102. Barth, A.J.; Bennert, V.N.; Canalizo, G.; Filippenko, A.V.; Gates, E.L.; Greene, J.E.; Li, W.; Malkan, M.A.; Pancoast, A.; Sand, D.J.; et al. The Lick AGN Monitoring Project 2011: Spectroscopic Campaign and Emission-line Light Curves. *Astrophys. J. Suppl.* **2015**, *217*, 26. [CrossRef]
103. Zu, Y.; Kochanek, C.S.; Kozłowski, S.; Peterson, B.M. Application of Stochastic Modeling to Analysis of Photometric Reverberation Mapping Data. *Astrophys. J.* **2016**, *819*, 122. [CrossRef]
104. Pei, L.; Fausnaugh, M.M.; Barth, A.J.; Peterson, B.M.; Bentz, M.C.; De Rosa, G.; Denney, K.D.; Goad, M.R.; Kochanek, C.S.; Korista, K.T.; et al. Space Telescope and Optical Reverberation Mapping Project. V. Optical Spectroscopic Campaign and Emission-line Analysis for NGC 5548. *Astrophys. J.* **2017**, *837*, 131. [CrossRef]
105. Fausnaugh, M.M.; Grier, C.J.; Bentz, M.C.; Denney, K.D.; De Rosa, G.; Peterson, B.M.; Kochanek, C.S.; Pogge, R.W.; Adams, S.M.; Barth, A.J.; et al. Reverberation Mapping of Optical Emission Lines in Five Active Galaxies. *Astrophys. J.* **2017**, *840*, 97. [CrossRef]
106. Du, P.; Zhang, Z.X.; Wang, K.; Huang, Y.K.; Zhang, Y.; Lu, K.X.; Hu, C.; Li, Y.R.; Bai, J.M.; Bian, W.H.; et al. Supermassive Black Holes with High Accretion Rates in Active Galactic Nuclei. IX. 10 New Observations of Reverberation Mapping and Shortened Hβ Lags. *Astrophys. J.* **2018**, *856*, 6. [CrossRef]
107. Williams, P.R.; Pancoast, A.; Treu, T.; Brewer, B.J.; Barth, A.J.; Bennert, V.N.; Buehler, T.; Canalizo, G.; Cenko, S.B.; Clubb, K.I.; et al. The Lick AGN Monitoring Project 2011: Dynamical Modeling of the Broad-line Region. *Astrophys. J.* **2018**, *866*, 75. [CrossRef]
108. Penton, A.; Malik, U.; Davis, T.M.; Martini, P.; Yu, Z.; Sharp, R.; Lidman, C.; Tucker, B.E.; Hoormann, J.K.; Aguena, M.; et al. OzDES reverberation mapping program: Lag recovery reliability for 6-yr C IV analysis. *Mon. Not. R. Astron. Soc.* **2022**, *509*, 4008–4023. [CrossRef]
109. Malik , U.; Sharp, R.; Penton, A.; Yu, Z.; Martini, P.; Lidman, C.; Tucker, B.E.; Davis, T.M.; Lewis, G.F.; Aguena, M.; et al. OzDES Reverberation Mapping Program: Hβ lags from the 6-yr survey. *Mon. Not. R. Astron. Soc.* **2023**, *520*, 2009–2023. [CrossRef]
110. Yu, Z.; Martini, P.; Penton, A.; Davis, T.M.; Kochanek, C.S.; Lewis, G.F.; Lidman, C.; Malik, U.; Sharp, R.; Tucker, B.E.; et al. OzDES Reverberation Mapping Programme: Mg II lags and R-L relation. *Mon. Not. R. Astron. Soc.* **2023**, *522*, 4132–4147. [CrossRef]
111. Shen, Y.; Grier, C.J.; Horne, K.; Stone, Z.; Li, J.I.; Yang, Q.; Homayouni, Y.; Trump, J.R.; Anderson, S.F.; Brandt, W.N.; et al. The Sloan Digital Sky Survey Reverberation Mapping Project: Key Results. *arXiv* **2023**, arXiv:2305.01014. [CrossRef]
112. Hu, C.; Du, P.; Lu, K.X.; Li, Y.R.; Wang, F.; Qiu, J.; Bai, J.M.; Kaspi, S.; Ho, L.C.; Netzer, H.; et al. Supermassive Black Holes with High Accretion Rates in Active Galactic Nuclei. III. Detection of Fe II Reverberation in Nine Narrow-line Seyfert 1 Galaxies. *Astrophys. J.* **2015**, *804*, 138. [CrossRef]
113. Shen, Y.; Hall, P.B.; Horne, K.; Zhu, G.; McGreer, I.; Simm, T.; Trump, J.R.; Kinemuchi, K.; Brandt, W.N.; Green, P.J.; et al. The Sloan Digital Sky Survey Reverberation Mapping Project: Sample Characterization. *Astrophys. J. Suppl.* **2019**, *241*, 34. [CrossRef]
114. Panda, S.; Marziani, P. High Eddington quasars as discovery tools: Current state and challenges. *Front. Astron. Space Sci.* **2023**, *10*, 1130103. [CrossRef]
115. Martínez-Aldama, M.L.; Czerny, B.; Kawka, D.; Karas, V.; Panda, S.; Zajaček, M.; Życki, P.T. Can Reverberation-measured Quasars Be Used for Cosmology? *Astrophys. J.* **2019**, *883*, 170. [CrossRef]
116. Shen, Y.; Ho, L.C. The diversity of quasars unified by accretion and orientation. *Nature* **2014**, *513*, 210–213. [CrossRef] [PubMed]
117. Martínez-Aldama, M.L.; Panda, S.; Czerny, B.; Marinello, M.; Marziani, P.; Dultzin, D. The CaFe Project: Optical Fe II and Near-infrared Ca II Triplet Emission in Active Galaxies. II. The Driver(s) of the Ca II and Fe II and Its Potential Use as a Chemical Clock. *Astrophys. J.* **2021**, *918*, 29. [CrossRef]
118. Wang, J.M.; Qiu, J.; Du, P.; Ho, L.C. Self-shadowing Effects of Slim Accretion Disks in Active Galactic Nuclei: The Diverse Appearance of the Broad-line Region. *Astrophys. J.* **2014**, *797*, 65. [CrossRef]
119. Abramowicz, M.A.; Czerny, B.; Lasota, J.P.; Szuszkiewicz, E. Slim accretion disks. *Astrophys. J.* **1988**, *332*, 646–658. [CrossRef]
120. Sadowski, A. Slim accretion disks around black holes. *arXiv* **2011**, arXiv:1108.0396.
121. Panda, S. The CaFe Project: Optical Fe II and Near-Infrared Ca II triplet emission in active galaxies: (II) synthetic EWs, co-dependence between cloud sizes and metal content. *arXiv* **2020**, arXiv:2004.13113.
122. Panda, S.; Marziani, P. Modelling the quasar spectra for super-Eddington sources – The What, the Why and the How. *arXiv* **2023**, arXiv:2308.05830. [CrossRef]

123. Lusso, E.; Risaliti, G. Quasars as standard candles. I. The physical relation between disc and coronal emission. *Astron. Astrophys.* **2017**, *602*, A79. [CrossRef]
124. Ferland, G.J.; Done, C.; Jin, C.; Landt, H.; Ward, M.J. State-of-the-art AGN SEDs for photoionization models: BLR predictions confront the observations. *Mon. Not. R. Astron. Soc.* **2020**, *494*, 5917–5922. [CrossRef]
125. Panda, S. Parameterizing the AGN Radius–Luminosity Relation from the Eigenvector 1 Viewpoint. *Front. Astron. Space Sci.* **2022**, *9*, 850409. [CrossRef]
126. Wills, B.J.; Netzer, H.; Wills, D. Broad emission features in QSOs and active galactic nuclei. II—New observations and theory of Fe II and H I emission. *Astrophys. J.* **1985**, *288*, 94–116. [CrossRef]
127. Davidson, K. On photoionization analyses of emission spectra of quasars. *Astrophys. J.* **1977**, *218*, 20–32. [CrossRef]
128. Osterbrock, D.E.; Ferland, G.J. *Astrophysics of Gaseous Nebulae and Active Galactic Nuclei*; University Science Books: Herndon, VA, USA, 2006.
129. Shen, Y.; Richards, G.T.; Strauss, M.A.; Hall, P.B.; Schneider, D.P.; Snedden, S.; Bizyaev, D.; Brewington, H.; Malanushenko, V.; Malanushenko, E.; et al. A Catalog of Quasar Properties from Sloan Digital Sky Survey Data Release 7. *Astrophys. J. Suppl.* **2011**, *194*, 45. [CrossRef]
130. Panda, S.; Martínez-Aldama, M.L.; Zajaček, M. Current and future applications of Reverberation-mapped quasars in Cosmology. *Front. Astron. Space Sci.* **2019**, *6*, 75. [CrossRef]
131. Haas, M.; Chini, R.; Ramolla, M.; Pozo Nuñez, F.; Westhues, C.; Watermann, R.; Hoffmeister, V.; Murphy, M. Photometric AGN reverberation mapping—an efficient tool for BLR sizes, black hole masses, and host-subtracted AGN luminosities. *Astron. Astrophys.* **2011**, *535*, A73. [CrossRef]
132. Pozo Nuñez, F.; Ramolla, M.; Westhues, C.; Bruckmann, C.; Haas, M.; Chini, R.; Steenbrugge, K.; Murphy, M. Photometric reverberation mapping of 3C 120. *Astron. Astrophys.* **2012**, *545*, A84. [CrossRef]
133. Ma, Q.; Wu, X.B.; Gu, H.; Wen, Y.; Fu, Y. The Hα Broadband Photometric Reverberation Mapping of Four Seyfert 1 Galaxies. *Astrophys. J.* **2023**, *949*, 22. [CrossRef]
134. Watson, D.; Denney, K.D.; Vestergaard, M.; Davis, T.M. A New Cosmological Distance Measure Using Active Galactic Nuclei. *Astrophys. J. Lett.* **2011**, *740*, L49. [CrossRef]
135. Czerny, B.; Hryniewicz, K.; Maity, I.; Schwarzenberg-Czerny, A.; Życki, P.T.; Bilicki, M. Towards equation of state of dark energy from quasar monitoring: Reverberation strategy. *Astron. Astrophys.* **2013**, *556*, A97. [CrossRef]
136. Sulentic, J.W.; Marziani, P.; del Olmo, A.; Dultzin, D.; Perea, J.; Alenka Negrete, C. GTC spectra of z $\approx$ 2.3 quasars: Comparison with local luminosity analogs. *Astron. Astrophys.* **2014**, *570*, A96. [CrossRef]
137. Cavaliere, A.; Vittorini, V. The Fall of the Quasar Population. *Astrophys. J.* **2000**, *543*, 599–610. [CrossRef]
138. Hopkins, P.F.; Hernquist, L.; Cox, T.J.; Di Matteo, T.; Robertson, B.; Springel, V. A Unified, Merger-driven Model of the Origin of Starbursts, Quasars, the Cosmic X-Ray Background, Supermassive Black Holes, and Galaxy Spheroids. *Astrophys. J. Suppl.* **2006**, *163*, 1–49. [CrossRef]
139. Davidson, K.; Netzer, H. The emission lines of quasars and similar objects. *Rev. Mod. Phys.* **1979**, *51*, 715–766. [CrossRef]
140. Lipari, S.; Macchetto, F. Coronal Lines and Starburst Features in Two New IRAS Active Galactic Nuclei: IRAS 04493-6441 and IRAS 22419-6049. *Astrophys. J.* **1992**, *387*, 522. [CrossRef]
141. Reunanen, J.; Kotilainen, J.K.; Prieto, M.A. Near-infrared spectroscopy of nearby Seyfert galaxies—I. First results. *Mon. Not. R. Astron. Soc.* **2002**, *331*, 154–168. [CrossRef]
142. Molina, M.; Reines, A.E.; Latimer, L.J.; Baldassare, V.; Salehirad, S. A Sample of Massive Black Holes in Dwarf Galaxies Detected via [Fe X] Coronal Line Emission: Active Galactic Nuclei and/or Tidal Disruption Events. *Astrophys. J.* **2021**, *922*, 155. [CrossRef]
143. Kynoch, D.; Landt, H.; Dehghanian, M.; Ward, M.J.; Ferland, G.J. Multiple locations of near-infrared coronal lines in NGC 5548. *Mon. Not. R. Astron. Soc.* **2022**, *516*, 4397–4416. [CrossRef]
144. Reefe, M.; Sexton, R.O.; Doan, S.M.; Satyapal, S.; Secrest, N.J.; Cann, J.M. CLASS Survey Description: Coronal-line Needles in the SDSS Haystack. *Astrophys. J. Suppl.* **2023**, *265*, 21. [CrossRef]
145. Short, P.; Lawrence, A.; Nicholl, M.; Ward, M.; Reynolds, T.M.; Mattila, S.; Yin, C.; Arcavi, I.; Carnall, A.; Charalampopoulos, P.; et al. Delayed appearance and evolution of coronal lines in the TDE AT2019qiz. *Mon. Not. R. Astron. Soc.* **2023**, *525*, 1568–1587. [CrossRef]
146. Marconi, A.; Moorwood, A.F.M.; Salvati, M.; Oliva, E. A [SiVI] (1.962 μm) coronal line survey of galactic nuclei. *Astron. Astrophys.* **1994**, *291*, 18–28.
147. Müller-Sánchez, F.; Prieto, M.A.; Hicks, E.K.S.; Vives-Arias, H.; Davies, R.I.; Malkan, M.; Tacconi, L.J.; Genzel, R. Outflows from Active Galactic Nuclei: Kinematics of the Narrow-line and Coronal-line Regions in Seyfert Galaxies. *Astrophys. J.* **2011**, *739*, 69. [CrossRef]
148. Rodríguez-Ardila, A.; Prieto, M.A.; Mazzalay, X.; Fernández-Ontiveros, J.A.; Luque, R.; Müller-Sánchez, F. Powerful outflows in the central parsecs of the low-luminosity active galactic nucleus NGC 1386. *Mon. Not. R. Astron. Soc.* **2017**, *470*, 2845–2860. [CrossRef]
149. Gravity Collaboration; Amorim, A.; Bauböck, M.; Brandner, W.; Clénet, Y.; Davies, R.; de Zeeuw, P.T.; Dexter, J.; Eckart, A.; Eisenhauer, F.; et al. The spatially resolved broad line region of IRAS 09149-6206. *Astron. Astrophys.* **2020**, *643*, A154. [CrossRef]
150. Reunanen, J.; Kotilainen, J.K.; Prieto, M.A. Near-infrared spectroscopy of nearby Seyfert galaxies—II. Molecular content and coronal emission. *Mon. Not. R. Astron. Soc.* **2003**, *343*, 192–208. [CrossRef]

151. Rodríguez-Ardila, A.; Prieto, M.A.; Portilla, J.G.; Tejeiro, J.M. The Near-infrared Coronal Line Spectrum of 54 nearby Active Galactic Nuclei. *Astrophys. J.* **2011**, *743*, 100. [CrossRef]
152. Prieto, A.; Rodríguez-Ardila, A.; Panda, S.; Marinello, M. A novel black hole mass scaling relation based on coronal gas, and its dependence with the accretion disc. *Mon. Not. R. Astron. Soc.* **2022**, *510*, 1010–1030. [CrossRef]
153. Bentz, M.C.; Katz, S. The AGN Black Hole Mass Database. *Publ. Astron. Soc. Pac.* **2015**, *127*, 67. [CrossRef]
154. Riffel, R.; Rodríguez-Ardila, A.; Pastoriza, M.G. A 0.8-2.4 μm spectral atlas of active galactic nuclei. *Astron. Astrophys.* **2006**, *457*, 61–70. [CrossRef]
155. Landt, H.; Bentz, M.C.; Ward, M.J.; Elvis, M.; Peterson, B.M.; Korista, K.T.; Karovska, M. The Near-Infrared Broad Emission Line Region of Active Galactic Nuclei. I. The Observations. *Astrophys. J. Suppl.* **2008**, *174*, 282–312. [CrossRef]
156. Cappellari, M.; Scott, N.; Alatalo, K.; Blitz, L.; Bois, M.; Bournaud, F.; Bureau, M.; Crocker, A.F.; Davies, R.L.; Davis, T.A.; et al. The ATLAS^{3D} project—XV. Benchmark for early-type galaxies scaling relations from 260 dynamical models: Mass-to-light ratio, dark matter, Fundamental Plane and Mass Plane. *Mon. Not. R. Astron. Soc.* **2013**, *432*, 1709–1741. [CrossRef]
157. Gültekin, K.; Richstone, D.O.; Gebhardt, K.; Lauer, T.R.; Tremaine, S.; Aller, M.C.; Bender, R.; Dressler, A.; Faber, S.M.; Filippenko, A.V.; et al. The M-σ and M-L Relations in Galactic Bulges, and Determinations of Their Intrinsic Scatter. *Astrophys. J.* **2009**, *698*, 198–221. [CrossRef]
158. Kaspi, S.; Maoz, D.; Netzer, H.; Peterson, B.M.; Vestergaard, M.; Jannuzi, B.T. The Relationship between Luminosity and Broad-Line Region Size in Active Galactic Nuclei. *Astrophys. J.* **2005**, *629*, 61–71. [CrossRef]
159. Ferland, G.J.; Chatzikos, M.; Guzmán, F.; Lykins, M.L.; van Hoof, P.A.M.; Williams, R.J.R.; Abel, N.P.; Badnell, N.R.; Keenan, F.P.; Porter, R.L.; et al. The 2017 Release Cloudy. *Rev. Mex. Astron. Astrofísica* **2017**, *53*, 385–438.
160. Mathews, W.G.; Ferland, G.J. What heats the hot phase in active nuclei? *Astrophys. J.* **1987**, *323*, 456–467. [CrossRef]
161. Frank, J.; King, A.; Raine, D.J. *Accretion Power in Astrophysics*, 3rd ed.; Cambridge University Press: Cambridge, UK, 2002.
162. Lasota, J.P. Black Hole Accretion Discs. In *Proceedings of the Astrophysics of Black Holes: From Fundamental Aspects to Latest Developments*; Bambi, C., Ed.; Astrophysics and Space Science Library; Springer: Berlin/Heidelberg, Germany, 2016; Volume 440, p. 1. [CrossRef]
163. Reynolds, C.S. Observing black holes spin. *Nat. Astron.* **2019**, *3*, 41–47. [CrossRef]
164. Campitiello, S.; Celotti, A.; Ghisellini, G.; Sbarrato, T. Black hole mass and spin estimates of the most distant quasars. *Astron. Astrophys.* **2019**, *625*, A23. [CrossRef]
165. Ivezić, Ž.; Kahn, S.M.; Tyson, J.A.; Abel, B.; Acosta, E.; Allsman, R.; Alonso, D.; AlSayyad, Y.; Anderson, S.F.; Andrew, J.; et al. LSST: From Science Drivers to Reference Design and Anticipated Data Products. *Astrophys. J.* **2019**, *873*, 111. [CrossRef]
166. Bianco, F.B.; Ivezić, Ž.; Jones, R.L.; Graham, M.L.; Marshall, P.; Saha, A.; Strauss, M.A.; Yoachim, P.; Ribeiro, T.; Anguita, T.; et al. Optimization of the Observing Cadence for the Rubin Observatory Legacy Survey of Space and Time: A Pioneering Process of Community-focused Experimental Design. *Astrophys. J. Suppl.* **2022**, *258*, 1. [CrossRef]
167. LSST Science Collaboration; Abell, P.A.; Allison, J.; Anderson, S.F.; Andrew, J.R.; Angel, J.R.P.; Armus, L.; Arnett, D.; Asztalos, S.J.; Axelrod, T.S.; et al. LSST Science Book, Version 2.0. *arXiv* **2009**, arXiv:0912.0201. [CrossRef]
168. LSST Dark Energy Science Collaboration. Large Synoptic Survey Telescope: Dark Energy Science Collaboration. *arXiv* **2012**, arXiv:1211.0310. [CrossRef]
169. LSST Science Collaboration; Marshall, P.; Anguita, T.; Bianco, F.B.; Bellm, E.C.; Brandt, N.; Clarkson, W.; Connolly, A.; Gawiser, E.; Ivezic, Z.; et al. Science-Driven Optimization of the LSST Observing Strategy. *arXiv* **2017**, arXiv:1708.04058. [CrossRef]
170. Narayan, G.; Zaidi, T.; Soraisam, M.D.; Wang, Z.; Lochner, M.; Matheson, T.; Saha, A.; Yang, S.; Zhao, Z.; Kececioglu, J.; et al. Machine-learning-based Brokers for Real-time Classification of the LSST Alert Stream. *Astrophys. J. Suppl.* **2018**, *236*, 9. [CrossRef]
171. LSST Dark Energy Science Collaboration (LSST DESC); Abolfathi, B.; Alonso, D.; Armstrong, R.; Aubourg, É.; Awan, H.; Babuji, Y.N.; Bauer, F.E.; Bean, R.; Beckett, G.; et al. The LSST DESC DC2 Simulated Sky Survey. *Astrophys. J. Suppl.* **2021**, *253*, 31. [CrossRef]
172. Shen, X.; Hopkins, P.F.; Faucher-Giguère, C.A.; Alexander, D.M.; Richards, G.T.; Ross, N.P.; Hickox, R.C. The bolometric quasar luminosity function at z = 0-7. *Mon. Not. R. Astron. Soc.* **2020**, *495*, 3252–3275. [CrossRef]
173. Kovačević, A.B.; Radović, V.; Ilić, D.; Popović, L.Č.; Assef, R.J.; Sánchez-Sáez, P.; Nikutta, R.; Raiteri, C.M.; Yoon, I.; Homayouni, Y.; et al. The LSST Era of Supermassive Black Hole Accretion Disk Reverberation Mapping. *Astrophys. J. Suppl.* **2022**, *262*, 49. [CrossRef]
174. Gravity Collaboration; Sturm, E.; Dexter, J.; Pfuhl, O.; Stock, M.R.; Davies, R.I.; Lutz, D.; Clénet, Y.; Eckart, A.; Eisenhauer, F.; et al. Spatially resolved rotation of the broad-line region of a quasar at sub-parsec scale. *Nature* **2018**, *563*, 657–660. [CrossRef]
175. GRAVITY Collaboration; Amorim, A.; Bauböck, M.; Bentz, M.C.; Brandner, W.; Bolzer, M.; Clénet, Y.; Davies, R.; de Zeeuw, P.T.; Dexter, J.; et al. A geometric distance to the supermassive black Hole of NGC 3783. *Astron. Astrophys.* **2021**, *654*, A85. [CrossRef]
176. Brandt, W.N.; Ni, Q.; Yang, G.; Anderson, S.F.; Assef, R.J.; Barth, A.J.; Bauer, F.E.; Bongiorno, A.; Chen, C.T.; De Cicco, D.; et al. Active Galaxy Science in the LSST Deep-Drilling Fields: Footprints, Cadence Requirements, and Total-Depth Requirements. *arXiv* **2018**, arXiv:1811.06542. [CrossRef]
177. Aihara, H.; AlSayyad, Y.; Ando, M.; Armstrong, R.; Bosch, J.; Egami, E.; Furusawa, H.; Furusawa, J.; Harasawa, S.; Harikane, Y.; et al. Third data release of the Hyper Suprime-Cam Subaru Strategic Program. *Publ. Astron. Soc. Jpn.* **2022**, *74*, 247–272. [CrossRef]

178. Zou, F.; Brandt, W.N.; Chen, C.T.; Leja, J.; Ni, Q.; Yan, W.; Yang, G.; Zhu, S.; Luo, B.; Nyland, K.; et al. Spectral Energy Distributions in Three Deep-drilling Fields of the Vera C. Rubin Observatory Legacy Survey of Space and Time: Source Classification and Galaxy Properties. *Astrophys. J. Suppl.* **2022**, *262*, 15. [CrossRef]
179. Scolnic, D.M.; Lochner, M.; Gris, P.; Regnault, N.; Hložek, R.; Aldering, G.; Allam, Tarek, J.; Awan, H.; Biswas, R.; Blazek, J.; et al. Optimizing the LSST Observing Strategy for Dark Energy Science: DESC Recommendations for the Deep Drilling Fields and other Special Programs. *arXiv* **2018**, arXiv:1812.00516. [CrossRef]
180. Sheng, X.; Ross, N.; Nicholl, M. Legacy Survey of Space and Time cadence strategy evaluations for active galactic nucleus time-series data in Wide-Fast-Deep field. *Mon. Not. R. Astron. Soc.* **2022**, *512*, 5580–5600. [CrossRef]
181. Raiteri, C.M.; Carnerero, M.I.; Balmaverde, B.; Bellm, E.C.; Clarkson, W.; D'Ammando, F.; Paolillo, M.; Richards, G.T.; Villata, M.; Yoachim, P.; et al. Blazar Variability with the Vera C. Rubin Legacy Survey of Space and Time. *Astrophys. J. Suppl.* **2022**, *258*, 3. [CrossRef]
182. Czerny, B.; Panda, S.; Prince, R.; Kumar Jaiswal, V.; Zajaček, M.; Martinez Aldama, M.L.; Kozłowski, S.; Kovacevic, A.B.; Ilic, D.; Popović, L.Č.; et al. Expectations for time-delay measurements in active galactic nuclei with the Vera Rubin Observatory. *Astron. Astrophys.* **2023**, *675*, A163. [CrossRef]
183. Fausnaugh, M.M.; Starkey, D.A.; Horne, K.; Kochanek, C.S.; Peterson, B.M.; Bentz, M.C.; Denney, K.D.; Grier, C.J.; Grupe, D.; Pogge, R.W.; et al. Continuum Reverberation Mapping of the Accretion Disks in Two Seyfert 1 Galaxies. *Astrophys. J.* **2018**, *854*, 107. [CrossRef]
184. Mudd, D.; Martini, P.; Zu, Y.; Kochanek, C.; Peterson, B.M.; Kessler, R.; Davis, T.M.; Hoormann, J.K.; King, A.; Lidman, C.; et al. Quasar Accretion Disk Sizes from Continuum Reverberation Mapping from the Dark Energy Survey. *Astrophys. J.* **2018**, *862*, 123. [CrossRef]
185. Sun, M.; Xue, Y.; Trump, J.R.; Gu, W.M. Winds can 'blow up' AGN accretion disc sizes. *Mon. Not. R. Astron. Soc.* **2019**, *482*, 2788–2794. [CrossRef]
186. Chan, J.H.H.; Millon, M.; Bonvin, V.; Courbin, F. Twisted quasar light curves: Implications for continuum reverberation mapping of accretion disks. *Astron. Astrophys.* **2020**, *636*, A52. [CrossRef]
187. Vincentelli, F.M.; McHardy, I.; Cackett, E.M.; Barth, A.J.; Horne, K.; Goad, M.; Korista, K.; Gelbord, J.; Brandt, W.; Edelson, R.; et al. On the multiwavelength variability of Mrk 110: Two components acting at different time-scales. *Mon. Not. R. Astron. Soc.* **2021**, *504*, 4337–4353. [CrossRef]
188. Cackett, E.M.; Zoghbi, A.; Ulrich, O. Frequency-resolved Lags in UV/Optical Continuum Reverberation Mapping. *Astrophys. J.* **2022**, *925*, 29. [CrossRef]
189. Guo , W.J.; Li, Y.R.; Zhang, Z.X.; Ho, L.C.; Wang, J.M. Accretion Disk Size Measurements of Active Galactic Nuclei Monitored by the Zwicky Transient Facility. *Astrophys. J.* **2022**, *929*, 19. [CrossRef]
190. Jha, V.K.; Joshi, R.; Chand, H.; Wu, X.B.; Ho, L.C.; Rastogi, S.; Ma, Q. Accretion disc sizes from continuum reverberation mapping of AGN selected from the ZTF survey. *Mon. Not. R. Astron. Soc.* **2022**, *511*, 3005–3016. [CrossRef]
191. Hagen, S.; Done, C. Modelling continuum reverberation in active galactic nuclei: A spectral-timing analysis of the ultraviolet variability through X-ray reverberation in Fairall 9. *Mon. Not. R. Astron. Soc.* **2023**, *521*, 251–268. [CrossRef]
192. Fian, C.; Chelouche, D.; Kaspi, S.; Sobrino Figaredo, C.; Lewis, T.; Catalan, S. Continuum reverberation mapping of MCG 08-11-011. *Astron. Astrophys.* **2023**, *672*, A132. [CrossRef]
193. Miller, J.A.; Cackett, E.M.; Goad, M.R.; Horne, K.; Barth, A.J.; Romero-Colmenero, E.; Fausnaugh, M.; Gelbord, J.; Korista, K.T.; Landt, H.; et al. Continuum Reverberation Mapping of Mrk 876 over Three Years with Remote Robotic Observatories. *Astrophys. J.* **2023**, *953*, 137. [CrossRef] [PubMed]
194. Collier, S.; Horne, K.; Wanders, I.; Peterson, B.M. A new direct method for measuring the Hubble constant from reverberating accretion discs in active galaxies. *Mon. Not. R. Astron. Soc.* **1999**, *302*, L24–L28. [CrossRef]
195. Oknyanskij, V.L. On the Possibility for Measuring the Hubble Constant from Optical-to-NIR Variability Time Delay in AGNs. *Odessa Astron. Publ.* **1999**, *12*, 99.
196. Panda, S.; Czerny, B.; Adhikari, T.P.; Hryniewicz, K.; Wildy, C.; Kuraszkiewicz, J.; Śniegowska, M. Modeling of the Quasar Main Sequence in the Optical Plane. *Astrophys. J.* **2018**, *866*, 115. [CrossRef]
197. Cackett, E.M.; Horne, K.; Winkler, H. Testing thermal reprocessing in active galactic nuclei accretion discs. *Mon. Not. R. Astron. Soc.* **2007**, *380*, 669–682. [CrossRef]
198. Pozo Nuñez, F.; Gianniotis, N.; Blex, J.; Lisow, T.; Chini, R.; Polsterer, K.L.; Pott, J.U.; Esser, J.; Pietrzyński, G. Optical continuum photometric reverberation mapping of the Seyfert-1 galaxy Mrk509. *Mon. Not. R. Astron. Soc.* **2019**, *490*, 3936–3951. [CrossRef]
199. Lobban, A.P.; Zola, S.; Pajdosz-Śmierciak, U.; Braito, V.; Nardini, E.; Bhatta, G.; Markowitz, A.; Bachev, R.; Carosati, D.; Caton, D.B.; et al. X-ray, UV, and optical time delays in the bright Seyfert galaxy Ark 120 with co-ordinated Swift and ground-based observations. *Mon. Not. R. Astron. Soc.* **2020**, *494*, 1165–1179. [CrossRef]
200. González-Buitrago, D.H.; García-Díaz, M.T.; Pozo Nuñez, F.; Guo, H. On the nature of the continuum reverberation of X-ray/UV and optical emission of IRAS 09149-6206. *Mon. Not. R. Astron. Soc.* **2023**, *525*, 4524–4539. [CrossRef]
201. Weaver, J.R.; Horne, K. Dust and the intrinsic spectral index of quasar variations: Hints of finite stress at the innermost stable circular orbit. *Mon. Not. R. Astron. Soc.* **2022**, *512*, 899–916. [CrossRef]
202. Gaskell, C.M. The case for cases B and C: Intrinsic hydrogen line ratios of the broad-line region of active galactic nuclei, reddenings, and accretion disc sizes. *Mon. Not. R. Astron. Soc.* **2017**, *467*, 226–238. [CrossRef]

203. Gaskell, C.M.; Anderson, F.C.; Birmingham, S.Á.; Ghosh, S. Estimating reddening of the continuum and broad-line region of active galactic nuclei: The mean reddening of NGC 5548 and the size of the accretion disc. *Mon. Not. R. Astron. Soc.* **2023**, *519*, 4082–4093. [CrossRef]
204. Kammoun, E.S.; Papadakis, I.E.; Dovčiak, M. Modelling the UV/optical continuum time-lags in AGN. *Mon. Not. R. Astron. Soc.* **2021**, *503*, 4163–4171. [CrossRef]
205. Kammoun, E.S.; Robin, L.; Papadakis, I.E.; Dovčiak, M.; Panagiotou, C. Revisiting UV/optical continuum time lags in AGN. *Mon. Not. R. Astron. Soc.* **2023**, *526*, 138–151. [CrossRef]
206. Edelson, R.; Gelbord, J.M.; Horne, K.; McHardy, I.M.; Peterson, B.M.; Arévalo, P.; Breeveld, A.A.; De Rosa, G.; Evans, P.A.; Goad, M.R.; et al. Space Telescope and Optical Reverberation Mapping Project. II. Swift and HST Reverberation Mapping of the Accretion Disk of NGC 5548. *Astrophys. J.* **2015**, *806*, 129. [CrossRef]
207. Kammoun, E.S.; Papadakis, I.E.; Dovčiak, M. A Hard Look at Thermal Reverberation and Optical/Ultraviolet Lags in NGC 5548. *Astrophys. J. Lett.* **2019**, *879*, L24. [CrossRef]
208. McHardy, I.M.; Connolly, S.D.; Horne, K.; Cackett, E.M.; Gelbord, J.; Peterson, B.M.; Pahari, M.; Gehrels, N.; Goad, M.; Lira, P.; et al. X-ray/UV/optical variability of NGC 4593 with Swift: Reprocessing of X-rays by an extended reprocessor. *Mon. Not. R. Astron. Soc.* **2018**, *480*, 2881–2897. [CrossRef]
209. Cackett, E.M.; Chiang, C.Y.; McHardy, I.; Edelson, R.; Goad, M.R.; Horne, K.; Korista, K.T. Accretion Disk Reverberation with Hubble Space Telescope Observations of NGC 4593: Evidence for Diffuse Continuum Lags. *Astrophys. J.* **2018**, *857*, 53. [CrossRef]
210. Hernández Santisteban, J.V.; Edelson, R.; Horne, K.; Gelbord, J.M.; Barth, A.J.; Cackett, E.M.; Goad, M.R.; Netzer, H.; Starkey, D.; Uttley, P.; et al. Intensive disc-reverberation mapping of Fairall 9: First year of Swift and LCO monitoring. *Mon. Not. R. Astron. Soc.* **2020**, *498*, 5399–5416. [CrossRef]
211. Cackett, E.M.; Gelbord, J.; Li, Y.R.; Horne, K.; Wang, J.M.; Barth, A.J.; Bai, J.M.; Bian, W.H.; Carroll, R.W.; Du, P.; et al. Supermassive Black Holes with High Accretion Rates in Active Galactic Nuclei. XI. Accretion Disk Reverberation Mapping of Mrk 142. *Astrophys. J.* **2020**, *896*, 1. [CrossRef]
212. Chelouche, D.; Pozo Nuñez, F.; Kaspi, S. Direct evidence of non-disk optical continuum emission around an active black hole. *Nat. Astron.* **2019**, *3*, 251–257. [CrossRef]
213. Korista, K.T.; Goad, M.R. The Variable Diffuse Continuum Emission of Broad-Line Clouds. *Astrophys. J.* **2001**, *553*, 695–708. [CrossRef]
214. Pozo Nuñez, F.; Bruckmann, C.; Deesamutara, S.; Czerny, B.; Panda, S.; Lobban, A.P.; Pietrzyński, G.; Polsterer, K.L. Modelling photometric reverberation mapping data for the next generation of big data surveys. Quasar accretion discs sizes with the LSST. *Mon. Not. R. Astron. Soc.* **2023**, *522*, 2002–2018. [CrossRef]
215. Netzer, H. Continuum reverberation mapping and a new lag-luminosity relationship for AGN. *Mon. Not. R. Astron. Soc.* **2021**, 509, 2637–2646. [CrossRef]
216. Lawther, D.; Goad, M.R.; Korista, K.T.; Ulrich, O.; Vestergaard, M. Quantifying the diffuse continuum contribution of BLR Clouds to AGN Continuum Inter-band Delays. *Mon. Not. R. Astron. Soc.* **2018**, *481*, 533–554. [CrossRef]
217. Jaiswal, V.K.; Prince, R.; Panda, S.; Czerny, B. Modeling time delays from two reprocessors in active galactic nuclei. *Astron. Astrophys.* **2023**, *670*, A147. [CrossRef]
218. Choloniewski, J. The Shape and Variability of the Nonthermal Component of the Optical Spectra of Active Galaxies. *Acta Astron.* **1981**, *31*, 293.
219. Winkler, H.; Glass, I.S.; van Wyk, F.; Marang, F.; Jones, J.H.S.; Buckley, D.A.H.; Sekiguchi, K. Variability studies of seyfert galaxies—I. Broad-band optical photometry. *Mon. Not. R. Astron. Soc.* **1992**, *257*, 659–676. [CrossRef]
220. Pozo Nuñez, F.; Haas, M.; Ramolla, M.; Bruckmann, C.; Westhues, C.; Chini, R.; Steenbrugge, K.; Lemke, R.; Murphy, M.; Kollatschny, W. Modelling photometric reverberation data: A disk-like broad-line region and a potentially larger black hole mass for 3C 120. *Astron. Astrophys.* **2014**, *568*, A36. [CrossRef]
221. Gaskell, C.M.; Goosmann, R.W.; Antonucci, R.R.J.; Whysong, D.H. The Nuclear Reddening Curve for Active Galactic Nuclei and the Shape of the Infrared to X-Ray Spectral Energy Distribution. *Astrophys. J.* **2004**, *616*, 147–156. [CrossRef]
222. Gaskell, C.M.; Benker, A.J. AGN Reddening and Ultraviolet Extinction Curves from Hubble Space Telescope Spectra. *arXiv* **2007**, arXiv:0711.1013. [CrossRef]
223. Korista, K.T.; Goad, M.R. Quantifying the impact of variable BLR diffuse continuum contributions on measured continuum interband delays. *Mon. Not. R. Astron. Soc.* **2019**, *489*, 5284–5300. [CrossRef]
224. Vanden Berk, D.E.; Richards, G.T.; Bauer, A.; Strauss, M.A.; Schneider, D.P.; Heckman, T.M.; York, D.G.; Hall, P.B.; Fan, X.; Knapp, G.R.; et al. Composite Quasar Spectra from the Sloan Digital Sky Survey. *Astron. J.* **2001**, *122*, 549–564. [CrossRef]
225. Glikman, E.; Helfand, D.J.; White, R.L. A Near-Infrared Spectral Template for Quasars. *Astrophys. J.* **2006**, *640*, 579–591. [CrossRef]
226. Wang, S.; Guo, H.; Woo, J.H. Estimating AGN Black Hole Masses via Continuum Reverberation Mapping in the Era of LSST. *Astrophys. J. Lett.* **2023**, *948*, L23. [CrossRef]
227. Elvis, M.; Karovska, M. Quasar Parallax: A Method for Determining Direct Geometrical Distances to Quasars. *Astrophys. J. Lett.* **2002**, *581*, L67–L70. [CrossRef]
228. Horne, K.; Korista, K.T.; Goad, M.R. Quasar tomography: Unification of echo mapping and photoionization models. *Mon. Not. R. Astron. Soc.* **2003**, *339*, 367–386. [CrossRef]

229. Marshall, J.; Bolton, A.; Bullock, J.; Burgasser, A.; Chambers, K.; DePoy, D.; Dey, A.; Flagey, N.; Hill, A.; Hillenbrand, L.; et al. The Maunakea Spectroscopic Explorer. In *Proceedings of the Bulletin of the American Astronomical Society*; IOP Publishing: Bristol, UK, 2019; Volume 51, p. 126,
230. Evans, C.; Puech, M.; Afonso, J.; Almaini, O.; Amram, P.; Aussel, H.; Barbuy, B.; Basden, A.; Bastian, N.; Battaglia, G.; et al. The Science Case for Multi-Object Spectroscopy on the European ELT. *arXiv* **2015**, arXiv:1501.04726.
231. Gardner, J.P.; Mather, J.C.; Clampin, M.; Doyon, R.; Greenhouse, M.A.; Hammel, H.B.; Hutchings, J.B.; Jakobsen, P.; Lilly, S.J.; Long, K.S.; et al. The James Webb Space Telescope. *Space Sci. Rev.* **2006**, *123*, 485–606. [CrossRef]
232. Jakobsen, P.; Ferruit, P.; Alves de Oliveira, C.; Arribas, S.; Bagnasco, G.; Barho, R.; Beck, T.L.; Birkmann, S.; Böker, T.; Bunker, A.J.; et al. The Near-Infrared Spectrograph (NIRSpec) on the James Webb Space Telescope. I. Overview of the instrument and its capabilities. *Astron. Astrophys.* **2022**, *661*, A80. [CrossRef]
233. Spergel, D.; Gehrels, N.; Breckinridge, J.; Donahue, M.; Dressler, A.; Gaudi, B.S.; Greene, T.; Guyon, O.; Hirata, C.; Kalirai, J.; et al. Wide-Field InfraRed Survey Telescope-Astrophysics Focused Telescope Assets WFIRST-AFTA Final Report. *arXiv* **2013**, arXiv:1305.5422.
234. Trindade Falcão, A.; Kraemer, S.B.; Crenshaw, D.M.; Melendez, M.; Revalski, M.; Fischer, T.C.; Schmitt, H.R.; Turner, T.J. Tracking X-ray outflows with optical/infrared footprint lines. *Mon. Not. R. Astron. Soc.* **2022**, *511*, 1420–1430. [CrossRef]
235. Backhaus, B.E.; Trump, J.R.; Cleri, N.J.; Simons, R.; Momcheva, I.; Papovich, C.; Estrada-Carpenter, V.; Finkelstein, S.L.; Matharu, J.; Ji, Z.; et al. CLEAR: Emission-line Ratios at Cosmic High Noon. *Astrophys. J.* **2022**, *926*, 161. [CrossRef]
236. U, V.; Lai, T.; Bianchin, M.; Remigio, R.P.; Armus, L.; Larson, K.L.; Díaz-Santos, T.; Evans, A.; Stierwalt, S.; Law, D.R.; et al. GOALS-JWST: Resolving the Circumnuclear Gas Dynamics in NGC 7469 in the Mid-infrared. *Astrophys. J. Lett.* **2022**, *940*, L5. [CrossRef]
237. Harikane, Y.; Ouchi, M.; Oguri, M.; Ono, Y.; Nakajima, K.; Isobe, Y.; Umeda, H.; Mawatari, K.; Zhang, Y. A Comprehensive Study of Galaxies at z 9-16 Found in the Early JWST Data: Ultraviolet Luminosity Functions and Cosmic Star Formation History at the Pre-reionization Epoch. *Astrophys. J. Suppl.* **2023**, *265*, 5. [CrossRef]
238. Bunker, A.J.; Saxena, A.; Cameron, A.J.; Willott, C.J.; Curtis-Lake, E.; Jakobsen, P.; Carniani, S.; Smit, R.; Maiolino, R.; Witstok, J.; et al. JADES NIRSpec Spectroscopy of GN-z11: Lyman-α emission and possible enhanced nitrogen abundance in a $z = 10.60$ luminous galaxy. *Astron. Astrophys.* **2023**, *677*, A88. [CrossRef]
239. Onoue, M.; Inayoshi, K.; Ding, X.; Li, W.; Li, Z.; Molina, J.; Inoue, A.K.; Jiang, L.; Ho, L.C. A Candidate for the Least-massive Black Hole in the First 1.1 Billion Years of the Universe. *Astrophys. J. Lett.* **2023**, *942*, L17. [CrossRef]
240. Armus, L.; Lai, T.; U, V.; Larson, K.L.; Diaz-Santos, T.; Evans, A.S.; Malkan, M.A.; Rich, J.; Medling, A.M.; Law, D.R.; et al. GOALS-JWST: Mid-infrared Spectroscopy of the Nucleus of NGC 7469. *Astrophys. J. Lett.* **2023**, *942*, L37. [CrossRef]
241. Larson, R.L.; Finkelstein, S.L.; Kocevski, D.D.; Hutchison, T.A.; Trump, J.R.; Arrabal Haro, P.; Bromm, V.; Cleri, N.J.; Dickinson, M.; Fujimoto, S.; et al. A CEERS Discovery of an Accreting Supermassive Black Hole 570 Myr after the Big Bang: Identifying a Progenitor of Massive $z > 6$ Quasars. *Astrophys. J. Lett.* **2023**, *953*, L29. [CrossRef]
242. Kocevski, D.D.; Onoue, M.; Inayoshi, K.; Trump, J.R.; Arrabal Haro, P.; Grazian, A.; Dickinson, M.; Finkelstein, S.L.; Kartaltepe, J.S.; Hirschmann, M.; et al. Hidden Little Monsters: Spectroscopic Identification of Low-mass, Broad-line AGNs at $z > 5$ with CEERS. *Astrophys. J. Lett.* **2023**, *954*, L4. [CrossRef]
243. Goulding, A.D.; Greene, J.E.; Setton, D.J.; Labbe, I.; Bezanson, R.; Miller, T.B.; Atek, H.; Bogdán, Á.; Brammer, G.; Chemerynska, I.; et al. UNCOVER: The Growth of the First Massive Black Holes from JWST/NIRSpec-Spectroscopic Redshift Confirmation of an X-Ray Luminous AGN at $z = 10.1$. *Astrophys. J. Lett.* **2023**, *955*, L24. [CrossRef]
244. Cresci, G.; Tozzi, G.; Perna, M.; Brusa, M.; Marconcini, C.; Marconi, A.; Carniani, S.; Brienza, M.; Giroletti, M.; Belfiore, F.; et al. Bubbles and outflows: The novel JWST/NIRSpec view of the $z = 1.59$ obscured quasar XID2028. *Astron. Astrophys.* **2023**, *672*, A128. [CrossRef]
245. Übler, H.; Maiolino, R.; Curtis-Lake, E.; Pérez-González, P.G.; Curti, M.; Perna, M.; Arribas, S.; Charlot, S.; Marshall, M.A.; D'Eugenio, F.; et al. GA-NIFS: A massive black hole in a low-metallicity AGN at $z \sim 5.55$ revealed by JWST/NIRSpec IFS. *Astron. Astrophys.* **2023**, *677*, A145. [CrossRef]
246. Cleri, N.J.; Yang, G.; Papovich, C.; Trump, J.R.; Backhaus, B.E.; Estrada-Carpenter, V.; Finkelstein, S.L.; Giavalisco, M.; Hutchison, T.A.; Ji, Z.; et al. CLEAR: High-ionization [Ne V] λ3426 Emission-line Galaxies at $1.4 < z < 2.3$. *Astrophys. J.* **2023**, *948*, 112. [CrossRef]
247. Cleri, N.J.; Olivier, G.M.; Hutchison, T.A.; Papovich, C.; Trump, J.R.; Amorín, R.O.; Backhaus, B.E.; Berg, D.A.; Fernández, V.; Finkelstein, S.L.; et al. Using [Ne V]/[Ne III] to Understand the Nature of Extreme-ionization Galaxies. *Astrophys. J.* **2023**, *953*, 10. [CrossRef]
248. Harikane, Y.; Zhang, Y.; Nakajima, K.; Ouchi, M.; Isobe, Y.; Ono, Y.; Hatano, S.; Xu, Y.; Umeda, H. A JWST/NIRSpec First Census of Broad-Line AGNs at z = 4–7: Detection of 10 Faint AGNs with M_BH∼10ˆ6-10ˆ8 M_sun and Their Host Galaxy Properties. *arXiv* **2023**, arXiv:2303.11946. [CrossRef]
249. Matthee, J.; Naidu, R.P.; Brammer, G.; Chisholm, J.; Eilers, A.C.; Goulding, A.; Greene, J.; Kashino, D.; Labbe, I.; Lilly, S.J.; et al. Little Red Dots: An abundant population of faint AGN at $z \sim 5$ revealed by the EIGER and FRESCO JWST surveys. *arXiv* **2023**, arXiv:2306.05448. [CrossRef]

250. Labbe, I.; Greene, J.E.; Bezanson, R.; Fujimoto, S.; Furtak, L.J.; Goulding, A.D.; Matthee, J.; Naidu, R.P.; Oesch, P.A.; Atek, H.; et al. UNCOVER: Candidate Red Active Galactic Nuclei at $3 < z < 7$ with JWST and ALMA. *arXiv* **2023**, arXiv:2306.07320. [CrossRef]
251. Greene, J.E.; Labbe, I.; Goulding, A.D.; Furtak, L.J.; Chemerynska, I.; Kokorev, V.; Dayal, P.; Williams, C.C.; Wang, B.; Setton, D.J.; et al. UNCOVER spectroscopy confirms a surprising ubiquity of AGN in red galaxies at $z > 5$. *arXiv* **2023**, arXiv:2309.05714. [CrossRef]

universe MDPI

Article

Deep Learning of Quasar Lightcurves in the LSST Era

Andjelka B. Kovačević [1,2,*], Dragana Ilić [1,3], Luka Č. Popović [1,2,4], Nikola Andrić Mitrović [5], Mladen Nikolić [1], Marina S. Pavlović [6], Iva Čvorović-Hajdinjak [1], Miljan Knežević [1] and Djordje V. Savić [4,7]

1 Faculty of Mathematics, University of Belgrade, Studentski trg 16, 11000 Belgrade, Serbia; dragana.ilic@matf.bg.ac.rs (D.I.); lpopovic@aob.rs (L.Č.P.); mladen.nikolic@matf.bg.ac.rs (M.N.)
2 PIFI Research Fellow, Key Laboratory for Particle Astrophysics, Institute of High Energy Physics, Chinese Academy of Sciences, 19B Yuquan Road, Beijing 100049, China
3 Humboldt Research Fellow, Hamburger Sternwarte, Universitat Hamburg, Gojenbergsweg 112, 21029 Hamburg, Germany
4 Astronomical Observatory, Volgina 7, 11000 Belgrade, Serbia
5 Department of Mathematics "Tullio Levi Civita", University of Padova, Via Trieste, 35121 Padova, Italy; andricmitrovicnikola@yahoo.com
6 Mathematical Institute of the Serbian Academy of Sciences and Arts, Kneza Mihaila 36, 11000 Belgrade, Serbia
7 Institut d'Astrophysique et de Géophysique, Université de Liège, Allée du 6 Août 19c, 4000 Liège, Belgium
* Correspondence: andjelka.kovacevic@matf.bg.ac.rs

Abstract: Deep learning techniques are required for the analysis of synoptic (multi-band and multi-epoch) light curves in massive data of quasars, as expected from the Vera C. Rubin Observatory Legacy Survey of Space and Time (LSST). In this follow-up study, we introduce an upgraded version of a conditional neural process (CNP) embedded in a multi-step approach for the analysis of large data of quasars in the LSST Active Galactic Nuclei Scientific Collaboration data challenge database. We present a case study of a stratified set of u-band light curves for 283 quasars with very low variability ∼0.03. In this sample, the CNP average mean square error is found to be ∼5% (∼0.5 mag). Interestingly, besides similar levels of variability, there are indications that individual light curves show flare-like features. According to the preliminary structure–function analysis, these occurrences may be associated with microlensing events with larger time scales of 5–10 years.

Keywords: high-energy astrophysics; quasars; astrostatistics techniques; time series analysis; computational astronomy; astronomy data modeling; observatories; optical observatories

1. Introduction

The launch of the Legacy Survey of Space and Time (LSST), which will be conducted by the Vera C. Rubin Observatory, is currently scheduled to take place in the first half of 2024. The cadences of the LSST, in concert with its large observational coverage, will capture a wide variety of intriguing time domain events, some of which are periodic signals of interest [1]. LSST should probe time series with cadences ranging from one minute to ten years across not only a vast portion of the sky but also across five photometric bands (see Figure 1).

Such synoptic (multi-band and multi-epoch) cadences combined with the large coverage will enable us to detect very short-lived events such as eclipses in ultracompact double-degenerate binary systems [2], fast faint transients such as optical phenomena associated with gamma-ray bursts [3], and electromagnetic counterparts to gravitational wave sources [4,5]. In contrast, the LSST decadal data catalogs will make it possible to investigate long-period variables, intermediate-mass black holes (IMBH), and quasars (QSO) [6–10].

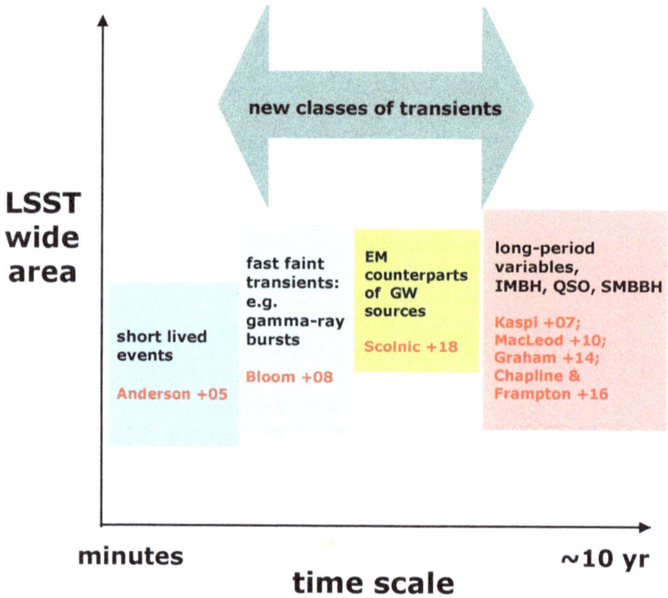

Figure 1. Schematic view of the LSST third scientific pillar. Exploring the transient optical sky as defined in Ivezić et al. [1]. The broad range of probed time scales and sky area may allow for the search for general variability properties across different types of objects, similar to the study by Burke et al. [10], which suggests a common process for all accretion disks. The references Anderson et al. [2], Bloom et al. [3], Scolnic et al. [4], Kaspi et al. [6], MacLeod et al. [7], Graham et al. [8], Chapline and Frampton [9] given in boxes are further explained in the text.

Quasars are an important population to study in order to have a better grasp of the physics behind the accretion of matter when it is subjected to extremely harsh conditions. Moreover, studies show that they may be used as cosmological probes (e.g., [11,12]). Up to this point, several hundred thousand quasars have been spectroscopically confirmed, and numerous efforts have been undertaken to identify the properties of their temporal flux variability [13]. The proposed physical mechanisms underlying the optical/UV variability range from the superposition of supernovae (e.g., [14]), microlensing [15,16], and thermal fluctuations from magnetic field turbulence [17] to instabilities in the accretion disk [14]. The amplitude of quasar observed optical variability is typically a few tenths of a magnitude (e.g., [18] found the SDSS quasars variability is $\sim$0.03 mag with a characteristic time-scale of several months) but it can also show larger variations over longer time-scales (see [19,20]), with statistical description via a damped random walk (DRW) model (e.g., [17,20,21]). The long lasting flare-like events (extreme tails of the variability distribution) are less clearly defined and represent a distinct kind of modeling problem (see, e.g., [22] and references therein). On top of this, the light curves have different topologies, which are superimposed on different types of cadences, which imposes many difficulties onto their modeling and the process of extracting knowledge from them [1].

Specifically, the LSST will provide a breakthrough in quasar observations in survey area and depth [23] as well as variability information of light curves sampled at a relatively high cadence (the order of days), over the course of a decade of the operations. Because of these new qualities, the LSST will be able to search even for supermassive black hole (SMBH) binaries having shorter periods (<5 years), which are significantly more uncommon. For instance, Xin and Haiman [23] suggested that it might be possible to identify ultra-short-period SMBH binaries (periods < 3 days) in the LSST quasar catalog. These binaries are thought to be so compact that they will 'chirp', or evolve in frequency, into the gravitational wave band, where the Laser Interferometer Space Antenna (LISA [24]) will be able to detect them in the middle of the 2030s. The expectations of such exciting discoveries are probable [23], as massive binary SMBHs are predicted to spend $O(10^5)$ years in orbits with periods of the order of a year if orbital decay is caused by either gravitational wave emission or negative torques exerted on the viscous time-scale by the surrounding gas disc [25].

Even while the LSST have cadences that are unprecedented in contrast to those of its predecessors, these cadences will primarily take the shape of more or less regular samplings that are separated by various short or lengthy periods (seasons) in which there are no observations.

When attempting to evaluate the variable features of quasar light curves, these frequent gaps represent one of the most challenging obstacles to overcome (along with the obviously irregular cadences, see [21]). In quasar time domain analysis, there are two main ways to deal with sampling that is not even in stochastic light curves [21]: the first approach is to use Monte Carlo simulations to forward the model in the frequency domain (see, e.g., [26]), whereas the second approach is to fit the light curve in the time domain mostly using Gaussian Processes (GP, see [27]).

Both approaches are viable, although they can be computationally expensive (see details in [21] and the references therein). If the cadences are similar to those found in the LSST with seasonal gaps, the first method is a costly one to compute because it necessitates either creating a highly dense light curve at the optimal sampling rate or segmenting the light curve and computing the periodograms of each segment separately. The likelihood function of GP, on the other hand, is computationally expensive (scales $\sim O(n^3)$) for the second approach because it requires inverting the $n \times n$ covariance matrix of the light curve, where n is the number of data points. A first-order continuous-time autoregressive process (CAR(1)) or an Ornstein–Uhlenbeck process, which could represent quasar light curves, is a special class of GP for which the computational complexity only scales linearly with the length of the light curve [17]. The application of CAR(1) to modeling typical active galactic nuclei (AGNs) optical light curves is questioned (see [20]), as some studies have found evidence for deviations from the CAR(1) process for optical light curves of AGN (see, e.g., [8,28,29]). Due to the nature of this issue, it was necessary to develop more complex Gaussian random process models, such as the continuous auto-regressive moving-average models (CARMA, [21]). Moreover, it has been demonstrated by Yu et al. [30] that a second-order stochastic process, a damped harmonic oscillator (DHO), is a more accurate way to characterize the variability of AGNs. Based on previous examples, the evolution of algorithms used to model AGN light curves typically emphasizes an increase in the total number of needed parameters. All of these models, however, are based on information gathered before the LSST era, which has a tendency to favor more luminous and nearby AGNs. Therefore, in order to make use of the tens of millions of the LSST AGN multi-band light curves effectively, it is highly desirable to employ flexible data-driven machine learning algorithms.

At the moment, kernel methods (such as GPs) and deep neural networks are seen as two of the most remarkable machine learning techniques [31]. The relationship between these two approaches has been the subject of a great deal of research in recent years [31,32]. In this light, here we present the AGN light curve modeling unit, which was created as a preprocessing module of the SER-SAG [2] team's LSST in-kind contribution of the time-domain periodicity mining pipeline and combines the best of two machine learning worlds. The neural latent variable model [Neural Process-NP [33]] is at the heart of this modeling unit. NPs, such as GPs, create distributions over functions, can adapt quickly to new observations, and can assess the uncertainty in their predictions [33]. NPs, such as neural networks, are computationally efficient during training and assessment, but they also learn to adapt their priors to data as well [33]. The NP module has been trained on the quasar light curves found in a dedicated database that arose from a challenge focused on the future use of the LSST quasar data (`LSST_AGN_DC` [34]). During the course of the operation, we also came to the realization that it is possible to distinguish the variable properties of quasars.

In our previous work (see [35] hereafter Paper I), we adapted a conditional neural process (CNP) for modeling the general variability of quasar light curves on a smaller sample of tens of objects. In this work, we complement the study of Paper I with an upgraded version of CNP on a much larger database of ($\sim$4 $\times$ 10^5) quasars (`LSST_AGN_DC`) which demanded 'deep learning' through a multi-step process and provide a case study example of how a complex procedure of 'deep learning' of quasar variability may lead to surprising results, i.e., detection of a larger collection of quasars with flare-like events. These flare-like incidents occur over a longer time period, and the preliminary structure–function analysis suggests that they may be related to microlensing events.

It is anticipated that the LSST would lead to an increase of at least one order of magnitude in the number of lensed quasars that are known [36]. Thus, optimizing the analysis or selecting sub-samples of those systems is becoming important (e.g., [37]).

The structure of the paper is as follows. In Section 2, we describe the data used for the machine learning experiments. In Section 3, we provide a concise explanation of the machine learning methods that were applied; while in Section 4, we report and discuss in further detail the series of experiments that were carried out. In the final Section 5, we summarize our findings.

2. Materials

2.1. Description of Quasar Data in `LSST_AGN_DC`

The dataset of AGN light curves used for the demonstration of our neural process modeling unit is selected from the LSST AGN data challenge 2021 dataset [`LSST_AGN_DC`, see details in [34]). The `LSST_AGN_DC` mimics the future LSST data release catalogs as much as possible (see also [38]). Calculations were run on NVIDIA T4, 2560 CUDA cores, Compute capability 7.5, Memory 16 GB GDDR6, Max memory bandwidth 300 GB/s.

The total number of objects in the `LSST_AGN_DC` is $\sim$440,000, with stars, galaxies, and 39,173 quasars drawn from two main survey fields, an expanded Stripe 82 area, and the XMM-LSS region. The total number of epochs for all objects is $\sim 5 \times 10^6$ [34]. The quantity of features (parameters) of objects is 381, sorted as (see details in [34,38]) astrometry (celestial equatorial coordinates, proper motion, parallax); photometry (point and extended source photometry in AB magnitudes and fluxes (in nano-Jansky); color (derived from flux ratio between different photometric bands); morphology (continuous number in range $[0, 1]$, extended sources has morphology closer to 1, while point-like sources closer to 0); light curve features (extracted from SDSS); spectroscopic and photometric redshift; and class labels (Star/Galaxy/quasar). The time domain data of light curves include observation epochs, photometric magnitudes, and errors in `u,g,r,i,z` bands as well as periodic and non-periodic features (see description of features in [39]).

2.2. Sample Selection

We chose 1006 quasars, that are spectroscopically confirmed and which have $\geq$100 epochs in u-band light curves (see Figure 2) from this large initial database. We employ this criterion because earlier research has demonstrated that this number of points is acceptable for both modeling light curves and extracting periodicity (see, e.g., [40]). We selected to study u-band light curves since they are less deformed by photometric filters [41]. In terms of mean sampling, the chosen quasar subsample exhibits a stratification into three non-intersecting branches of mean sampling (Figure 2).

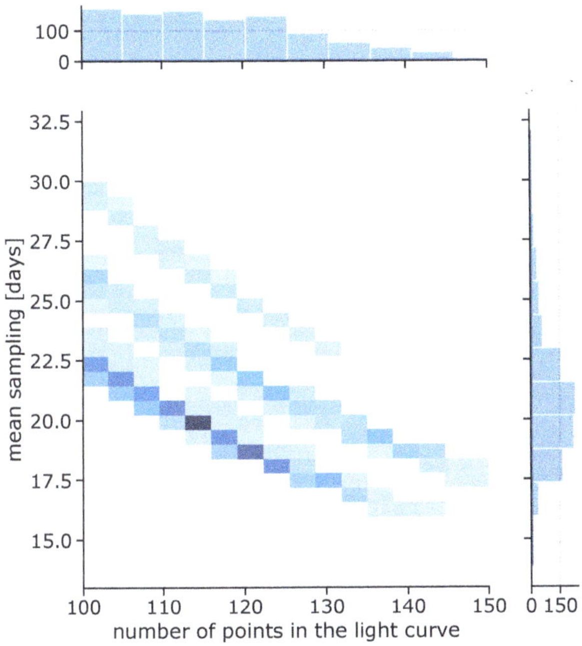

Figure 2. Statistical description of 1006 quasar light curves in the u band having ≥100 points. The primary joint distribution plot of the number of points in the light curves and mean sampling makes the stratification of mean sampling very obvious. The non-Gaussianity of the mean sample and the number of points in the light curves are brought to light by the marginal distributions that are presented on the side plots. The marginals are presented as counts in this instance. Upper marginal: y-axis counts a number of light curves having a number of points given in x-axis in the main plot. Right marginal: y-axis counts a number of light curves having mean sampling in y-axis in the main plot.

Both the non-Gaussianity of marginal distributions and stratification in the phase space of mean sampling and the number of points in the light curves indicate that we are encountering a highly diverse sample.

Kasliwal et al. [42] provided the first case study of the relevance of using modeling procedures individually on stratified light curves, as opposed to the "one-size-fits-all" approach, which allows us to account for the prevailing physical process's heterogeneity. The authors categorized the Kepler light curves of 20 objects based on visual similarities and found that the light curves fall into five broad strata: stochastic-looking, somewhat stochastic-looking+weak oscillatory features, oscillatory features dominant, flare features dominant, and not-variable. Certain light curves appear to change from one state of variability to another.

Motivated by Kasliwal et al. [42]'s example and a large number of our preselected objects (1006) that cannot be visually stratified, we employed the Self-Organizing Maps (SOM) algorithm [43] to stratify (cluster) light curves with similar topological patterns. From 36 clusters obtained from SOM, we chose interesting strata containing 310 light curves with apparent low variability (see Figure 3).

To thoroughly check the characteristics of 283 sources, we calculated the fractional root mean square (rms) variability amplitude F_{var} and the optical luminosities (L_{op}).

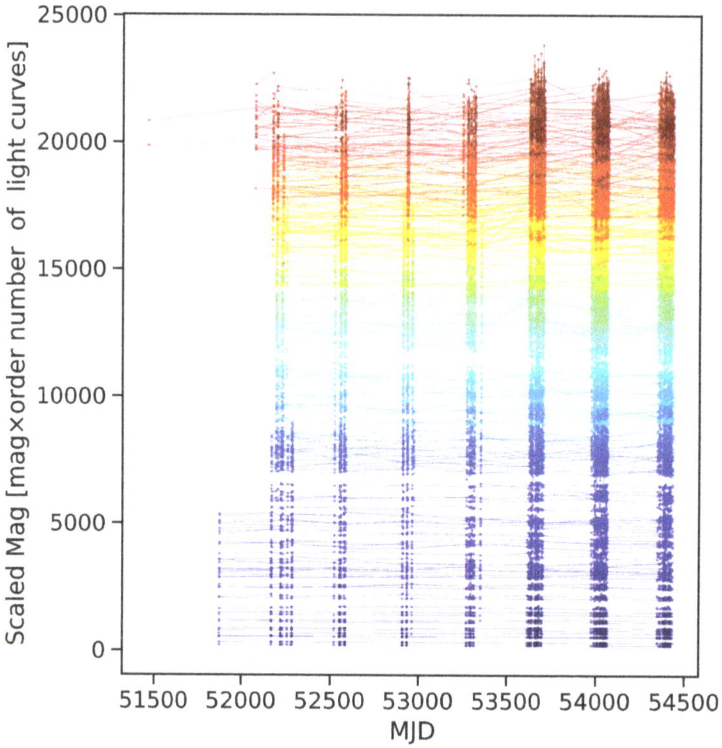

Figure 3. Cluster of 283 u-band light curves with low variability obtained via SOM algorithm. Observed magnitudes are given as points, while solid lines serve as an eye guide. The magnitudes of each light curve have been multiplied by their ordinal number for clarity. Warmer colors correspond to higher ordinal numbers.

The uncertainties of the individual magnitude measurements contribute an additional variance, which is captured in F_{var} (see [44,45]) as:

$$S^2 = \frac{1}{N-1}\sum_{i=1}^{N}(mag_i - < mag >)^2 \tag{1}$$

$$F_{var} = \sqrt{\frac{S^2 - < \sigma >^2}{< mag >^2}} \tag{2}$$

where N is the number of points in the light curves, $mag_i, i = 1, \ldots, N$ are observed magnitudes, $< mag >= \frac{1}{N}\sum_{i=1}^{N} mag_i$, $< \sigma >^2 = \frac{1}{N}\sum_{i=1}^{N}\sigma_i^2$ and $\sigma_i, i = 1, \ldots, N$ are measurement errors [3].

The black hole masses (M) were randomly assigned using the probability distribution based on absolute magnitude M_u by [7]:

$$P(\log_{10} M|M_u) = \frac{1}{\sqrt{2\pi\sigma_M^2}} exp^{-\frac{(\log_{10} M - \log_{10} < M >)^2}{2\sigma_M}} \tag{3}$$

where $\log_{10} < M >= 2.0 - 0.27M_u$, $\sigma_M = 0.58 + 0.011M_u$, and M_u is an absolute magnitude and is calculated using the known u-band magnitude and K-correction, $K(z) = -2.5(1+\delta)\log(1+z)$, with the canonical spectral index $\delta = -0.5$ as in Solomon and Stojkovic [46]. The assigned masses of black holes serve as proxies that complement other inferred quasar properties.

We approximate the optical luminosity (L_{op}) of the sampled objects via (see [13])

$$L_{op}[\mathrm{erg\,s^{-1}}] = 4\pi d_L^2 F_{0,\lambda}\lambda_{eff}10^{-2.5(<mag>-A)} \tag{4}$$

where d_L is the estimated luminosity distance for a flat universe (using `astropy` module, [47]) with standard cosmological parameters $H_0 = 67.4$ km s^{-1}Mpc^{-1} and $\Omega_m = 0.315$ [48], c is the light speed, z is the redshift of the object, $F_{0,\lambda} = 3.75079 \times 10^{-9}$ erg, cm^2 s^{-1}Å^{-1} is the zero point flux density, $\lambda_{eff} = 3608.04$ Å is the effective wavelength of the SDSS filter system (see [49,50]), $< mag >$ is a mean u-band magnitude, and A is the Galactic absorption at the effective wavelength along the line of sight. However, for our purposes we did not take into account A.

We present a corner plot of the four parameter distributions (M, F_{var}, L_{op}, z) in Figure 4. The selected objects are indeed characterized by small variability $F_{var} \sim [0, 0.03]$ and larger redshift $1 \leq z \leq 3$. Moreover, the two-dimensional plots reveal some scatter in the parameter distributions. In contrast, the 2D distribution of luminosity and redshift of objects is strongly nonlinear, as expected (see also [13]). Finally, after excluding objects with an exact number of 100 points in the light curves, our sample contained 283 light curves with > 100 data points that were used for NP modeling.

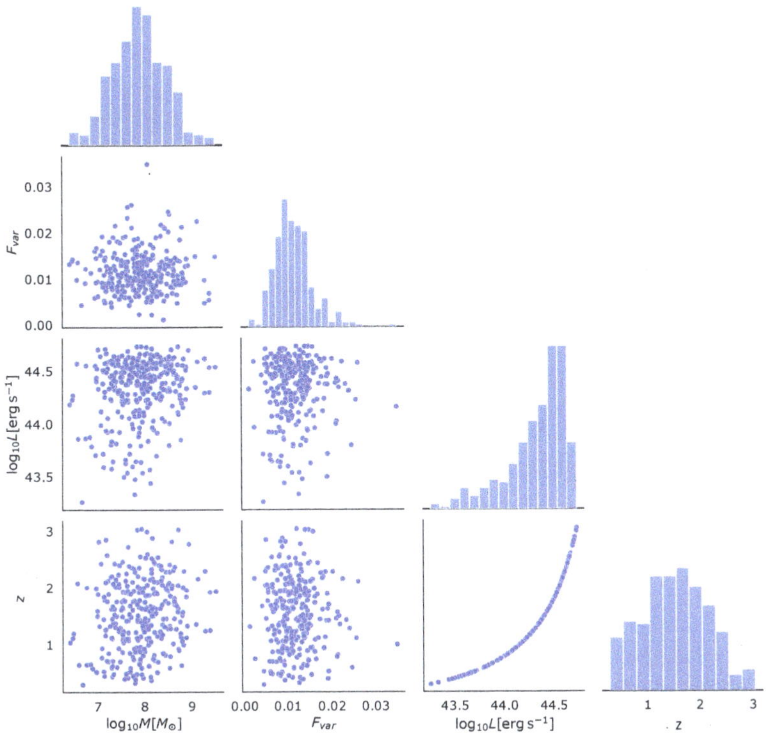

Figure 4. Corner plot of the probability distribution of quasars parameters in selected Cluster 36. Because the parameter distribution is multi-dimensional (only four are seen here), the information is unfolded into a succession of 1D and 2D representations. The histograms of the marginalized probability for each parameter are given on the diagonal. The scatter plots for all pairwise parameter combinations are given on the off-diagonal.

3. Methods

In this section, we provide the motivation and description of the computational model.

3.1. Motivation

The stochastic variability seen in quasar flux time series is thought to be caused by emissions from an accretion disc with local 'spots' that contribute more or less flux than the disc's mean flux level. These spots appear at random and dissipate over a specific physical time scale [51]. Because the spots do not dissipate instantly within the disc, some long-term correlations may exist, which can be described by power spectrum density (PSD $\sim f^{-2}$, where f is frequency) consistent with the Autoregressive (1) model (AR(1)), or damped random walk, or simplest form of Gaussian process characterized by the relaxation time, and the variability on timescales much shorter than relaxation time (see [17]). Some ground-based studies [8,52] show that AGN light curves could have PSD slopes steeper than AR(1) PSD on very short time scales, indicating that the damped random walk process oversimplifies optical quasar variability (see [8,53]). Moreover, Kasliwal et al. [42] developed the damped power-law (DPL) model by generalizing the PSD of AR(1) as $\sim 1/f^{\gamma}$. If $\gamma < 2$, the process exhibits weaker autocorrelation on short time scales than the AR(1), resulting in a less smooth time series. When $\gamma > 2$, the process exhibits stronger autocorrelation on short time scales than the AR(1), resulting in a smoother time series. However, the light curves show a wide variety of different types of behavior, even superposition of at least two features (e.g., stochastic+flare [42]). Ruan et al. [54] modeled blazar variability flare-like features with an AR(1) process using observed 101 blazar light curves from the Lincoln Near-Earth Asteroid Research (LINEAR) near-Earth asteroid survey. Moreover, Kasliwal et al. [42] tried to model flare-like features in Kepler light curves with DPL and AR(1) and point out that both models are unable to model flare-like features. Besides the variety of the strength of correlation in the AGN light curves, and different topologies of light curves, the next issue that should be taken into account is the cadence gaps (long time ranges without observation) of variable size. As we want to predict data in these gaps, it is important that we do not introduce any additional relation that could be reflected in PSD (see [29]). Various types of neural processes can be considered, such as attentive processes that can introduce attentive mechanisms for correlation among data points. However, due to large cadence gaps in quasar light curves, in Paper I, we presented the successful application of the conditional neural process (CNP, general Neural process which does not introduce correlation), combining the neural network and general Gaussian process's capabilities, to model tens of stochastic light curves with large gaps and without flares. Here, we show an upgraded version of the code, which is applied on the quasar light curves obtained from the largest database mimicking LSST survey (`LSST_AGN_DC`, containing about 40,000 quasars).

3.2. Conditional Neural Process

Here, we briefly summarize the conditional neural process description; for a detailed mathematical description, the reader is referred to the given literature. It is commonly accepted in the field of machine learning that models must be "trained" with a large number of examples before they can make meaningful predictions about data they have never seen before. However, there are several instances where we do not have enough data to meet this demand: acquiring a substantial volume of data may prove prohibitively expensive, if not impossible. For example, it is not possible to obtain homogeneous cadences of observations with any ground-based telescope, including the LSST. Nonetheless, there are compelling grounds to believe that this is not a side effect of learning. Humans are known to be particularly good at generalizing after only seeing a small number of examples, according to what we know ("few-shot estimate"). In current meta-learning terminology, NPs and GPs are examples of approaches for "few-shot function estimates" [33]. NPs, as opposed to GPs, are metalearners [4] (see [55]).

We assume the latent continuous-time light curve $(\mathbf{x},\mathbf{y})_{\mathcal{L}} = \left\{(x^l,y^l)\right\}_{l=1}^{L}$ with time instances $\mathbf{x}$ and fluxes $\mathbf{y}$, is a realization of a stochastic process, so that observed points are sampled from it at irregular instances $(\mathbf{x},\mathbf{y})_{\mathcal{O}} = \{(x^o,y^o)\}_{o=1}^{O})$ (see [56]), and which can be learned through neural processes, as it is a way to meta-learn a map from datasets to predictive stochastic processes using neural networks [55].

If we are given target inputs of time instances $(\mathbf{x}_{\mathcal{T}} = \{x^t\}_{t=1}^{T})$, and corresponding unknown fluxes $\mathbf{y}_{\mathcal{T}} = \{y^t\}^{T}{}_{t=1}$, we need a distribution over predictions $(p(\mathbf{y}_{\mathcal{T}}|\mathbf{x}_{\mathcal{T}};\mathcal{C}^{\sim}))$ [5], where $\mathcal{C}^{\sim} = \{(\mathbf{x}_{\mathcal{C}},\mathbf{y}_{\mathcal{C}}) = (x^c,y^c), c = 1,\dots C,\}$ are context points from the light curve (used for training the model). The $p(\mathbf{y}_{\mathcal{T}}|\mathbf{x}_{\mathcal{T}};\mathcal{C}^{\sim})$ distribution is called the stochastic process.

If we chose predictors at random from this distribution, each one would be a plausible way to fit the data, and the distribution of the samples would show how uncertain our predictions are. Therefore, the NP can be seen as using neural networks to meta-learn a map from datasets to predictive stochastic processes.

Generally, NPs are constructed to firstly map the entire context set to a representation R, $Enc_\theta : \mathcal{C}^{\sim} \to R$ [6], using an encoder $\text{Enc}_\theta(\mathcal{C}^{\sim}) = \rho\left(\sum_{c=1}^{C}\phi(x^{(c)},y^{(c)})\right)$, where ρ,ϕ are defined by neural network [55]. The sum operation in the encoder is a key as it ensures that the resulting R "resides" in the same space regardless of the number of context points $\mathcal{C}$. The predictive distribution at any set of target inputs $\mathbf{x}_{\mathcal{T}}$ is factorized and conditioned on lower dimensional representation R. Keeping this in mind, one can write that $p_\theta(\mathbf{y}_{\mathcal{T}}|\mathbf{x}_{\mathcal{T}};\mathcal{C}^{\sim}) = \prod_{t=1}^{T} p_\theta(y^{(t)}|x^{(t)},R)$. In the next step, NP calls the decoder, Dec_θ, which is the map parametrizing the predictive distribution using the target input $x^{(t)}$ and the encoding of the context set (R). Typically the predictive distribution is multivariate Gaussian, meaning that the decoder predicts a mean $\mu^{(t)}$ model and a variance $(\sigma^{2(t)})$ [55].

Specifically, the scheme of the particular member of NP called the conditional neural process (CNP [33]) is given in Figure 5. Each pair in the context set is locally encoded by a multilayer perceptron:

$$R^c = \text{Enc}_\theta(\mathcal{C}^{\sim}) = \text{MLP}\left([x^{(c)};y^{(c)}]\right) \tag{5}$$

Here, we discuss the main difference between the CNP and other NP: the local encodings R^c are then aggregated by a mean pooling to a global representation R:

$$R = \text{Enc}_\theta(\mathcal{C}^{\sim}) = \frac{1}{C}\sum_{c=1}^{C}\text{MLP}\left([x^{(c)};y^{(c)}]\right) \tag{6}$$

Finally, the global representation R is fed along with the target input x^t into a decoder MLP to yield the mean and variance of the predictive distribution of the target output:

$$(\mu^{(t)},\sigma^{2(t)}) = \text{Dec}_\theta(R,x^{(t)}) = \text{MLP}\left([R,x^{(t)}]\right) \tag{7}$$

The encoder consists of 1 hidden layer (dimension 2×128) that encodes the features of time instances, followed by 3 hidden layers (dimension 128×128 that locally encode each feature-value pair (time instances, magnitudes) with final activation ReLU layer. The decoder has a 4-hidden-layer MLP of the dimension (128×128) that predicts the distribution of the target value, with the last layer consisting of `softmax` activation function. We note that the number of layers, batch sizes, learning rates, and optimizers are chosen as a balance between the hyperparameters found in the literature where the multilayer approach is more feasible (see [13]) for our experimentation. The aim of the training CNP is to minimize the negative conditional log probability (or loss) (a more detailed explanation is given in [33,35]):

$$\mathcal{L} = -\frac{1}{N}\sum_{\mathbf{y}_{\mathcal{T}}\in\mathcal{T}}\log(p(\mathbf{y}_{\mathcal{T}}|\mathbf{x}_{\mathcal{T}};\mathcal{C}))$$

where $p(\mathbf{y}_\mathcal{T}|\mathbf{x}_\mathcal{T};\mathcal{C})$ defines a posterior distribution for target values [7] over functions that could be fitted through observed data points, and N is a cardinality of a randomly chosen subset of observations used for conditioning. An increase in the log-probability indicates that the predicted distribution better describes the data sample statistically [8].

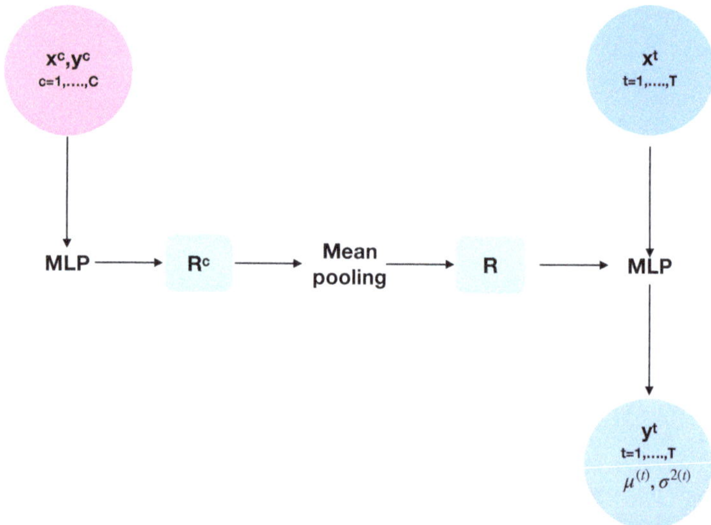

Figure 5. Forward pass (left to right) computational graph of the conditional neural process (see [55]). Contexts (observed points) are given in a red circle. MLP is a multilayer perceptron. R^c is local encoding of context points, whereas R is general encoding. For given target time instances x^t in a blue circle, the decoder, which is an MLP, produces the predictive distribution of the target output $(\mu^{(t)}, \sigma^{2(t)})$ and predicted values y^t.

The computational cost of prediction estimates for $\mathcal{T}$ target points conditioned on $\mathcal{C}$ context points with is $\mathcal{O}(\mathcal{T}+\mathcal{C})$, which is more efficient than for GP [33,55].

Our initial adaptation of the CNP for the purposes of the LSST quasars light curve modeling was described in Čvorović-Hajdinjak et al. [35], Breivik et al. [57]. Using the above concept, we fully developed an NP-module for LSST quasar light curve modeling, which is upgraded to `pytorch` and refactorized into 6 subunits:

1. Model architecture;
2. Definition of dataset class and collate function;
3. Metrics (loss and mean squared error—MSE);
4. Training and calculation of training and validation metrics (loss and MSE);
5. Saving model in predefined repository;
6. Upload of the trained model so that prediction can be performed anytime.

The following features (from the above list) are new in contrast to the earlier version of the CNP module (see [35]): (2), the MSE given in (3), as well as (5), and (6).

4. Results and Discussion

In this section, we present and discuss the results of our procedures for 'gaining the knowledge from large data' comprising the training of CNP on strata (Section 4.1), CNP modeling of variability of light curves in strata (Section 4.2) and modified structure–function analysis of the observed and modeled light curves (Section 4.3).

4.1. Training of CNP

Following the prescription for splitting the data set into training, testing, and validating subsamples (see, e.g., [13]), we randomly divided the strata of u-band light curves (seen in Figure 5), into a training dataset with 80% of the total number of objects, a test dataset with 10% of the total number of objects, and a validation dataset with 10% of the total number of objects. The training, target, and validation time instances, originally given as modified Julian dates (MJDs), are transformed to a $[-2, 2]$ range alongside corresponding magnitudes and measured errors [9]. The training, validating, and targeting light curves are given as tensors of size $128 \times N$, where 128 is a batch size and N is the corresponding number of epochs in the light curves [10]. We independently carried out the method of dividing the data and applying our algorithm to the training, validating, and testing data about 100 times.

During the training process, the training set was augmented by the addition of extra curves that were generated from the original by adding and subtracting measured uncertainty from observed points. The method of adding noise to neural network inputs during training has been known for a long time. Many theoretical studies have demonstrated that it allowed for increasing generalization capabilities of the network (e.g., [59,60]). Bishop [61] has shown that the use of this method was equivalent to Tikhonov regularization. Wang and Principe [62] showed that if this method was applied, the network also trained faster. Most often it is considered as one of the methods to avoid ANN overtraining (see [63]). Currently, this method is also used when training deep neural networks [64]. Noise can be introduced into a training neural network in four different phases: input data, model parameters, loss function, and sample labels (see [65,66]). For injecting the noise, it is necessary that the probability distribution from which noise is drawn corresponds to the real-world situation of observed data (see for application in astronomical light curves [67]). As photometric errors in observed light curves mostly follow Gaussian distribution (see, e.g., [19]), the added Gaussian noise ($N(0, \sigma_i)$) to the input training light curves magnitudes corresponds to estimated measurement error(σ_i) at each time step (see for application in astronomical light curves [67]). The performance metric values for the training dataset and the validation data set for 100 runs are shown in Figure 6. The lower training loss but higher training MSE compared to the validation loss and MSE can be attributed to the sensitivity of MSE to particularly the sharp peaks in the light curves. Switching to mean absolute error (MAE) led to a more typical behavior, with the training MAE lower than the validation MAE.

As already mentioned, as a balance between hyperparameters was found in the literature and our experimentation, we used the Adam optimization algorithm (see [35] and references therein) implemented in Python package `torch.optim` with a learning rate of 0.0001 (see [13,67]) and a batch size of 32. Adam is probably the most frequently used optimization algorithm for training deep learning models thanks to its adaptive step size, which in practice most often leads to decreased oscillations of the gradients and faster convergence. It combines the best aspects of the `AdaGrad` and `RMSProp` algorithms to provide an optimization that can handle sparse gradients on noisy problems as we encounter in quasar light curves (see [13,67]). Once again, we emphasize that the general model was introduced in Paper I, and here we provide its upgraded version, with application on quasar light curves strata (see [68]), such granular application can potentially help the understanding of physical properties of categories of objects (see [68]), as we also demonstrated here. Moreover, CNP is well generalizable on various data sets (strata) as it inherits GP's ability to determine the predictive distribution of data.

The left panel in Figure 6 shows that both the validation and train loss decrease rapidly until epoch 2000, after which both losses stabilize. The loss and MSE are expected to be high in the early epochs (since the network is initialized at random and the network's behavior differs from the desired one in the early epochs) and thus inconsequential. The larger extent of epochs is depicted for illustration purposes that justify the inclusion of early stopping criterion so that CNP is trained at epochs < 2000.

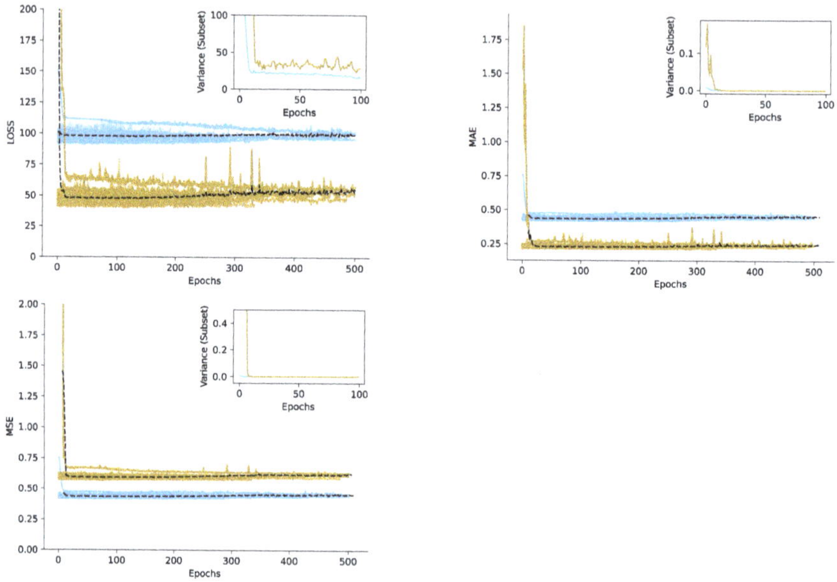

Figure 6. Loss (**the upper left panel**), mean absolute error (MAE, **the upper right panel**) and MSE (**the bottom left panel**) for the training (orange) and for the validation (blue) data set during 100 runs of training. The dashed black lines represent the mean values of metrics at each epoch over 100 runs. Training loss is estimated in each iteration within each epoch, but validation loss is obtained at the end of each epoch. **Inset plots show corresponding variances for training (orange) and validation (blue) data sets (a subset is shown for clarity).** The metrics are expected to be high in the early epochs (since the network is initialized at random and the network's behavior differs from the desired one in the early epochs). The behavior of metrics at a larger extent of epochs justifies the inclusion of early stopping criterion in the CNP.

Overfitting might be indicated by a decreasing training loss and an increasing or plateauing validation loss. However, our loss curves do not show this behavior. To prevent underfitting, we employed data augmentation with noise, and early stopping.

4.2. CNP Modeling of Quasar Variability

A detailed catalog of CNP models of light curves in our sample of 283 low variable and high redshift quasars ($F_{var} \sim 0.03$, $1 \leq z \leq 3$) is given in the Appendix A (see Figures A1–A32). Each plot shows the modeling performance of the CNP. The most notable quality is that the CNP catches the overall trends and major flare-like events. Both the autoencoder neural network constructed by (compare to Figure 7 in Tachibana et al. [13]) and CNP model the quasar temporal behavior purely based on the characteristics of the data without any prior assumptions. However, the main difference is that CNP also inherits the flexibility of stochastic process modeling such as GP. To assess the modeling accuracy of the CNP each plot presents MSE and loss values along the 95% confidence interval of the model. As we included observational errors in the CNP training process, the confidence bands for the regions of light curves dominated by points with larger errors are wider. Furthermore, we discovered that practically all light curves contain flare-like occurrences and even outliers, which also raise the obtained confidence band. We note that MSE is of comparable value to the MSE found in other studies (see, e.g., [67]). MSE is calculated on original (nontransformed data) and its value of $\sim$0.5 mag corresponds to $\sim$5%. Given that MSE represents variance, it is also more resistant to outliers (flares). Because loss is measured as the log of probability density, it is particularly susceptible to large gaps and outliers in our light curves. We emphasize that deep learning studies of astronomical time series report MSE (see [13,67]) frequently, so we will provide both MSE and loss.

The corner plot of mean square error, loss of each model fitting, and the corresponding mean magnitude and mean photometric error of observed light curves are given in Figure 7. According to the individual plots, a higher MSE (>0.5) is coupled with mean magnitudes in the range [20, 22] and mean photometric errors greater than 0.002. The tail of the marginal mean error distribution likewise contains these mean magnitudes.

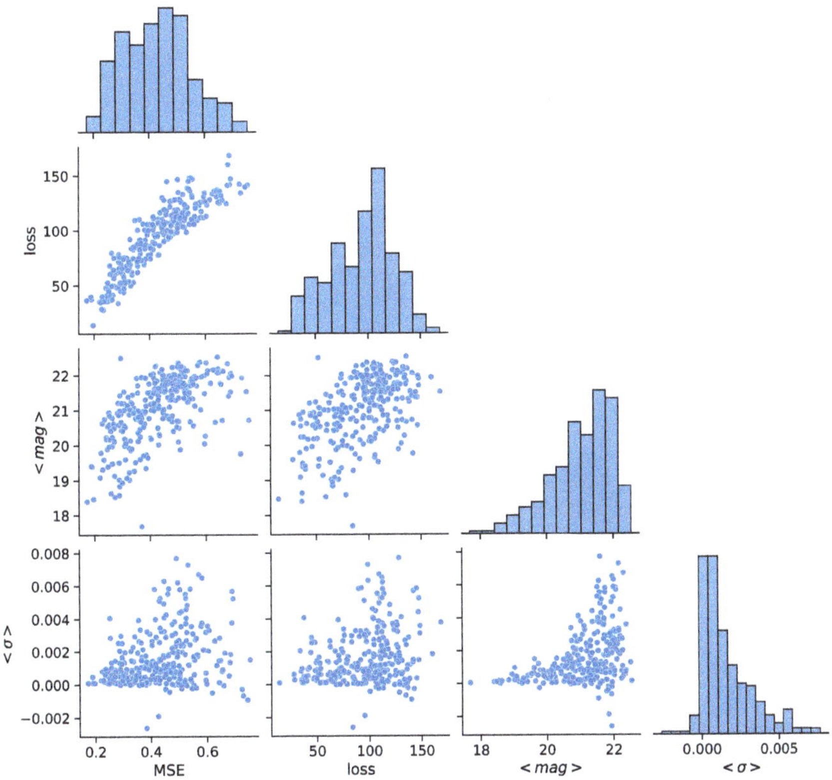

Figure 7. Corner plot of the probability distribution of modeling parameters (MSE, loss) for corresponding observed light curves parameters (mean magnitude $< mag >$, mean photometric error $< \sigma >$). Histograms of the marginalized probability for each parameter are given on the diagonal.

In our previous analysis, CNP was applied to the light curves with significant changes of gradients, having inhomogeneous cadences and infrequent flare-like features [35]. However, as demonstrated in [68] and here, deeper analysis of flare-like patterns is needed. We are testing additional CNP alternatives, but we are treading carefully to avoid the unintended introduction of relations that do not exist in data (see [29]).

4.3. Modified Structure–Function Analysis of Observed and Modeled Light Curves

The results from the neural network clustering suggest that there is no significant variability in the quasar light curves of the chosen sample. In order to test and further investigate this finding, we used the structure functions (SF, see, e.g., [14,68,69] and references therein). The SF could be defined as [70]:

$$S(\tau) = \frac{1}{N(\tau)} \sum_{i<j} (m(t_j) - m(t_i))^2 \tag{8}$$

where $m(t_i)$ is the magnitude measured at the epoch t_i, and summation runs across the $N(\tau)$ epochs for which is satisfied $t_j - t_i = \tau$. In addition to S defined above, we also use the two modified structure functions S_+ and S_- introduced by Kawaguchi et al. [14]. For S_+, the integration only includes pairings of magnitudes for which the flux increases $m(t_j) - m(t_i) > 0$, whereas for S_-, the integration only includes pairs of magnitudes for which the flux becomes dimmer $m(t_j) - m(t_i) < 0$.

Both modified structure functions measure the underlying asymmetry of the emission process as manifested in the light curves [14]. A comparison of modified structure functions could, in fact, disclose distinct mechanisms that are responsible for the variability of light curves, as was indicated in Hawkins [70]. To be more specific, the disc instability model shows itself in the form of an asymmetry in the light curves, such that the relation $S_- > S_+$ is observed at shorter time scales τ. On the other hand, transient events such as supernovae show up as the case that $S_+ > S_-$ as the time scale gets shorter. In the scenario of microlensing, which is a fundamentally symmetrical process in the setting of quasar variability, the two functions are indistinguishable, denoted by the equation, i.e., $S_- = S_+$ [70]. Even though it looks paradoxical for quasars, Hawkins [70] explained this with a model where the fluctuations in the light curves originating from the accretion disc become smaller with increasing luminosity, while the effects of microlensing become more pronounced at higher redshift, which for quasars typically means higher luminosity.

Figure 8 displays S_+ and S_-, and their relative difference normalized by standard SF (S, bottom panel) for observed quasar sample. Because of the regularity of large seasonal gaps in the observations, we were able to partition the data into five distinct time bins. In the top panel, both modified functions S_+ and S_- overlap. The normalized relative difference between them is also consistent with time symmetry.

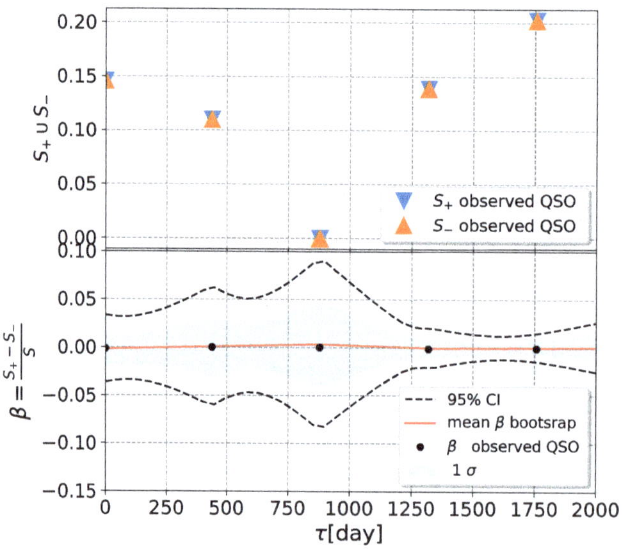

Figure 8. Time symmetry in the light curves of selected quasar samples. ***Top***: S_- and S_+ for the quasar sample are indicated by upward and downward pointing triangles, respectively. ***Bottom***: The normalized relative difference $\beta = \frac{S_+ - S_-}{S}$ for observed quasar sample (black dots), mean β (red line), 1σ(blue band), and 95% (dashed lines) confidence intervals as inferred from the bootstrapping quasar sample.

The striking feature of both modified structure functions is their zero value at the time lag of around 800 days. Going back to the overview plot of observed light curves in Figure 3, we could see that this time lag comprises combinations of vertical columns of data points after MJD = 52,000 such as the first and third vertical, the second and fourth, the third and the sixth, etc., which are more similar in variability than other combinations that could be made. Moreover, we note that in the range 52,000 $\leq MJD \leq$ 53,500 is an evidently smaller number of data points than for $MJD >$ 52,500, which affects the bootstrapping method as it has larger uncertainty.

We can further compare the CNP-modeled light curves of the selected quasar sample by testing for time asymmetries (see Figure 9). It is important to note that modeled light curves made it possible to construct structure functions using a finer bin grid that contained 20 bins. As we have seen, S_+ and S_- of modeled light curves are practically identical, which lends further support to the idea that microlensing is at play here. We can see that the CNP modeling does not modify the fundamental variability characteristics of the light curves that have been observed.

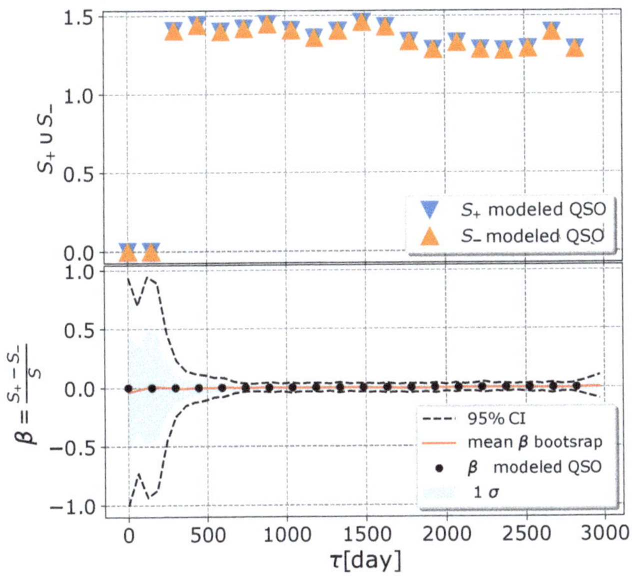

Figure 9. The same as Figure 8 but for CNP-modeled light curves.

For the structure functions of modeled light curves, the variability in the shortest timescale (<200 days) appears very different from that at a longer scale. This is an inevitable consequence of the fact that CNP is more uncertain about time lags corresponding to cadence gaps at epochs < 52,700 (see Figures 3 and A1–A32). We could also see that the structure function's confidence interval is rather large for these time lags.

We bring attention to the fact that our quasar sample was selected by our neural network algorithm in the absence of any other criteria (e.g., quasar parameter relationships), and that our clustering method based on the SOM could segregate objects with specific variability characteristics.

It is worth mentioning that microlensing can become obvious in multiple lensed quasar. This happens when some variations can only be observed in one image and appear to predominate over fluctuations that can be seen in all images over lengthy time scales. The light curves that are produced by microlensing can be simulated in a variety of different ways. Especially relevant to our collection of light curves are some interesting simulations by Lewis et al. [71]. In microlensing simulations, the source size has a significant impact on the look of the light curves, which become smoother and more rounded as the source size increases [70,71]. Light curves of our quasar sample given in Figures A1–A32 share a striking similarity with Figures 2a and 3b from Lewis et al. [71] regarding dominating non-smoothed flare-like patterns.

According to the findings of Tachibana et al. [13], modified structure functions and $\beta(\tau)$ for their sample of observed quasars exhibit asymmetry. They also noted that their simulated light curves as damped random walk (DRW) processes do not exhibit any substantial deviation from symmetric processes when compared to the bounds of confidence intervals. On the other hand, when we move deeper into the confidence interval, the values of the modified structure functions and $\beta(\tau)$ begin to fluctuate. Additionally, their simulated DRW light curves do not reflect the many flare-like events along the time baseline of light curves that are observed in our sample.

Based on the face value of these results, we find that the photometric variability of this sample of quasars could be explained with the microlensing model [70]. A justification that might be called is a model in which the fluctuations in the emission coming from the accretion disc becomes smaller as the luminosity of the object increases, but the effects of microlensing become more pronounced at higher redshifts, which for quasars typically coincide with higher luminosities (see [70]). We also note that low variability corresponds to high bolometric luminosity. The bolometric quasar luminosity is closely tied to the accretion rate of the SMBH. A bolometric quasar luminosity function (QLF) was constructed by Hopkins et al. [72] and updated for the bolometric QLF at $z = 0 - 7$ by Shen et al. [73].

Importantly, the flare-like patterns remained present in g and r curves (see an example in Figure 10). This is a hint that the flare-like events are achromatic since the gravitational bending of light is independent of the frequency of the radiation [74].

Nonetheless, it may be too soon to conclude that microlensing is the only plausible explanation. This is due to the fact that photometric monitoring for a period of four years is not yet sufficient to construct meaningful statistics based on the cadences that are available. Extreme optical extragalactic transients caused by explosion of supernovae, tidal disruption events around dormant black holes, rare blazars, stellar-mass black hole mergers in quasars disks, and intrinsic accretion outbursts in quasars may also be possible mechanisms behind observed flares (see [75]).

The discovery of such a distinct cluster of quasars in the LSST AGN data challenge database that makes ~0.8% could have interesting implications.

Assuming that the LSST will harvest $n \sim \mathcal{O}(2 \times 10^7 - 10^8)$ quasars (see [23] and references therein), a very simple extrapolation indicates that future LSST data releases may contain a population of microlensed quasars that is not negligible ($N \sim \mathcal{O}(10^5 - 10^6)$). Interestingly, the microlensing duration should be shorter in the X-ray (several months) than in the UV/optical emission range [several years, see [76]). Moreover, cosmologically distributed lens objects may contribute significantly to the X-ray variability of high-redshifted QSOs ($z > 2$, [16]). Moreover, we could estimate N assuming the microlensing rate Γ, i.e., the number of events expected to be detected per quasar and per year. Γ was evaluated for particular quasar J1249+3449 as $\Gamma \sim 2 \times 10^{-4}(m/0.1M_{\odot})^{-0.5}$ events per quasar per year, where m is the lens mass (see [74]). The value of parameter m could be in the range of $(0.01 - 0.5)M_{\odot}$, if the host galaxy of quasar has stars with velocities 200–400 km s^{-1}, and a source lens distance in the range 1–10 kpc. Even if we caution that Γ was derived for a particular AGN [74], we could use it as a proxy to estimate within a time interval Δt the total number N of expected microlensing events as simply $N \sim \epsilon \Gamma_{total} \Delta t$ (see [77]). For simplicity, we assume that for maximal survey efficiency $\epsilon \sim 1$, if n sources are monitored over a time interval Δt the total rate is $\Gamma_{total} \sim n \times \Gamma$. For Δt, we could expect to be within the range of weeks up to decades [78]. For example, Hawkins [15] found that if a lower limit of the time scale of ~24 years was supposed to be caused by microlensing, it would correspond to a minimum mass for microlensing bodies of $0.4M_{\odot}$ [15]. Taking into account that S_+ and S_- are more certain in our study for time scales >800 days, and since the LSST operation time is ~10 yr, we estimated that the number of microlensing quasars in the LSST data releases may be $N \sim \mathcal{O}(10^4 - 10^5)$. Because of these assumptions, our estimates are most likely an upper bound on the number of microlensing quasars produced by LSST.

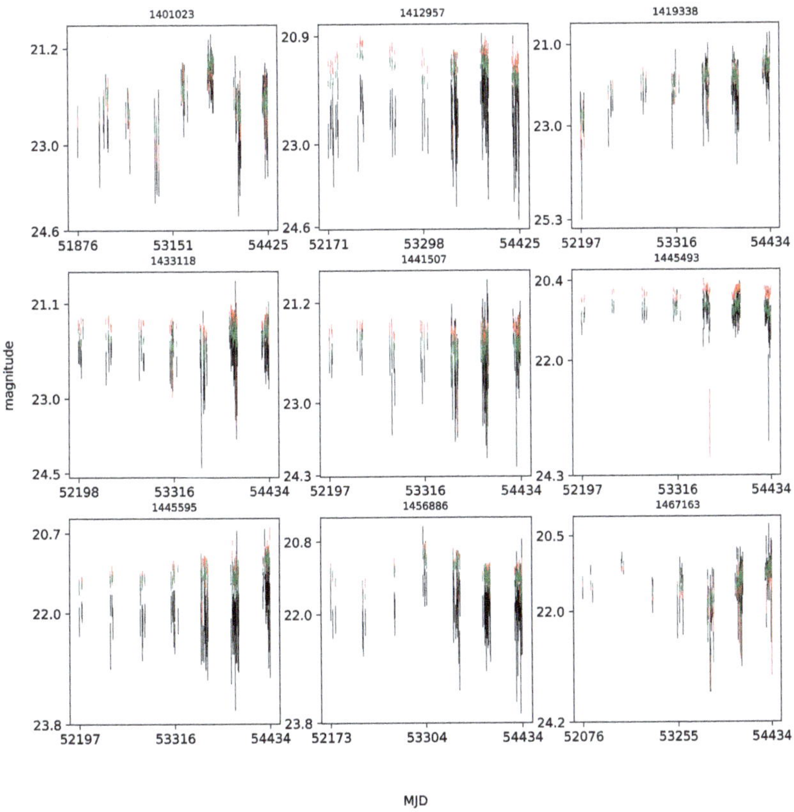

Figure 10. Comparison of presence of flare-like patterns in u (black), g (green) and r (red) bands. Object identification number is given in the corresponding plot title.

Nonetheless, in order to extract the physical properties of variability, the microlensing quasar population should be handled independently during the analysis process of the LSST quasar data, particularly the detection of binary candidates. Furthermore, it is probable that this population will have an influence on how quasars are classified in the LSST data pipelines.

5. Conclusions

In this follow-up study, we presented an improved version of a conditional neural process (CNP, Paper I) that was embedded in a multi-step approach for learning (stratification of light curves via neural network, deep learning of each stratum, statistical analysis of observed and modeled light curves in each stratum) from large amounts quasars contained in the LSST Active Galactic Nuclei Scientific Collaboration data challenge database. The main observations are:

- Individual light curves of 1006 quasars having more than 100 epochs in LSST Active Galactic Nuclei Scientific Collaboration data challenge database exhibit a variety of behaviors, which can be generally stratified via the neural network into 36 clusters.

- A case study of one of the stratified sets of u-band light curves for 283 quasars with very low variability $F_{var} \sim 0.03$ is presented here. The CNP model has an average mean square error of $\sim$5% (0.5 mag) on this stratum. Interestingly, all of the light curves in this stratum show features resembling the flares. An initial modified structure-function analysis suggests that these features may be linked to microlensing events that occur over longer time scales of five to ten years.
- As many of the light curves in the LSST AGN data challenge database could be modeled with CNP, there are still enough objects having interesting features in the light curves (as our case study suggests) to urge a more extensive investigation.

With the help of this scientific case, we were also able to demonstrate the importance of CNP (along with other deep learning methods) for data-driven modeling in contexts where considerable samples of objects may have variability patterns that differ from the DRW. In the future, investigations of this nature should receive a greater amount of attention.

Author Contributions: Conceptualization, A.B.K., D.I., L.Č.P. and M.N.; LSST SER-SAG team members, L.Č.P., D.I., A.B.K. and M.N.; supervising and conceptualization of machine learning methodology, M.N., software design, M.N., A.B.K., N.A.M., I.Č.-H. and M.S.P.; supervising of mathematical analysis of light curves, M.K.; writing—original draft preparation, A.B.K., D.I., L.Č.P., M.S.P., M.N., N.A.M., I.Č.-H., M.K. and D.V.S., writing—review and editing, A.B.K., D.I., L.Č.P., M.S.P., M.N., N.A.M., I.Č.-H., M.K. and D.V.S. All authors have read and agreed to the published version of the manuscript.

Funding: A.B.K., D.I., L.Č.P., M.N. and M.K. acknowledge funding provided by University of Belgrade-Faculty of Mathematics (the contract №451-03-47/2023-01/200104), through the grants by the Ministry of Science, Technological Development and Innovation of the Republic of Serbia. A.B.K. and L.Č.P. thank the support by Chinese Academy of Sciences President's International Fellowship Initiative (PIFI) for visiting scientist. L.Č.P. and D.V.S. acknowledges funding provided by Astronomical Observatory (the contract 451-03-68/2022-14/200002), through the grants by the Ministry of Education, Science, and Technological Development of the Republic of Serbia.

Data Availability Statement: The data are available from the corresponding author upon reasonable request and with the permission of the LSST AGN Scientific Collaboration.

Acknowledgments: We sincerely thank Gordon T. Richards, and Weixiang Yu for their essential efforts in the construction of LSST AGN data challenge within the Rubin-LSST Science Collaborations. This work was conducted as a joint action of the Rubin-LSST Active Galactic Nuclei (AGN) and Transients and Variable Stars (TVS) Science Collaborations. The authors express their gratitude to the Vera C. Rubin LSST AGN and TVS Science Collaborations for fostering cooperation and the interchange of ideas and knowledge during their numerous meetings.

Conflicts of Interest: The authors declare no conflict of interest. The funders had no role in the design of the study; in the collection, analyses, or interpretation of data; in the writing of the manuscript; or in the decision to publish the results.

Abbreviations

The following abbreviations are used in this manuscript:

AGN	active galactic nuclei
LSST	Legacy Survey of Space and Time
LSST_AGN_DC	LSST AGN data challenge database
SMBH	Supermassive black hole

Appendix A. Catalogue of CNP Models of u-Band Light Curves

Figures A1–A32 show a collection of plots of the observed u-band and corresponding predicted light curves for selected 283 objects using a training set for the CNP model.

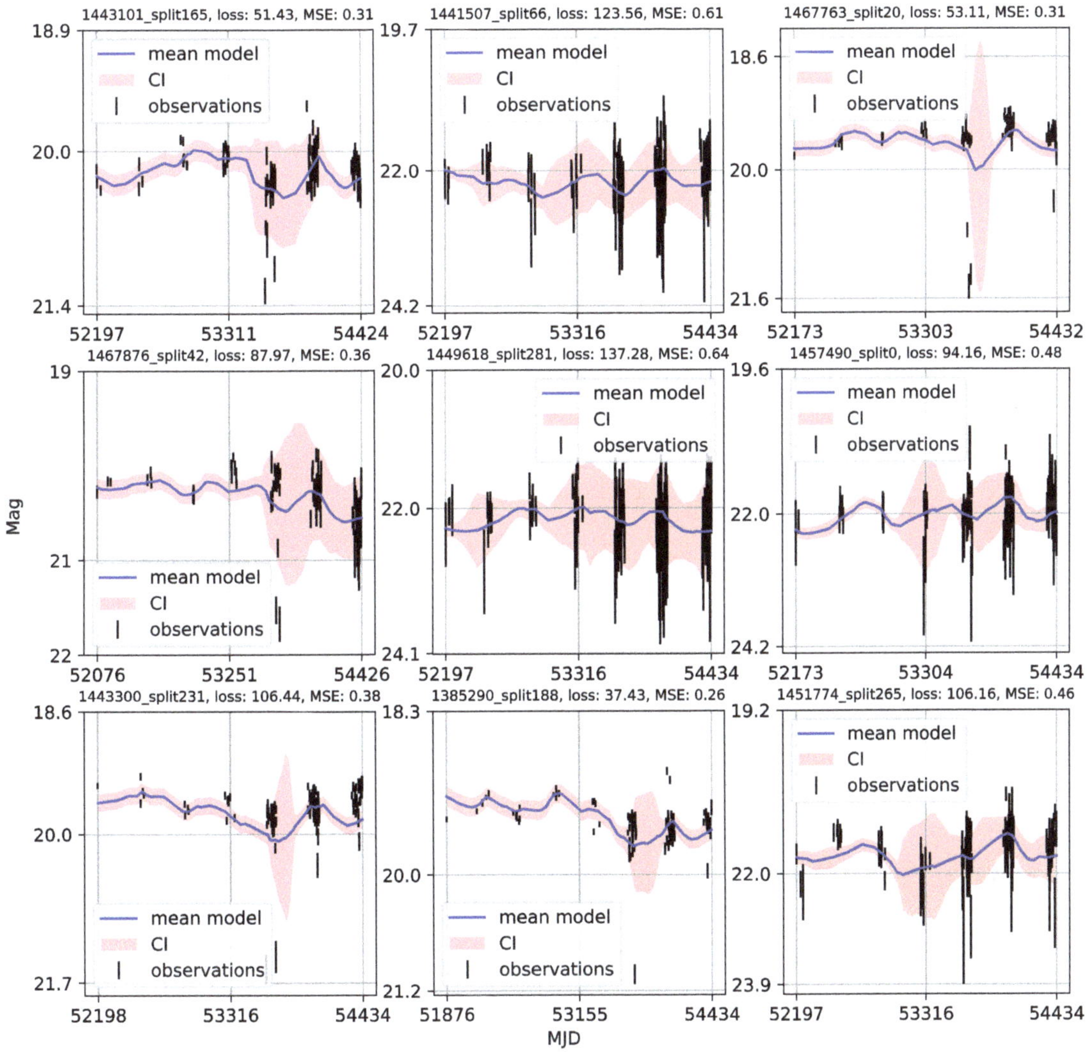

Figure A1. CNP modeling of light curves. The subtitle of each plot shows the object ID from the LSST AGN DC database with a tag indicating the iteration when it is selected for training, testing, or validation during the training process, the logarithmic value of probability loss, and the MSE value. Black error bars indicate observations with measurement uncertainty. Solid blue lines are model fits to the data. The red band represents the 1 σ confidence interval.

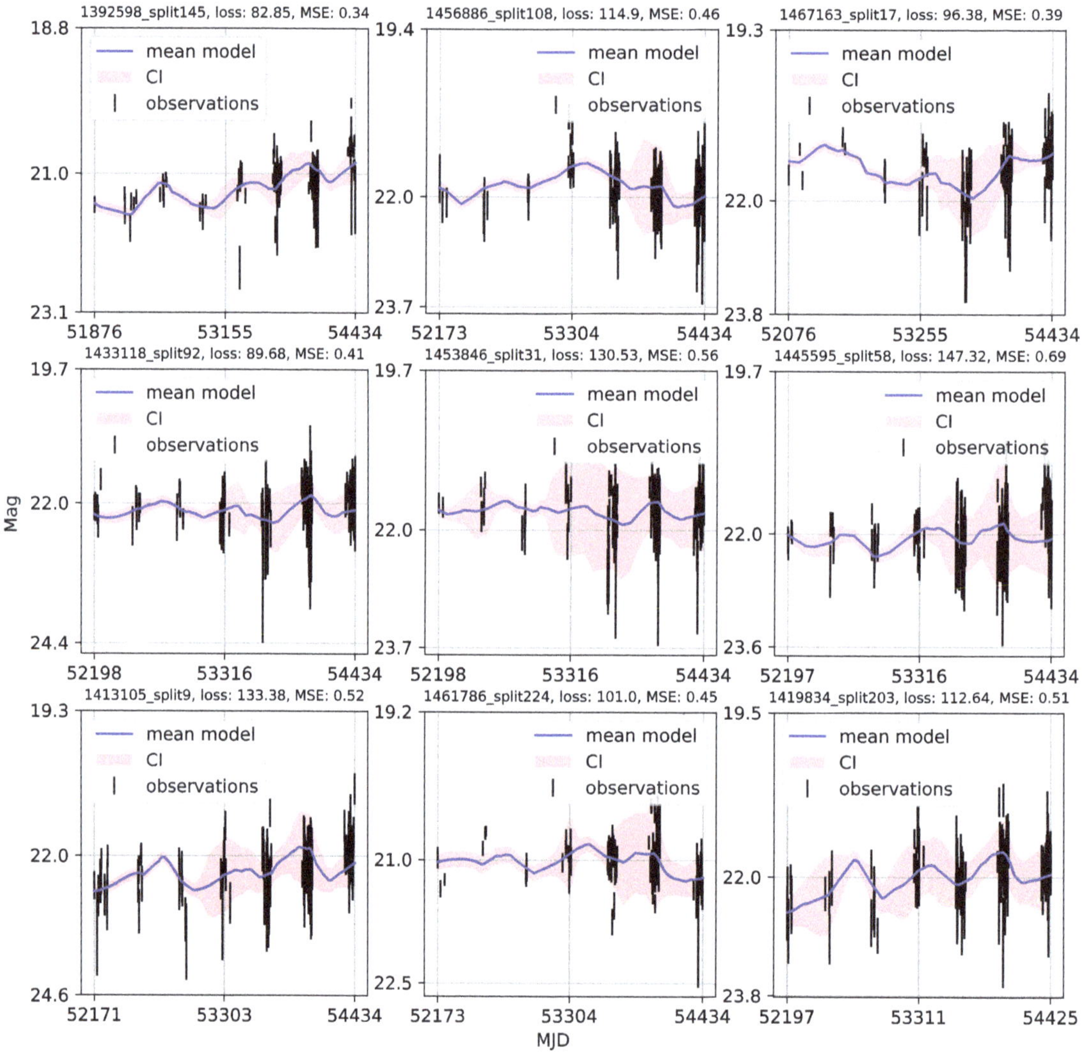

Figure A2. CNP models of light curves. The subtitle of each plot shows the object ID from the LSST AGN DC database with a tag indicating the iteration when it is selected for training, testing, or validation during the training process, the logarithmic value of probability loss, and the MSE value. Black error bars indicate observations with measurement uncertainty. Solid blue lines are model fits to the data. The red band represents the 1 σ confidence interval.

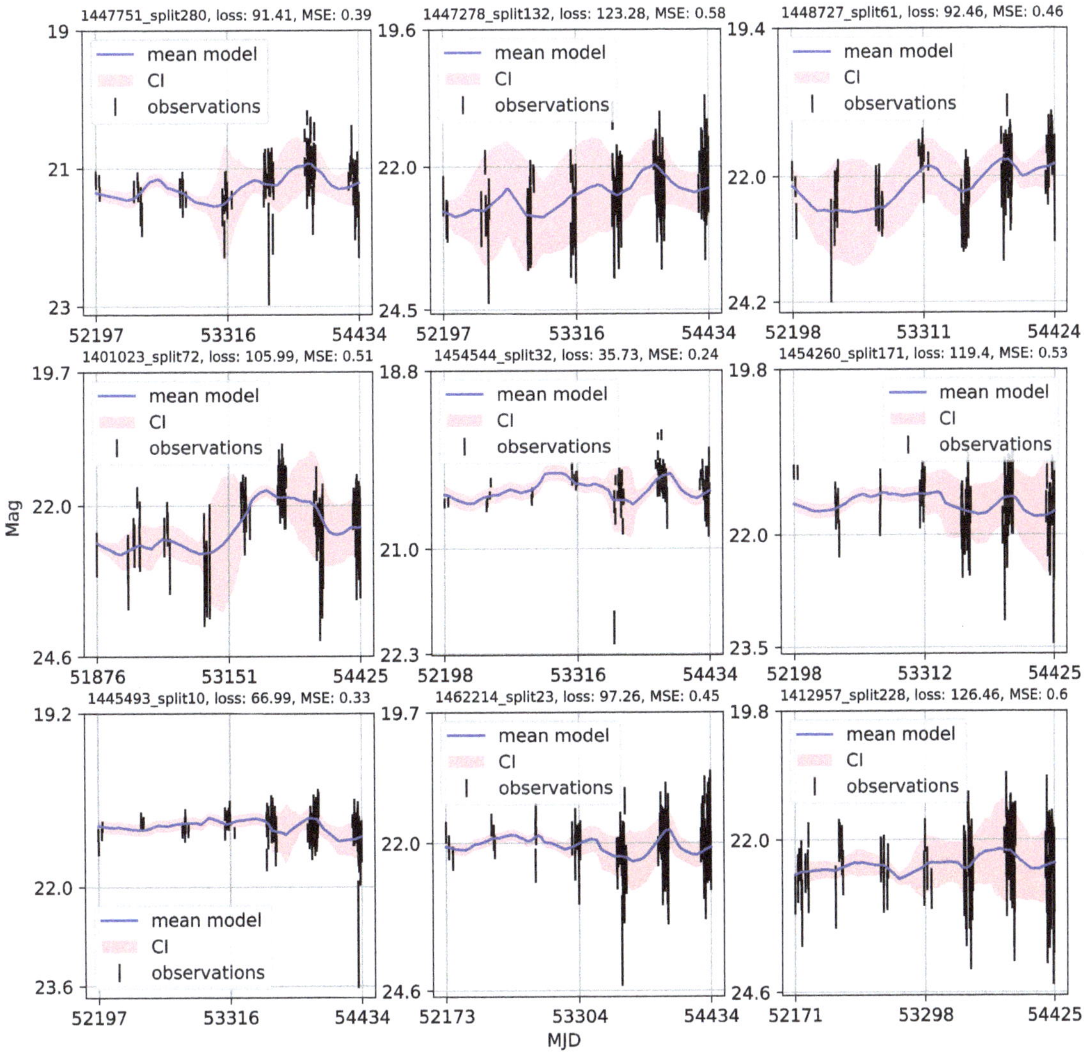

Figure A3. CNP models of light curves. The subtitle of each plot shows the object ID from the LSST AGN DC database with a tag indicating the iteration when it is selected for training, testing, or validation during the training process, the logarithmic value of probability loss, and the MSE value. Black error bars indicate observations with measurement uncertainty. Solid blue lines are model fits to the data. The red band represents the $1\,\sigma$ confidence interval.

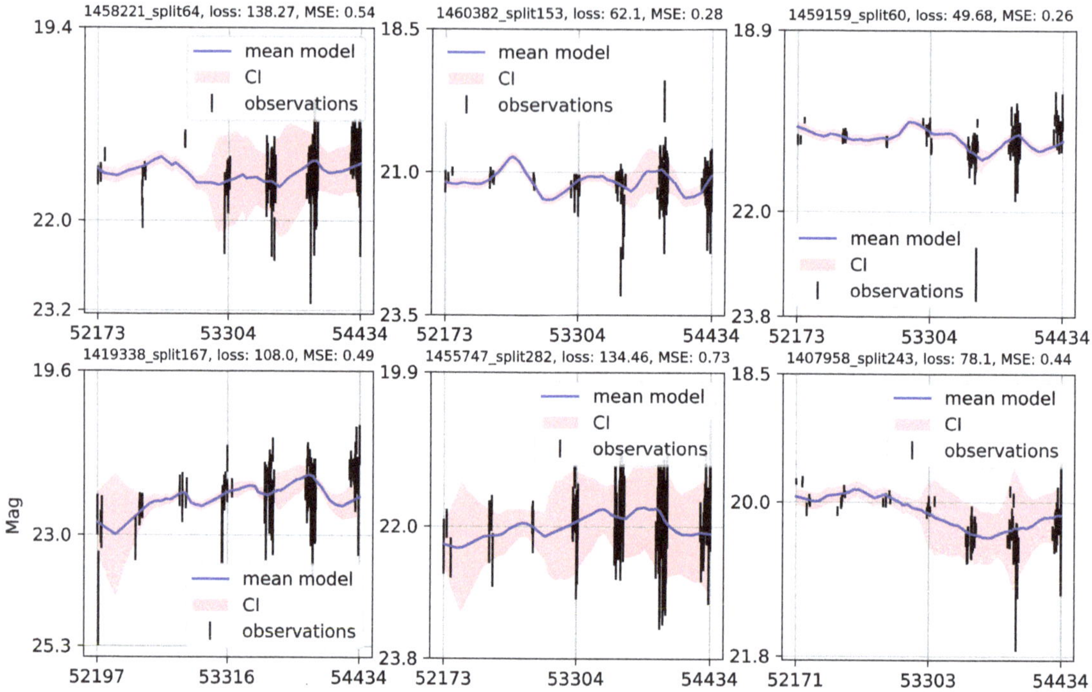

Figure A4. CNP models of light curves. The subtitle of each plot shows the object ID from the LSST AGN DC database with a tag indicating the iteration when it is selected for training, testing, or validation during the training process, the logarithmic value of probability loss, and the MSE value. Black error bars indicate observations with measurement uncertainty. Solid blue lines are model fits to the data. The red band represents the 1 σ confidence interval.

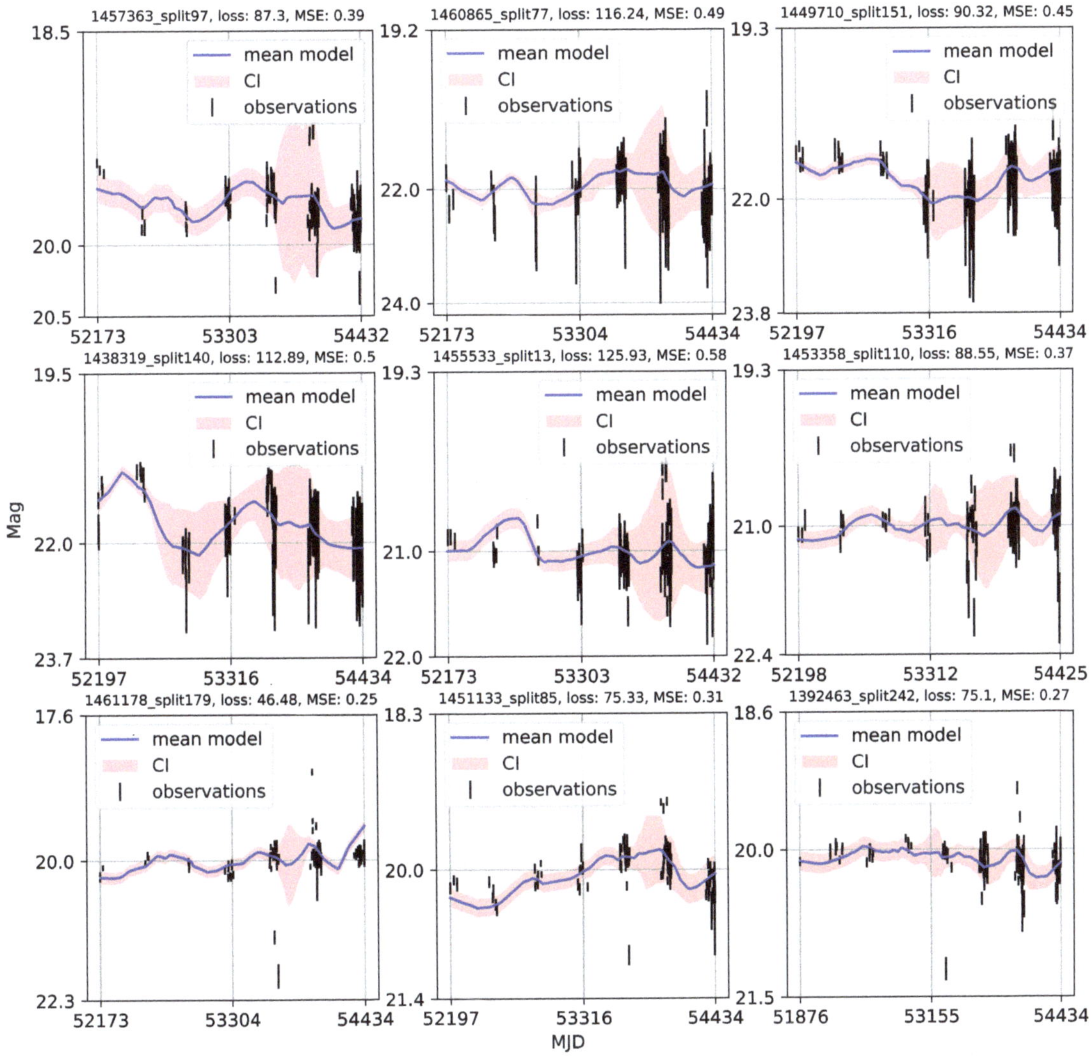

Figure A5. CNP models of light curves. The subtitle of each plot shows the object ID from the LSST AGN DC database with a tag indicating the iteration when it is selected for training, testing, or validation during the training process, the logarithmic value of probability loss, and the MSE value. Black error bars indicate observations with measurement uncertainty. Solid blue lines are model fits to the data. The red band represents the 1 σ confidence interval.

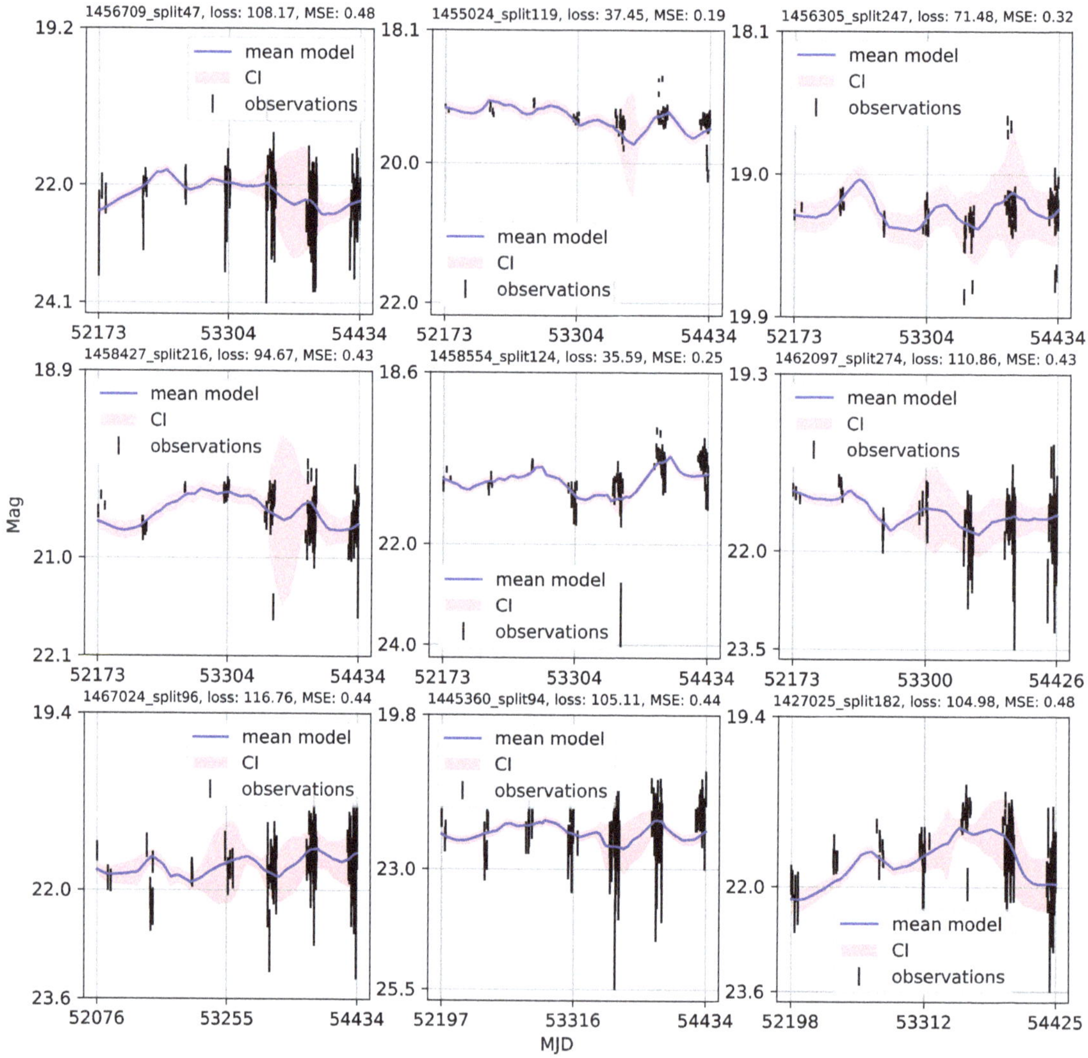

Figure A6. CNP models of light curves. The subtitle of each plot shows the object ID from the LSST AGN DC database with a tag indicating the iteration when it is selected for training, testing, or validation during the training process, the logarithmic value of probability loss, and the MSE value. Black error bars indicate observations with measurement uncertainty. Solid blue lines are model fits to the data. The red band represents the 1 σ confidence interval.

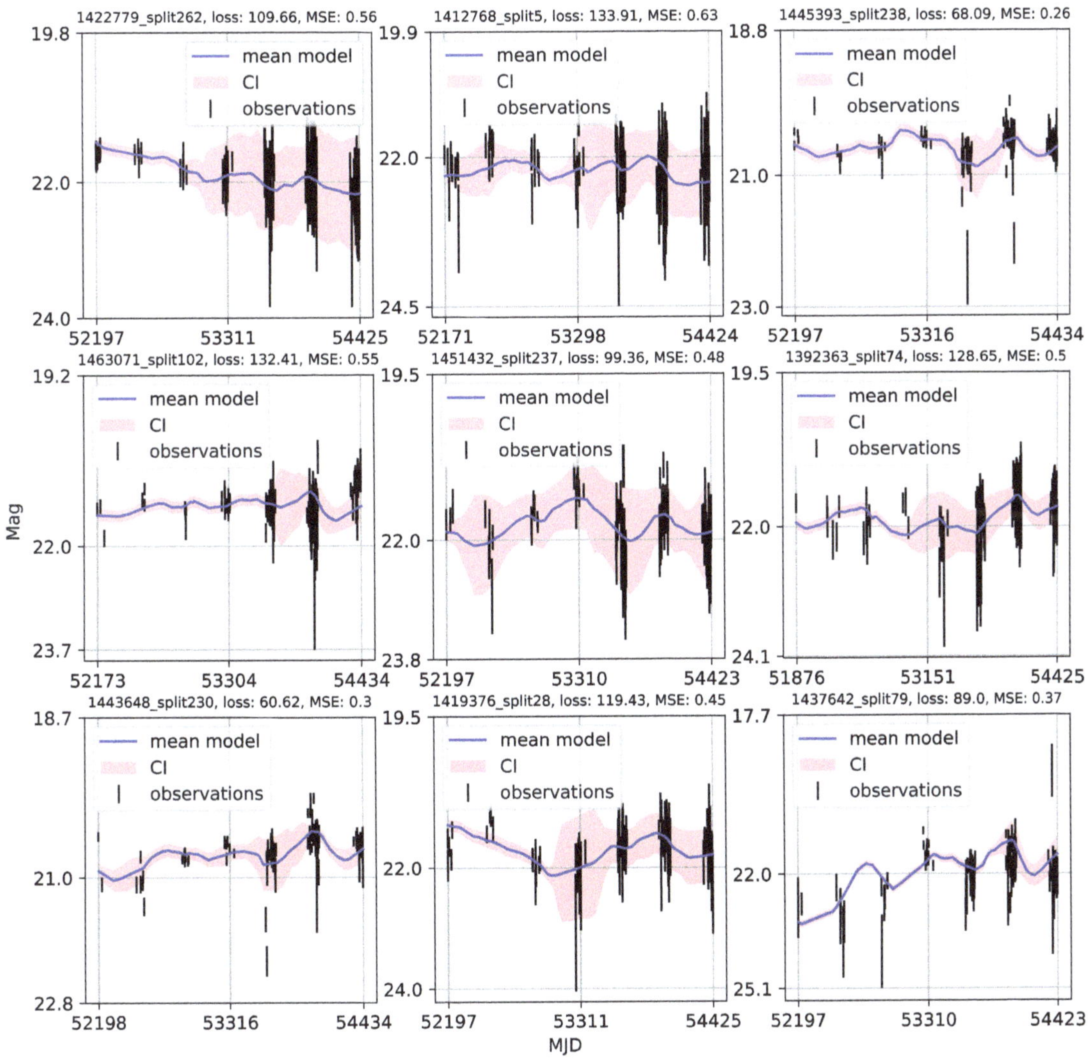

Figure A7. CNP models of light curves. The subtitle of each plot shows the object ID from the LSST AGN DC database with a tag indicating the iteration when it is selected for training, testing, or validation during the training process, the logarithmic value of probability loss, and the MSE value. Black error bars indicate observations with measurement uncertainty. Solid blue lines are model fits to the data. The red band represents the 1 σ confidence interval.

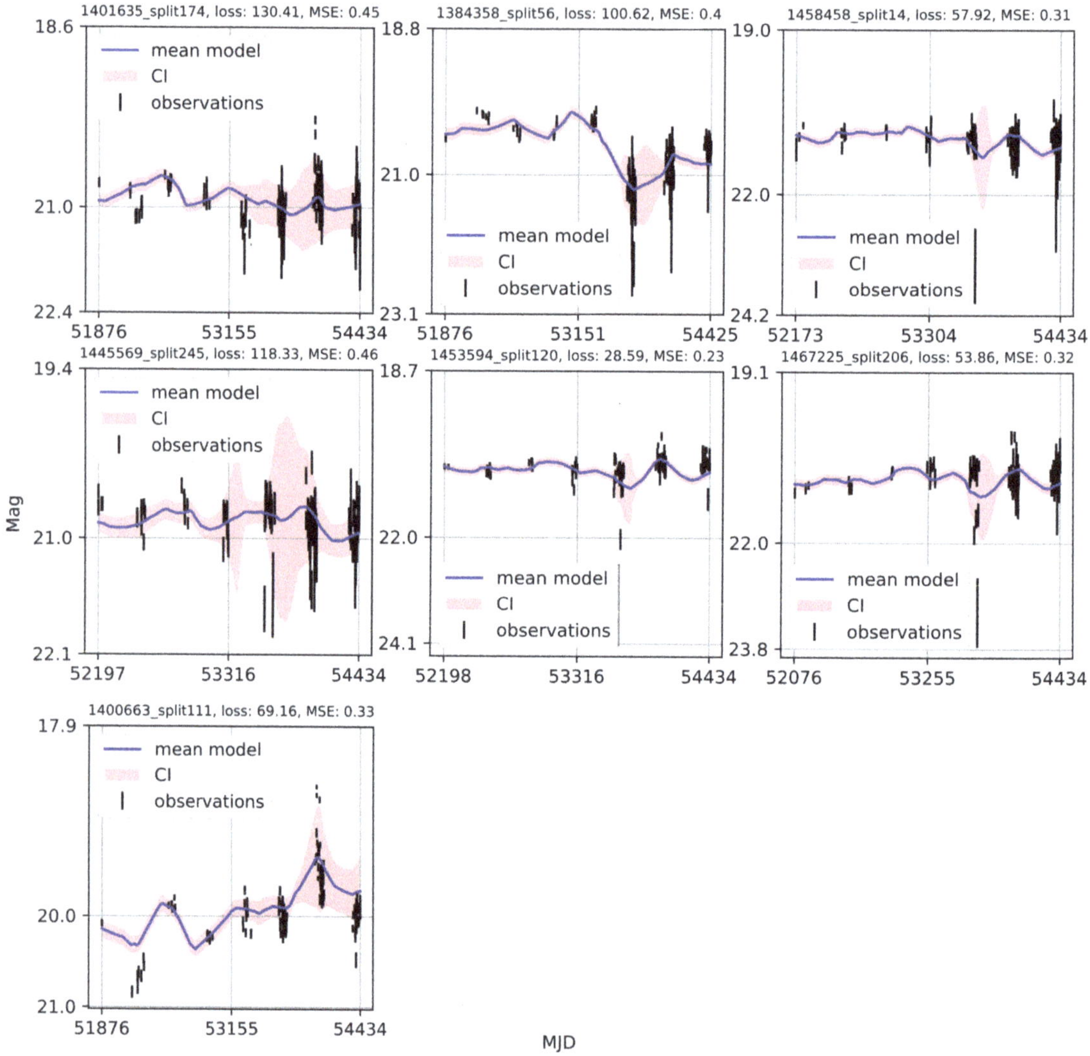

Figure A8. CNP models of light curves. The subtitle of each plot shows the object ID from the LSST AGN DC database with a tag indicating the iteration when it is selected for training, testing, or validation during the training process, the logarithmic value of probability loss, and the MSE value. Black error bars indicate observations with measurement uncertainty. Solid blue lines are model fits to the data. The red band represents the $1\,\sigma$ confidence interval.

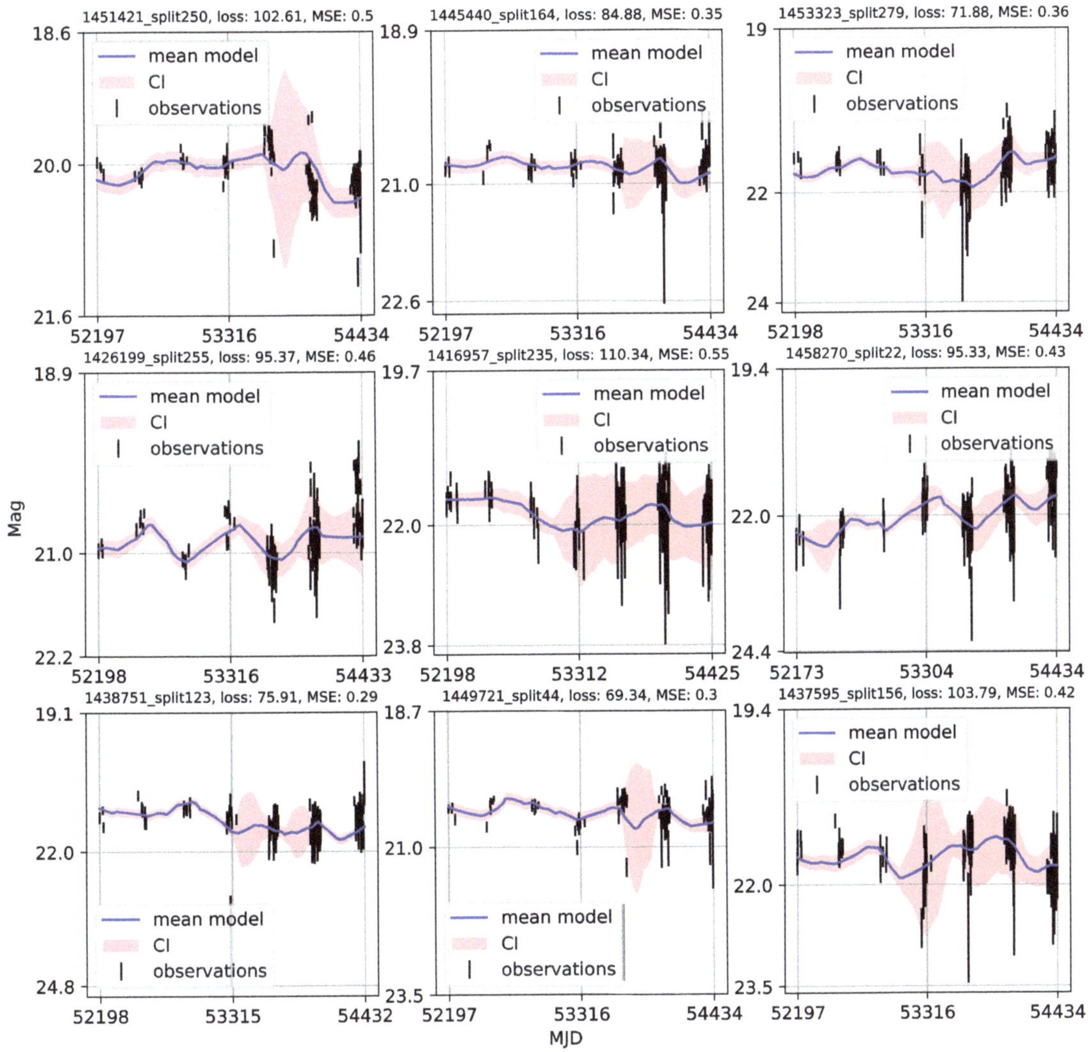

Figure A9. CNP models of light curves. The subtitle of each plot shows the object ID from the LSST AGN DC database with a tag indicating the iteration when it is selected for training, testing, or validation during the training process, the logarithmic value of probability loss, and the MSE value. Black error bars indicate observations with measurement uncertainty. Solid blue lines are model fits to the data. The red band represents the 1 σ confidence interval.

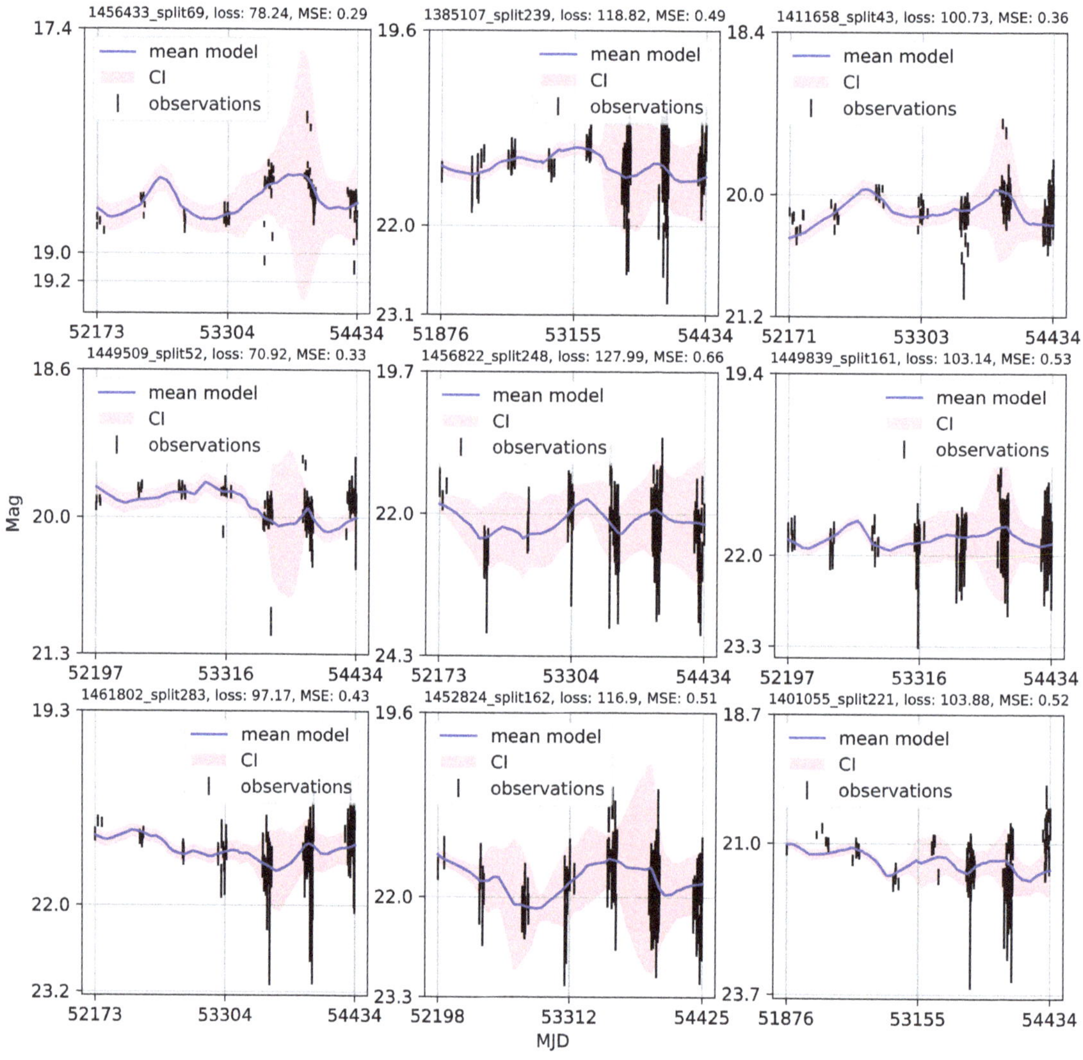

Figure A10. CNP models of light curves. The subtitle of each plot shows the object ID from the LSST AGN DC database with a tag indicating the iteration when it is selected for training, testing, or validation during the training process, the logarithmic value of probability loss, and the MSE value. Black error bars indicate observations with measurement uncertainty. Solid blue lines are model fits to the data. The red band represents the $1\,\sigma$ confidence interval.

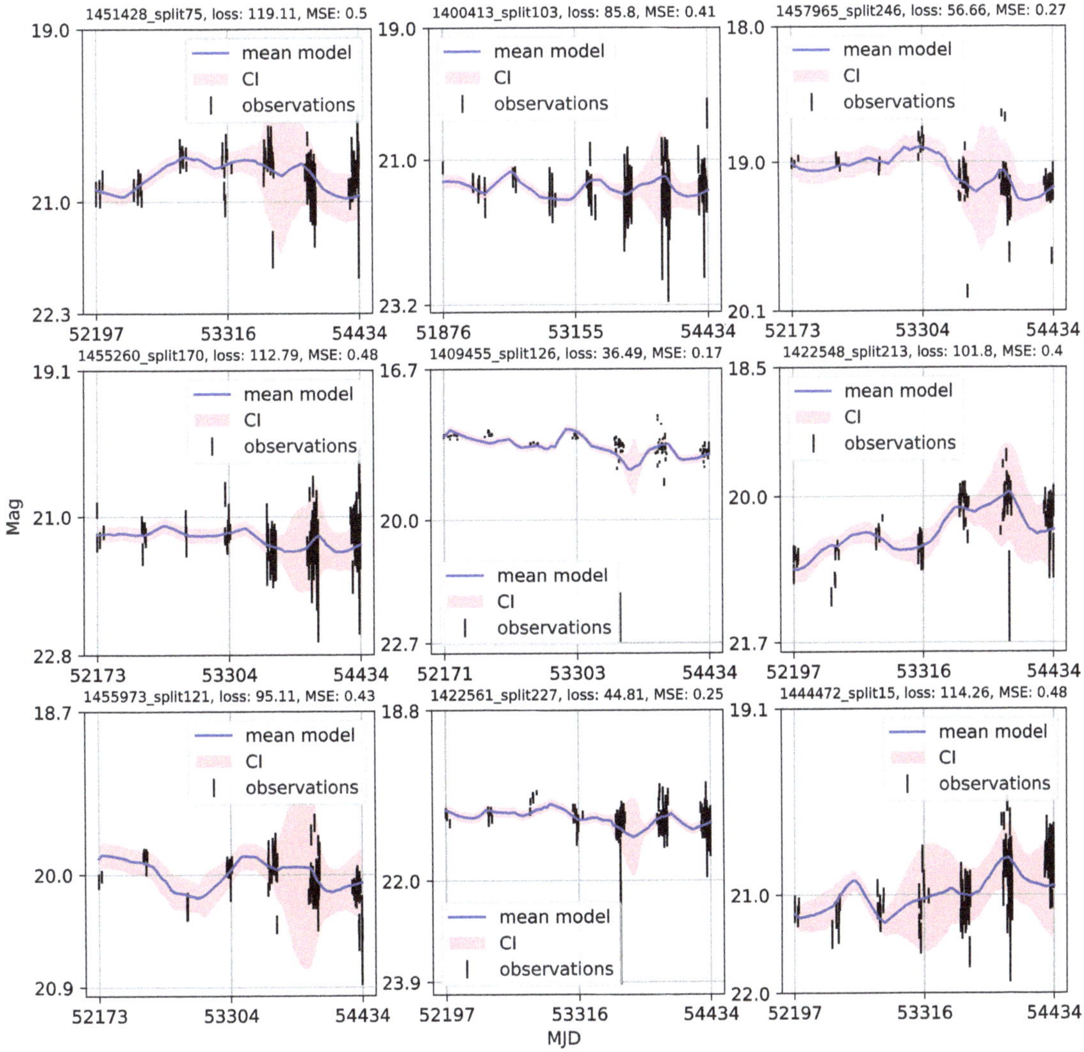

Figure A11. CNP models of light curves. The subtitle of each plot shows the object ID from the LSST AGN DC database with a tag indicating the iteration when it is selected for training, testing, or validation during the training process, the logarithmic value of probability loss, and the MSE value. Black error bars indicate observations with measurement uncertainty. Solid blue lines are model fits to the data. The red band represents the 1 σ confidence interval.

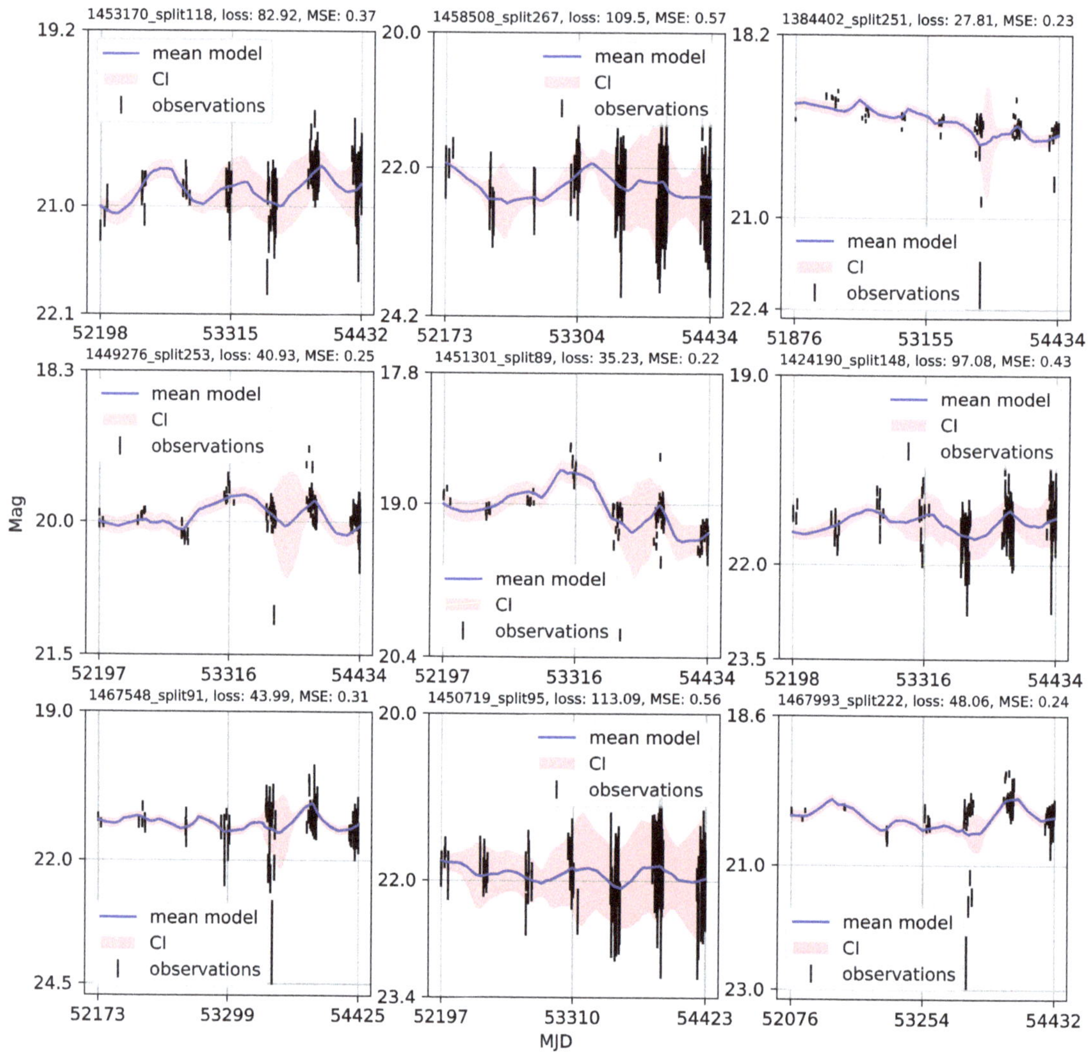

Figure A12. CNP models of light curves. The subtitle of each plot shows the object ID from the LSST AGN DC database with a tag indicating the iteration when it is selected for training, testing, or validation during the training process, the logarithmic value of probability loss, and the MSE value. Black error bars indicate observations with measurement uncertainty. Solid blue lines are model fits to the data. The red band represents the $1\,\sigma$ confidence interval.

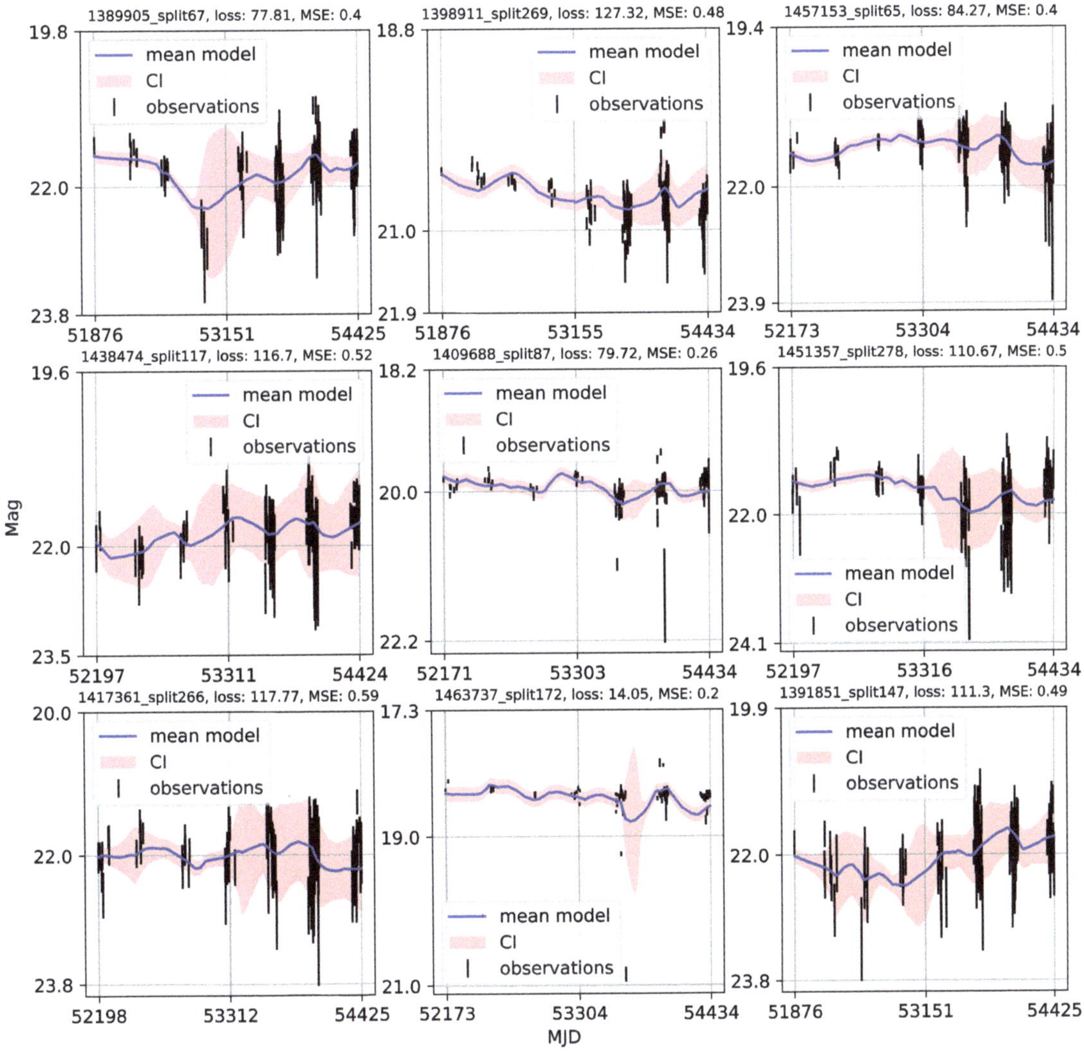

Figure A13. CNP models of light curves. The subtitle of each plot shows the object ID from the LSST AGN DC database with a tag indicating the iteration when it is selected for training, testing, or validation during the training process, the logarithmic value of probability loss, and the MSE value. Black error bars indicate observations with measurement uncertainty. Solid blue lines are model fits to the data. The red band represents the 1 σ confidence interval.

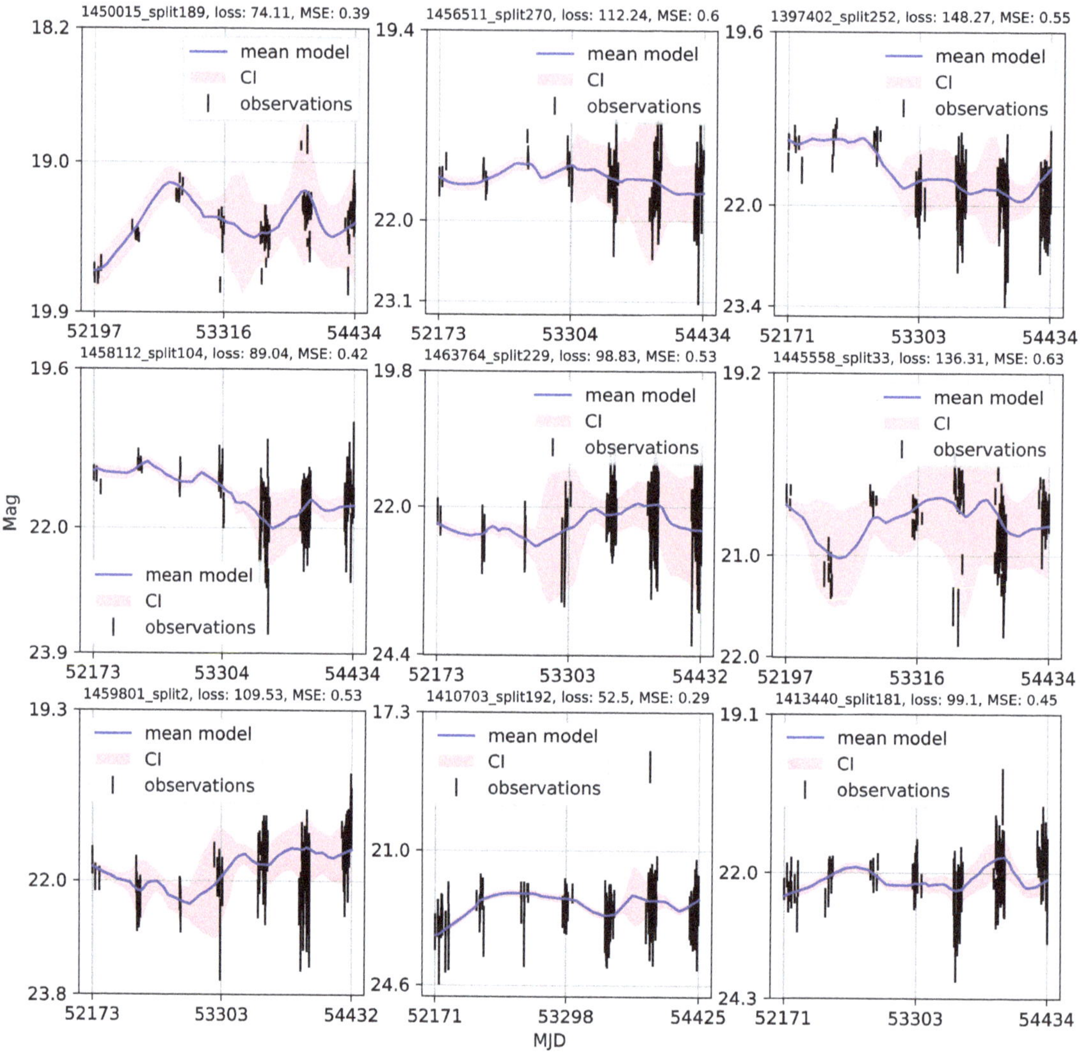

Figure A14. CNP models of light curves. The subtitle of each plot shows the object ID from the LSST AGN DC database with a tag indicating the iteration when it is selected for training, testing, or validation during the training process, the logarithmic value of probability loss, and the MSE value. Black error bars indicate observations with measurement uncertainty. Solid blue lines are model fits to the data. The red band represents the 1 σ confidence interval.

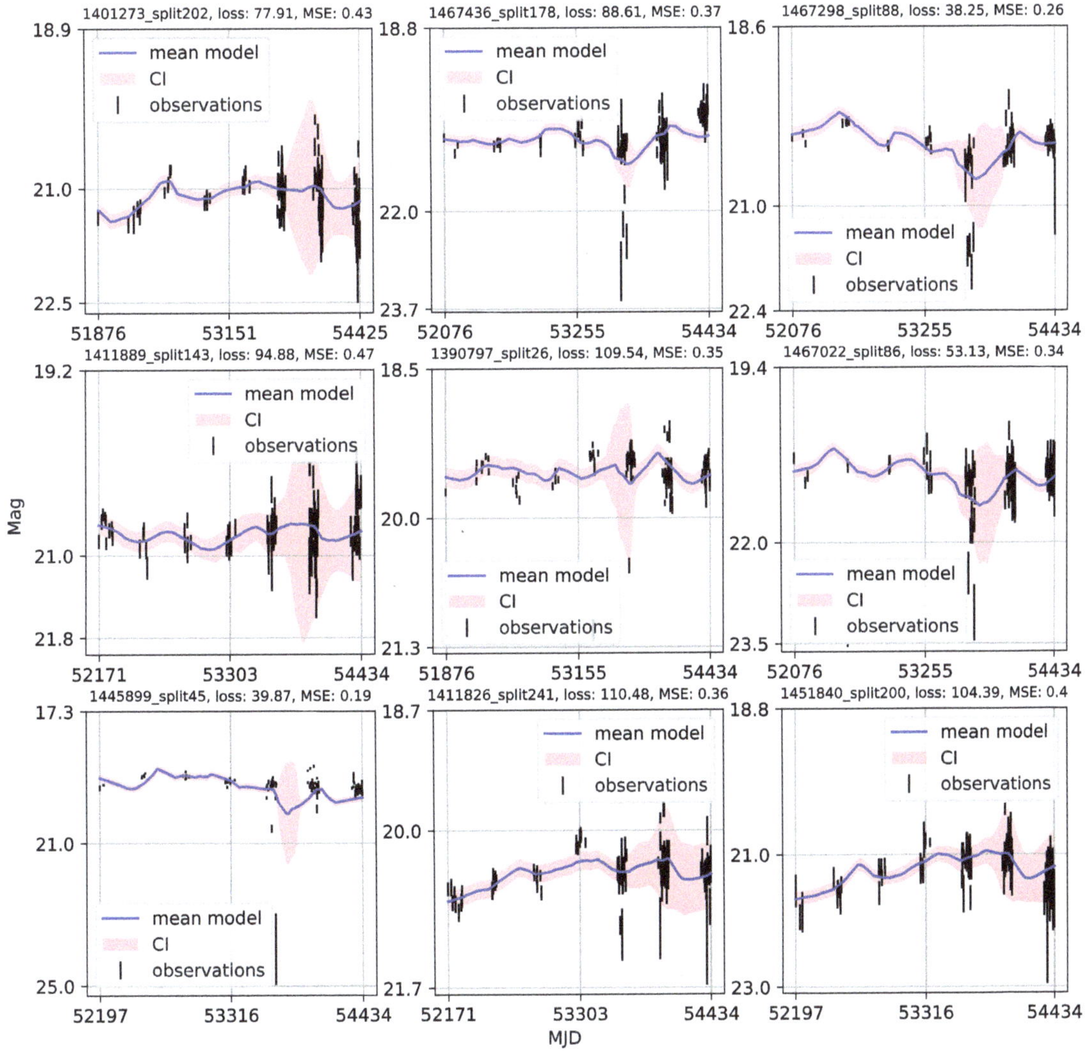

Figure A15. CNP models of light curves. The subtitle of each plot shows the object ID from the LSST AGN DC database with a tag indicating the iteration when it is selected for training, testing, or validation during the training process, the logarithmic value of probability loss, and the MSE value. Black error bars indicate observations with measurement uncertainty. Solid blue lines are model fits to the data. The red band represents the 1 σ confidence interval.

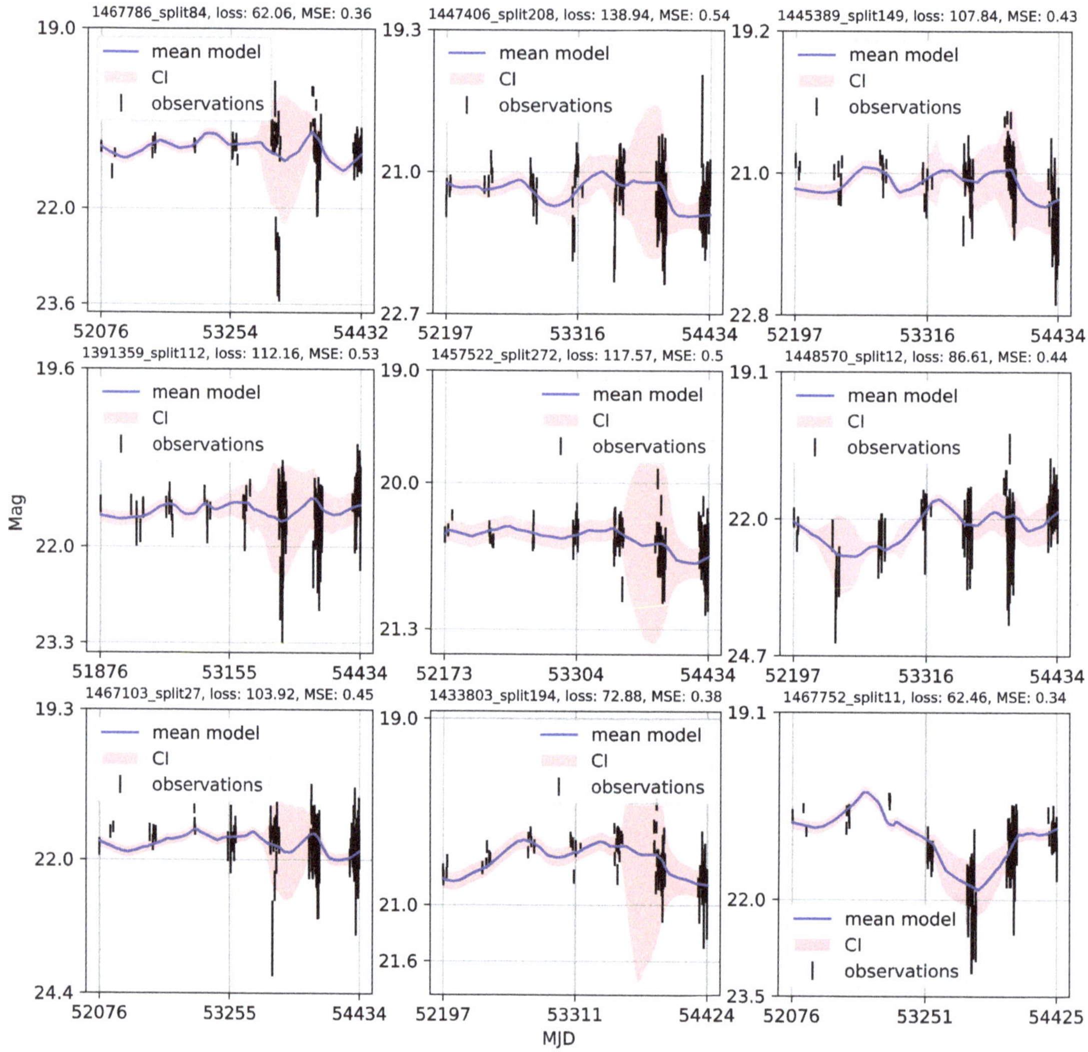

Figure A16. CNP models of light curves. The subtitle of each plot shows the object ID from the LSST AGN DC database with a tag indicating the iteration when it is selected for training, testing, or validation during the training process, the logarithmic value of probability loss, and the MSE value. Black error bars indicate observations with measurement uncertainty. Solid blue lines are model fits to the data. The red band represents the 1 σ confidence interval.

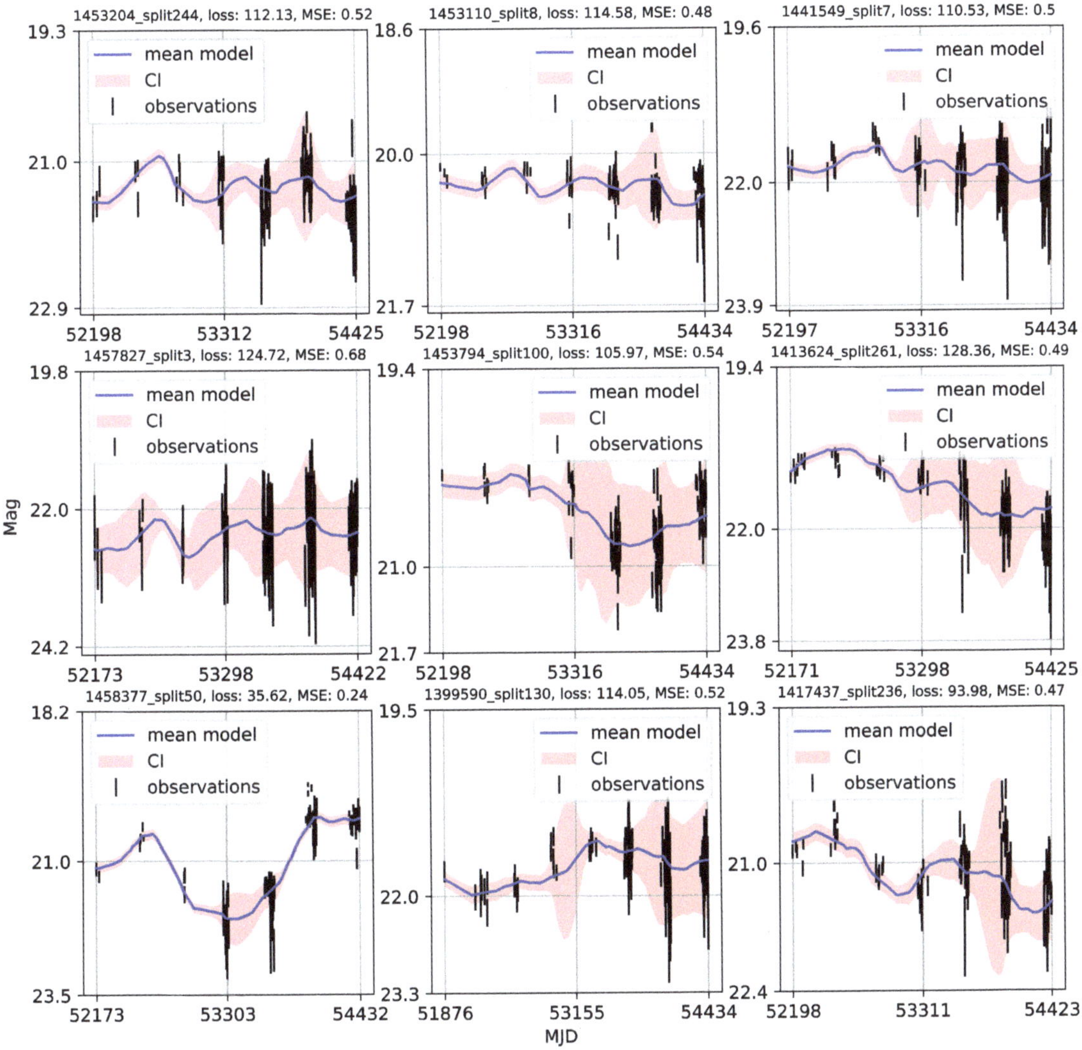

Figure A17. CNP models of light curves. The subtitle of each plot shows the object ID from the LSST AGN DC database with a tag indicating the iteration when it is selected for training, testing, or validation during the training process, the logarithmic value of probability loss, and the MSE value. Black error bars indicate observations with measurement uncertainty. Solid blue lines are model fits to the data. The red band represents the 1 σ confidence interval.

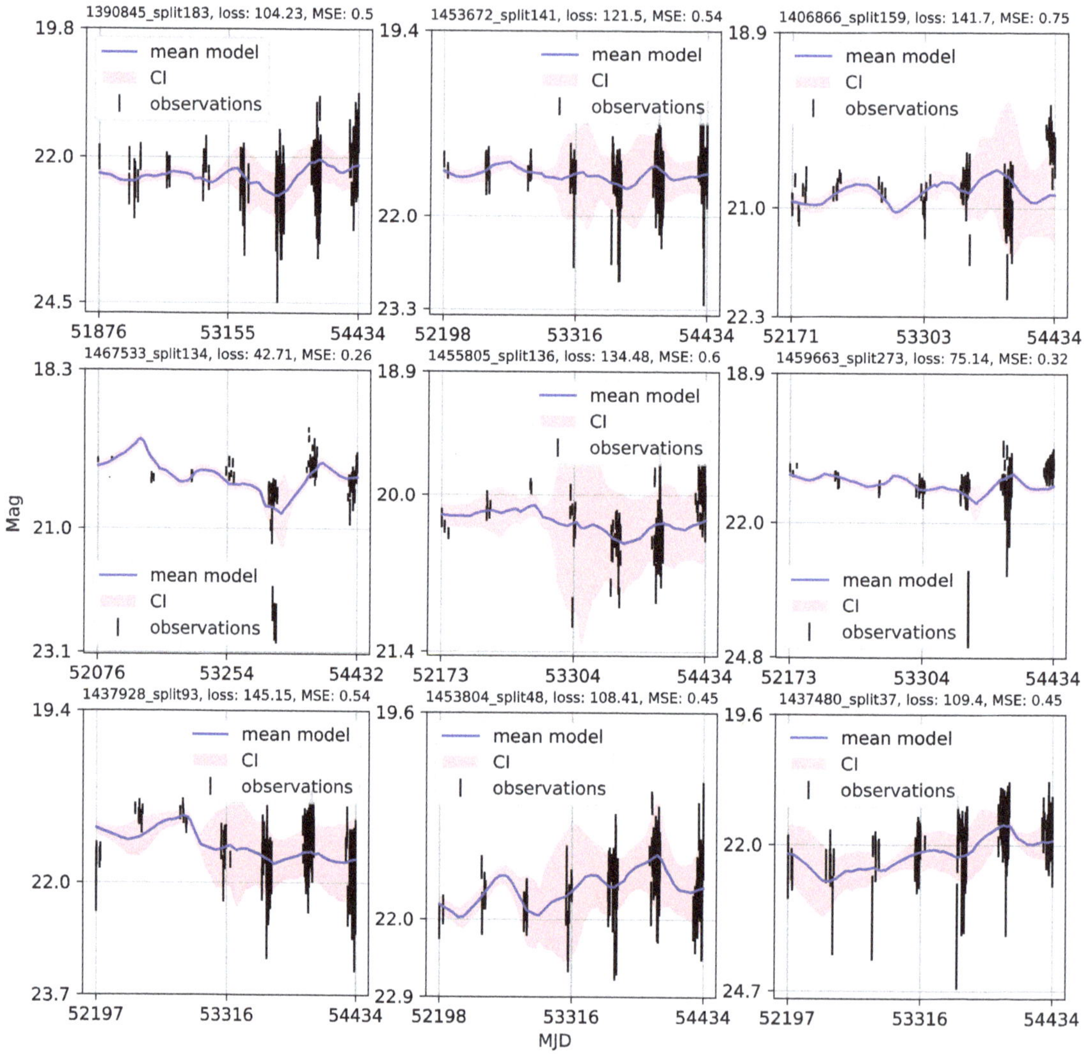

Figure A18. CNP models of light curves. The subtitle of each plot shows the object ID from the LSST AGN DC database with a tag indicating the iteration when it is selected for training, testing, or validation during the training process, the logarithmic value of probability loss, and the MSE value. Black error bars indicate observations with measurement uncertainty. Solid blue lines are model fits to the data. The red band represents the 1 σ confidence interval.

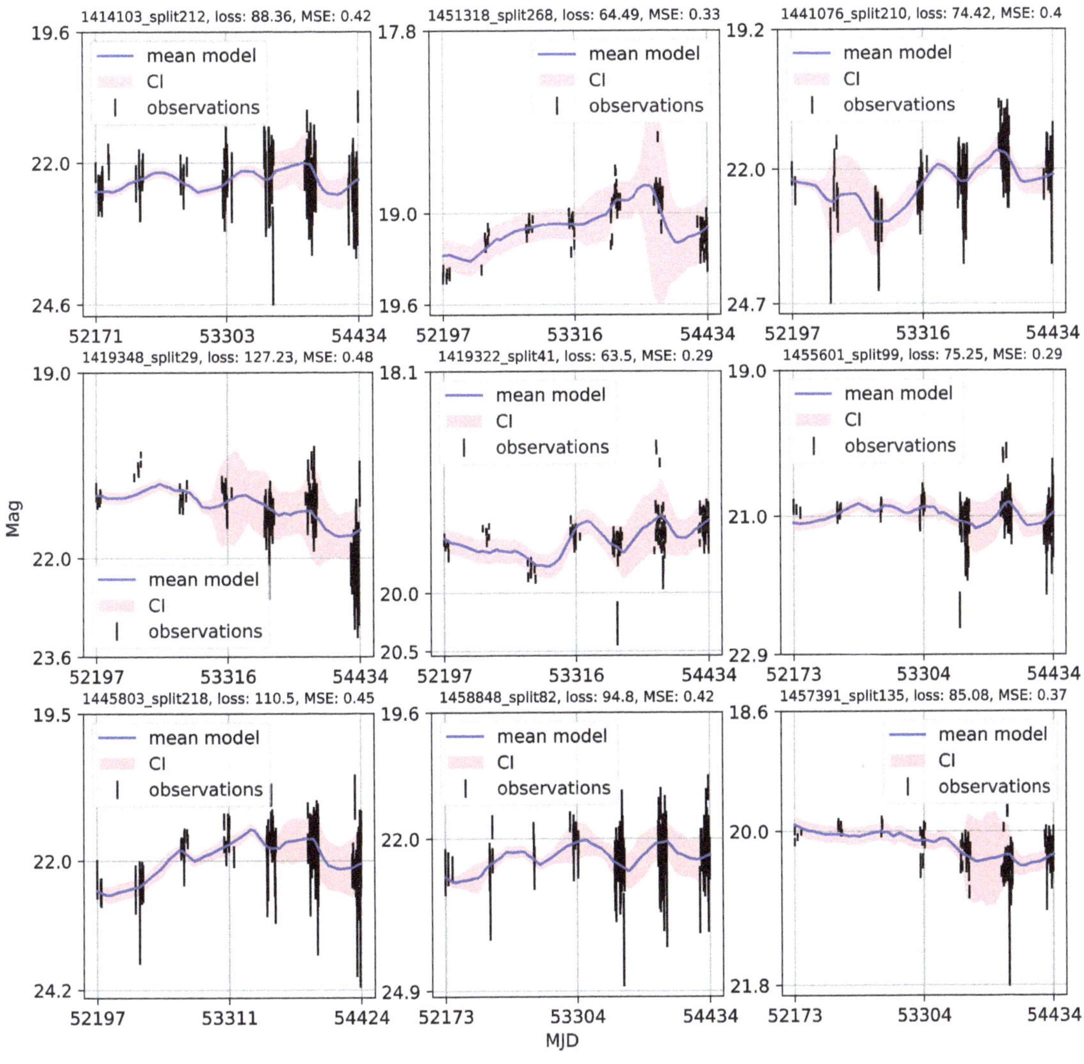

Figure A19. CNP models of light curves. The subtitle of each plot shows the object ID from the LSST AGN DC database with a tag indicating the iteration when it is selected for training, testing, or validation during the training process, the logarithmic value of probability loss, and the MSE value. Black error bars indicate observations with measurement uncertainty. Solid blue lines are model fits to the data. The red band represents the $1\,\sigma$ confidence interval.

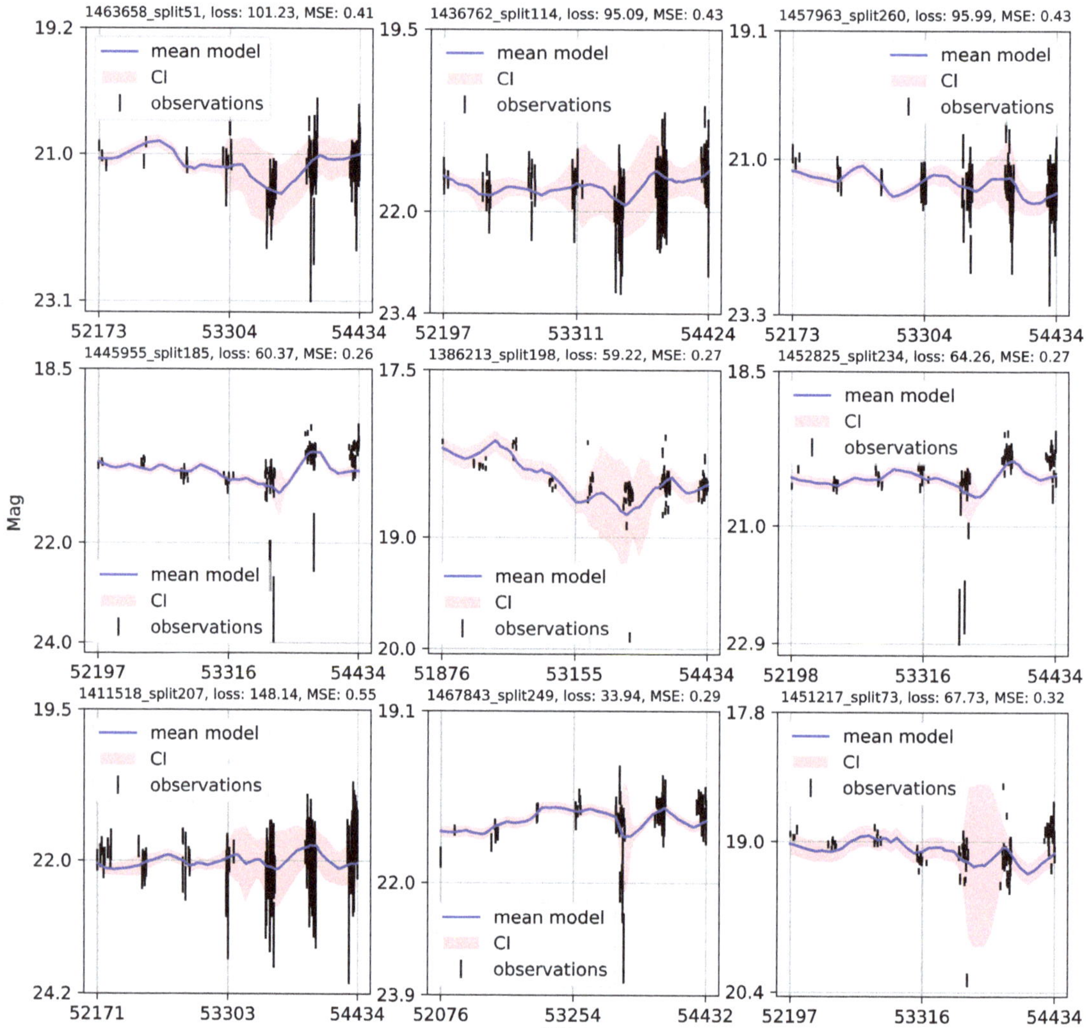

Figure A20. CNP models of light curves. The subtitle of each plot shows the object ID from the LSST AGN DC database with a tag indicating the iteration when it is selected for training, testing, or validation during the training process, the logarithmic value of probability loss, and the MSE value. Black error bars indicate observations with measurement uncertainty. Solid blue lines are model fits to the data. The red band represents the 1 σ confidence interval.

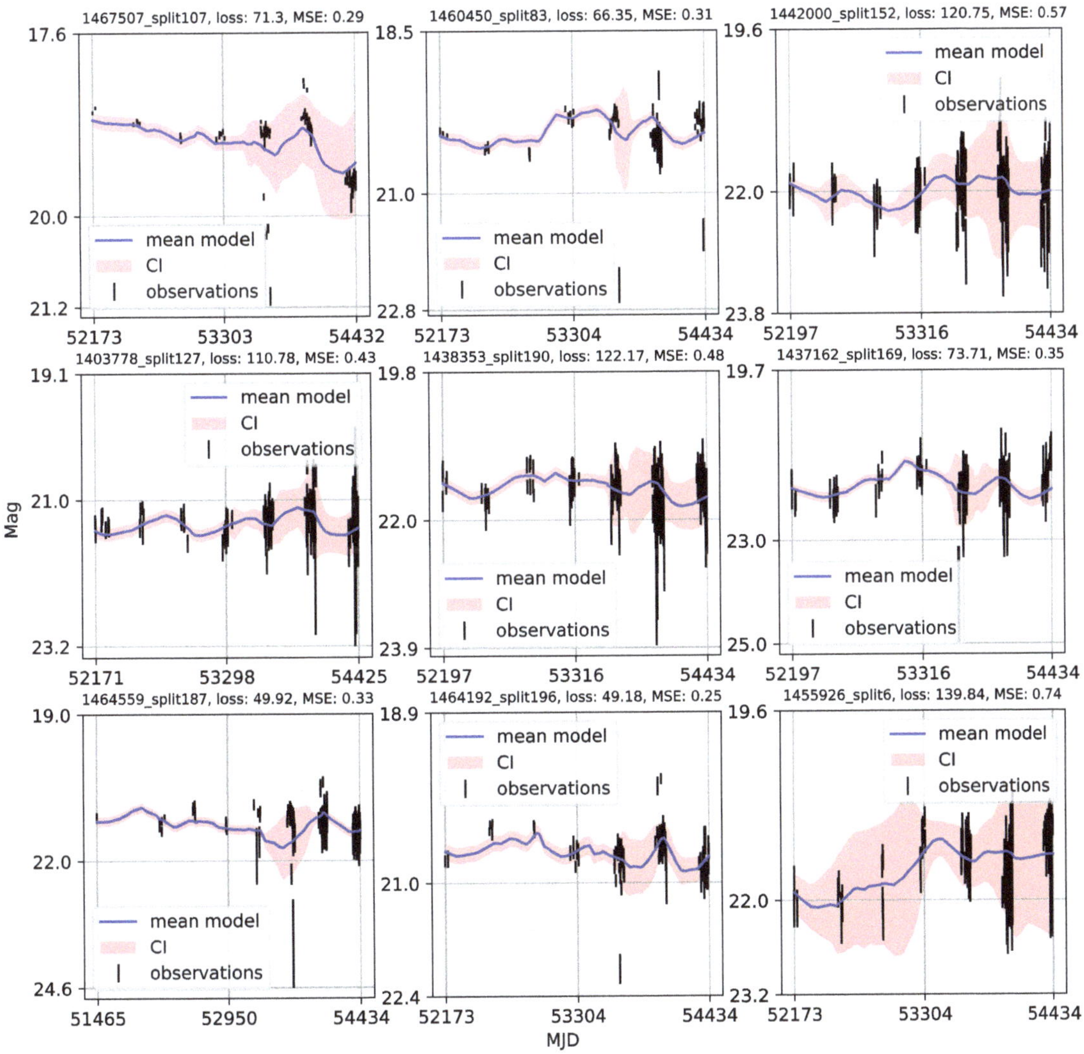

Figure A21. CNP models of light curves. The subtitle of each plot shows the object ID from the LSST AGN DC database with a tag indicating the iteration when it is selected for training, testing, or validation during the training process, the logarithmic value of probability loss, and the MSE value. Black error bars indicate observations with measurement uncertainty. Solid blue lines are model fits to the data. The red band represents the 1 σ confidence interval.

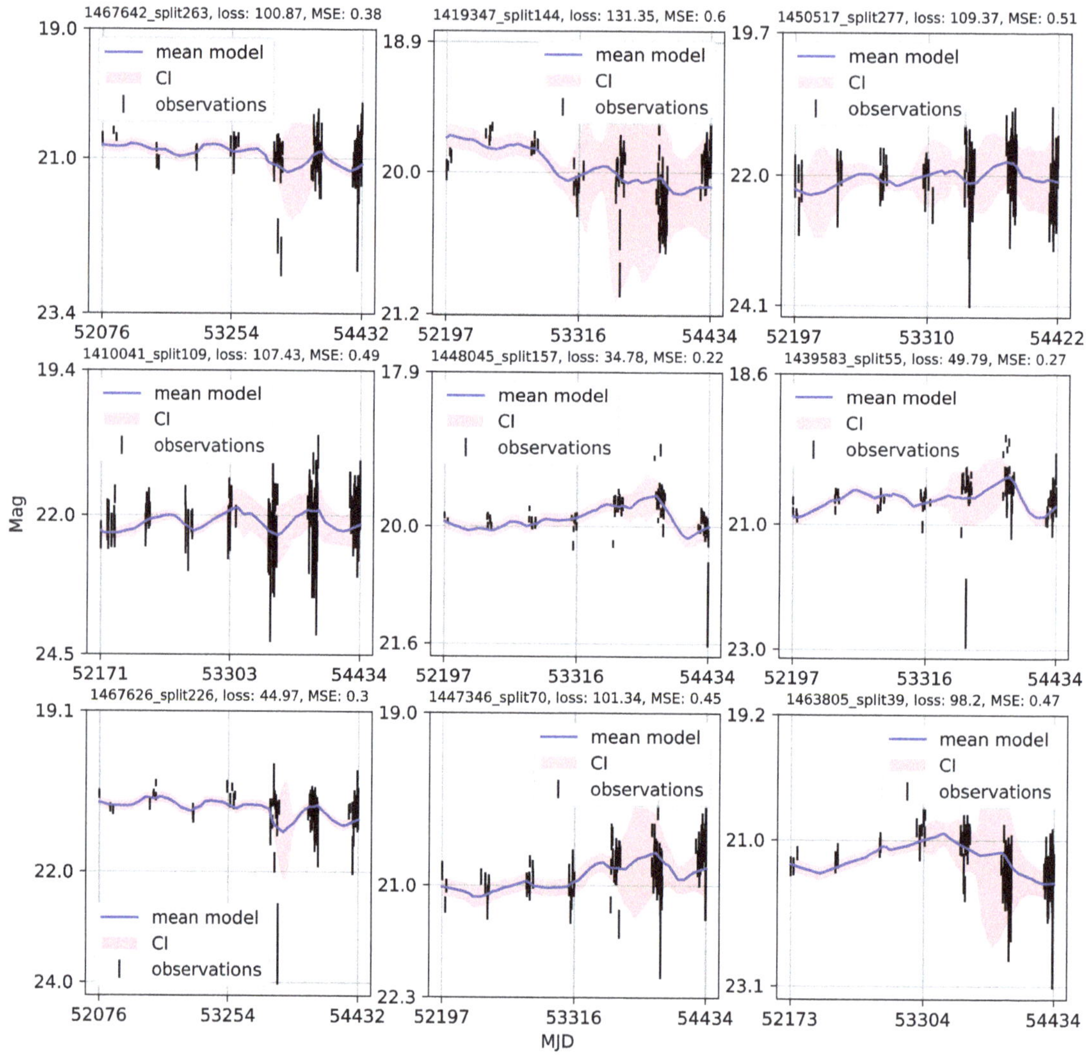

Figure A22. CNP models of light curves. The subtitle of each plot shows the object ID from the LSST AGN DC database with a tag indicating the iteration when it is selected for training, testing, or validation during the training process, the logarithmic value of probability loss, and the MSE value. Black error bars indicate observations with measurement uncertainty. Solid blue lines are model fits to the data. The red band represents the 1 σ confidence interval.

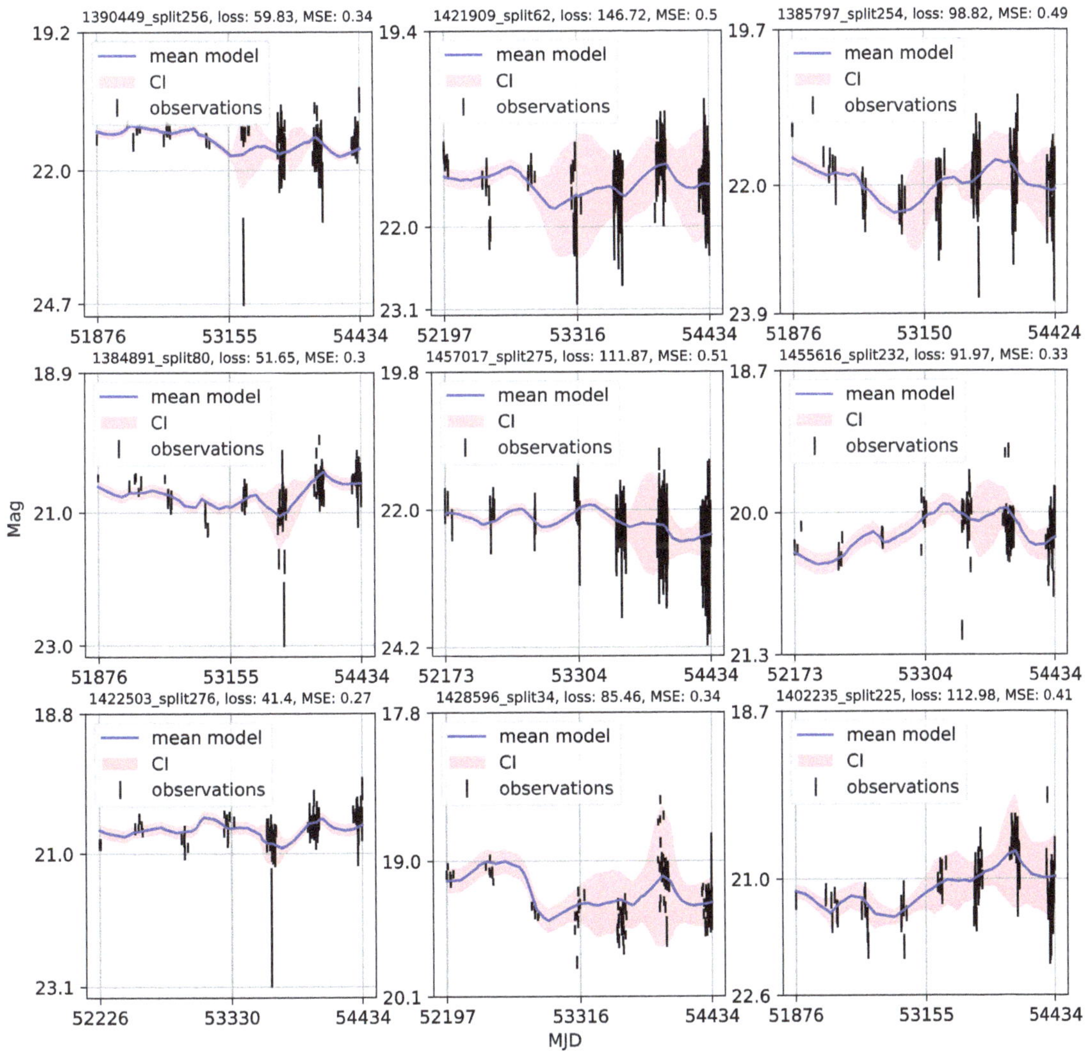

Figure A23. CNP models of light curves. The subtitle of each plot shows the object ID from the LSST AGN DC database with a tag indicating the iteration when it is selected for training, testing, or validation during the training process, the logarithmic value of probability loss, and the MSE value. Black error bars indicate observations with measurement uncertainty. Solid blue lines are model fits to the data. The red band represents the 1 σ confidence interval.

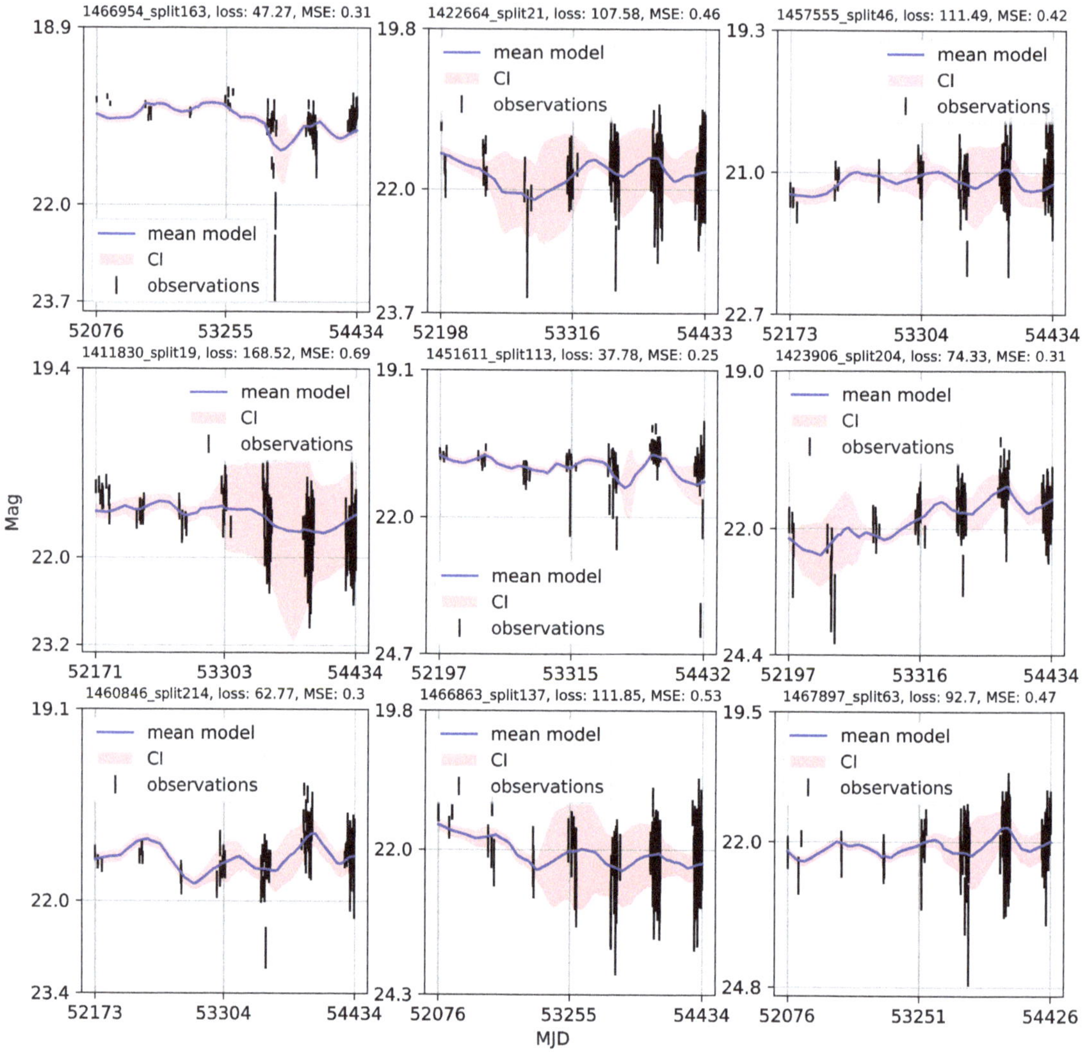

Figure A24. CNP models of light curves. The subtitle of each plot shows the object ID from the LSST AGN DC database with a tag indicating the iteration when it is selected for training, testing, or validation during the training process, the logarithmic value of probability loss, and the MSE value. Black error bars indicate observations with measurement uncertainty. Solid blue lines are model fits to the data. The red band represents the $1\,\sigma$ confidence interval.

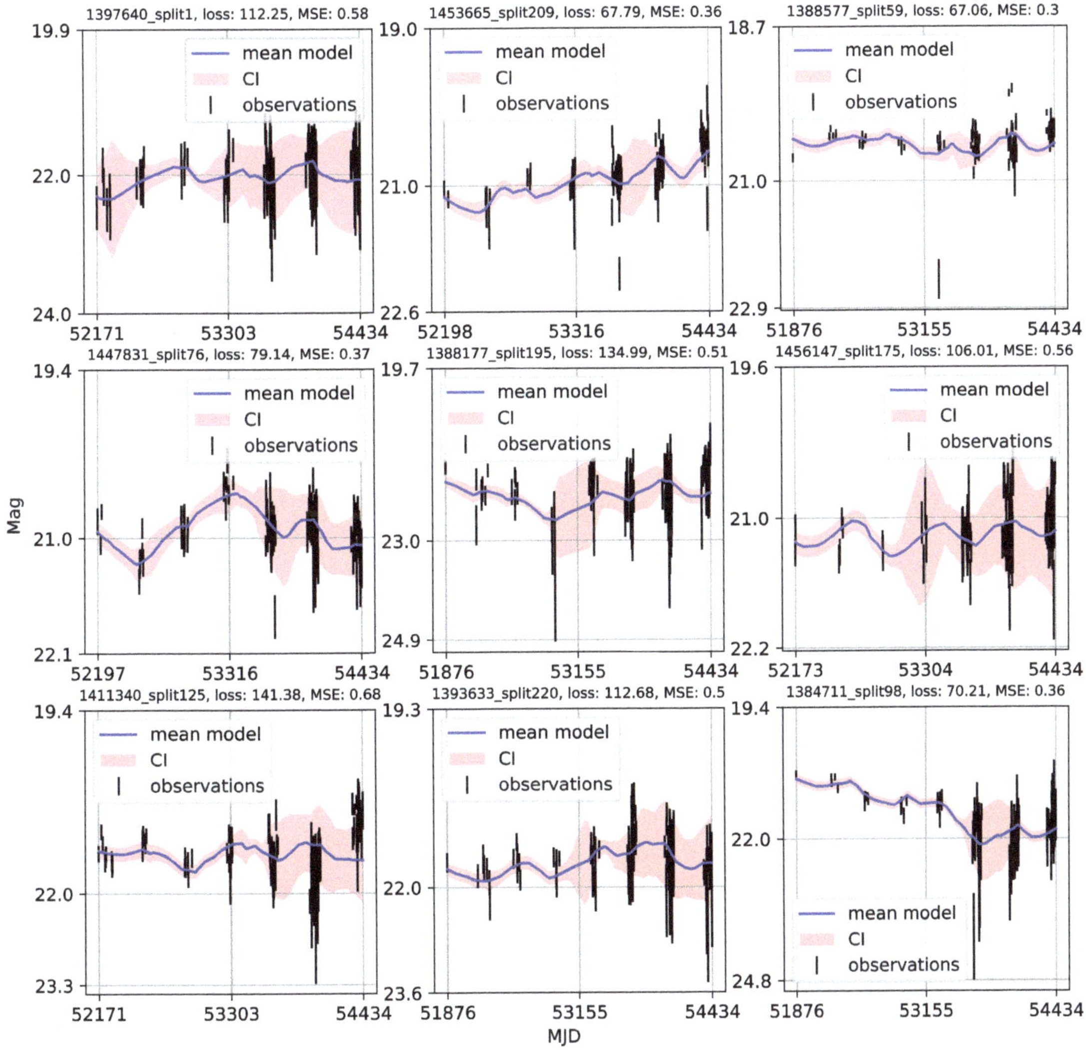

Figure A25. CNP models of light curves. The subtitle of each plot shows the object ID from the LSST AGN DC database with a tag indicating the iteration when it is selected for training, testing, or validation during the training process, the logarithmic value of probability loss, and the MSE value. Black error bars indicate observations with measurement uncertainty. Solid blue lines are model fits to the data. The red band represents the 1 σ confidence interval.

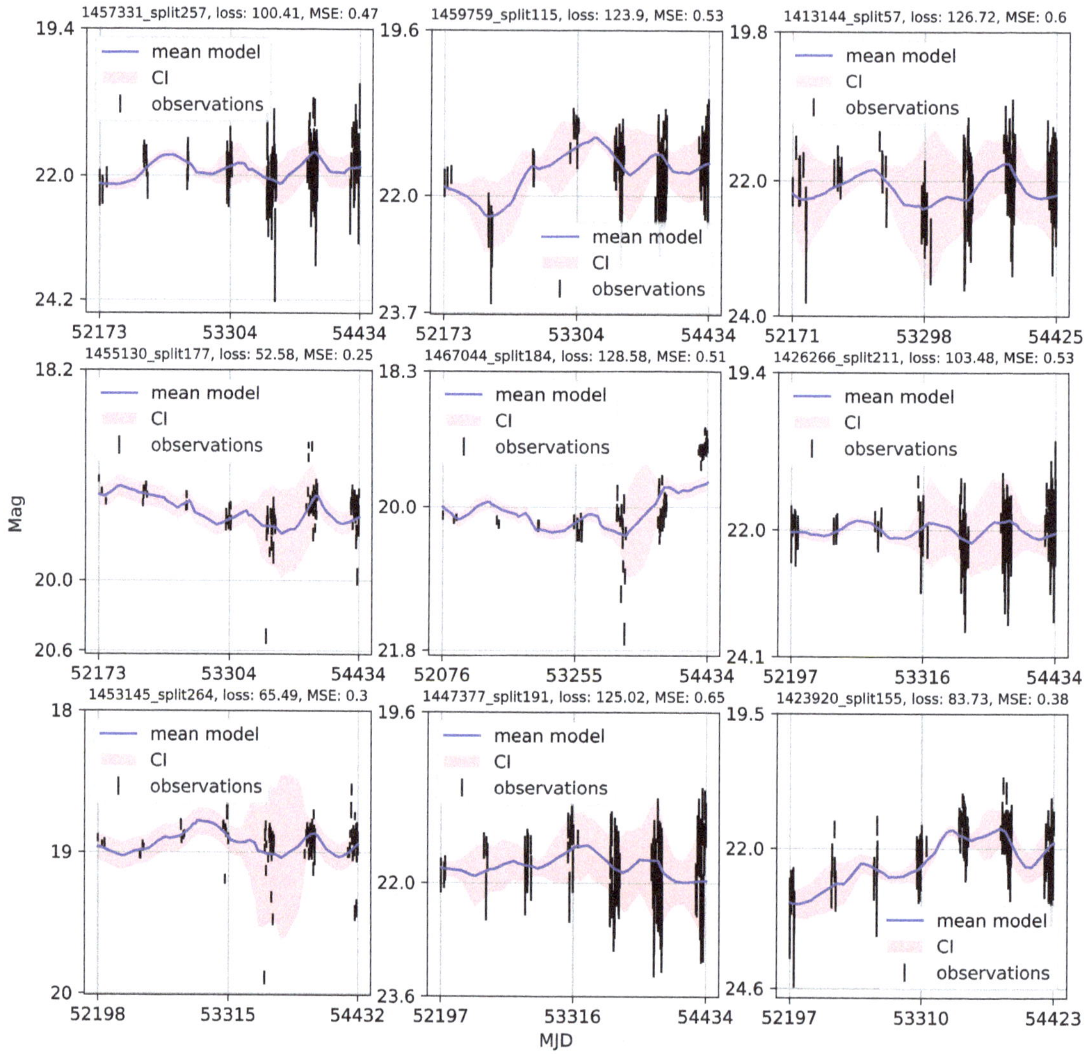

Figure A26. CNP models of light curves. The subtitle of each plot shows the object ID from the LSST AGN DC database with a tag indicating the iteration when it is selected for training, testing, or validation during the training process, the logarithmic value of probability loss, and the MSE value. Black error bars indicate observations with measurement uncertainty. Solid blue lines are model fits to the data. The red band represents the $1\,\sigma$ confidence interval.

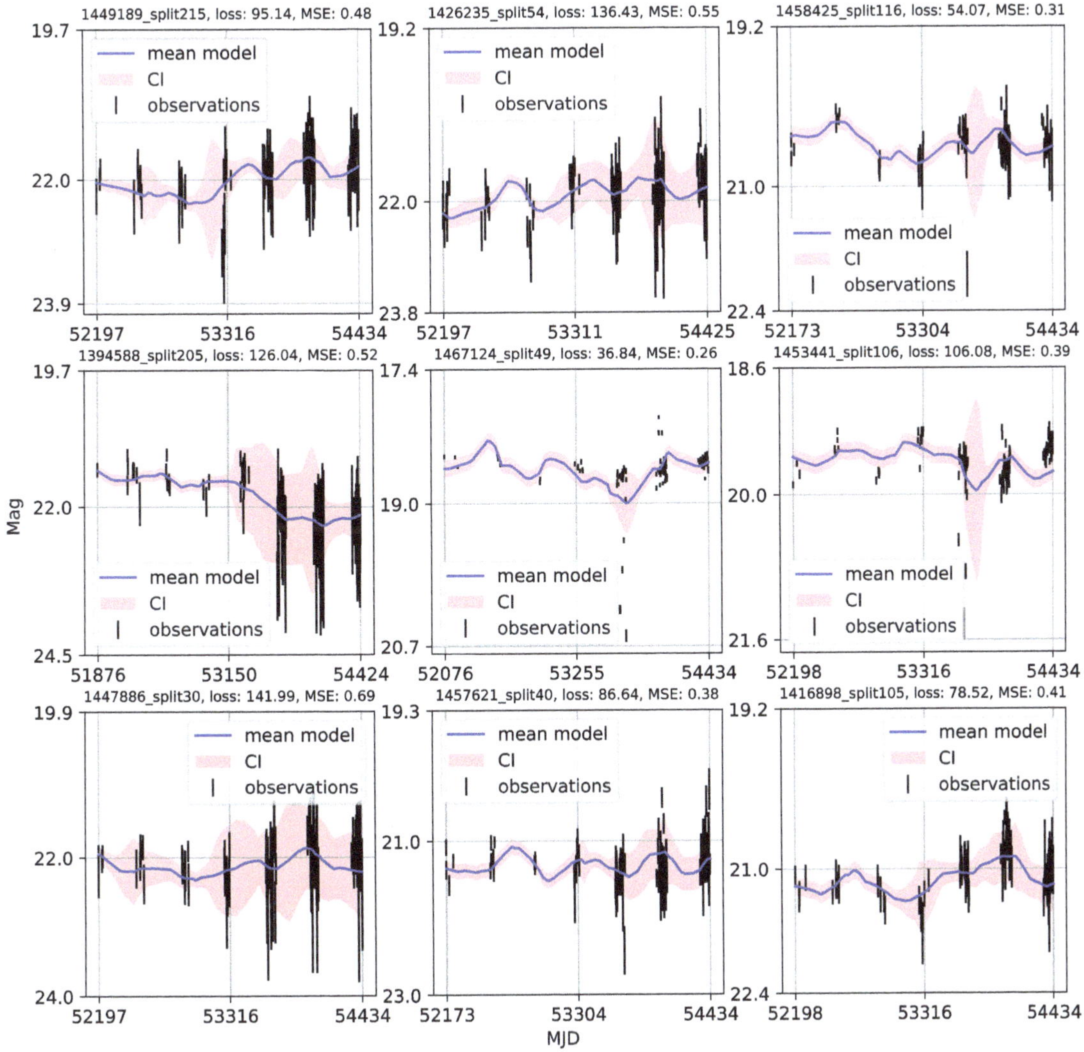

Figure A27. CNP models of light curves. The subtitle of each plot shows the object ID from the LSST AGN DC database with a tag indicating the iteration when it is selected for training, testing, or validation during the training process, the logarithmic value of probability loss, and the MSE value. Black error bars indicate observations with measurement uncertainty. Solid blue lines are model fits to the data. The red band represents the 1 σ confidence interval.

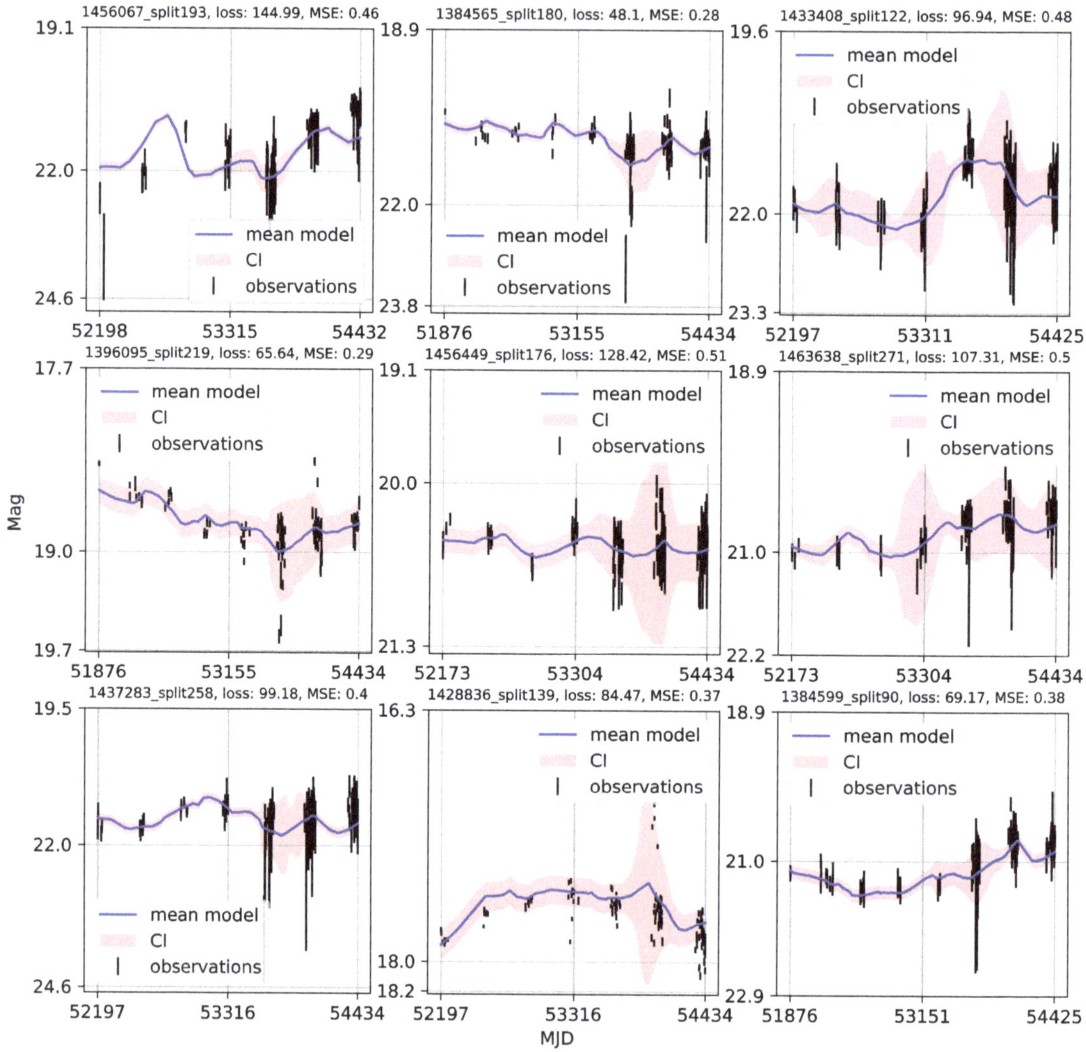

Figure A28. CNP models of light curves. The subtitle of each plot shows the object ID from the LSST AGN DC database with a tag indicating the iteration when it is selected for training, testing, or validation during the training process, the logarithmic value of probability loss, and the MSE value. Black error bars indicate observations with measurement uncertainty. Solid blue lines are model fits to the data. The red band represents the 1 σ confidence interval.

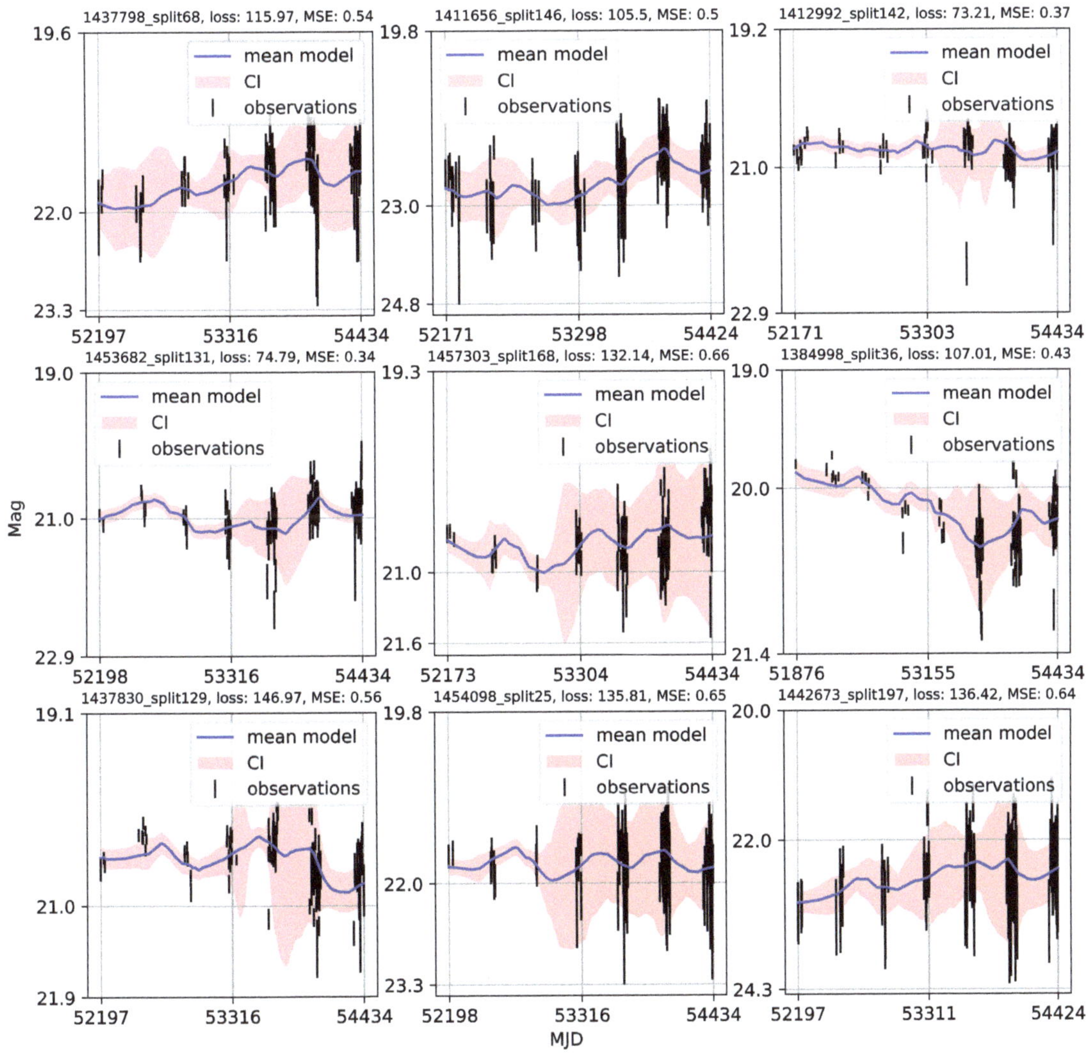

Figure A29. CNP models of light curves. The subtitle of each plot shows the object ID from the LSST AGN DC database with a tag indicating the iteration when it is selected for training, testing, or validation during the training process, the logarithmic value of probability loss, and the MSE value. Black error bars indicate observations with measurement uncertainty. Solid blue lines are model fits to the data. The red band represents the 1 σ confidence interval.

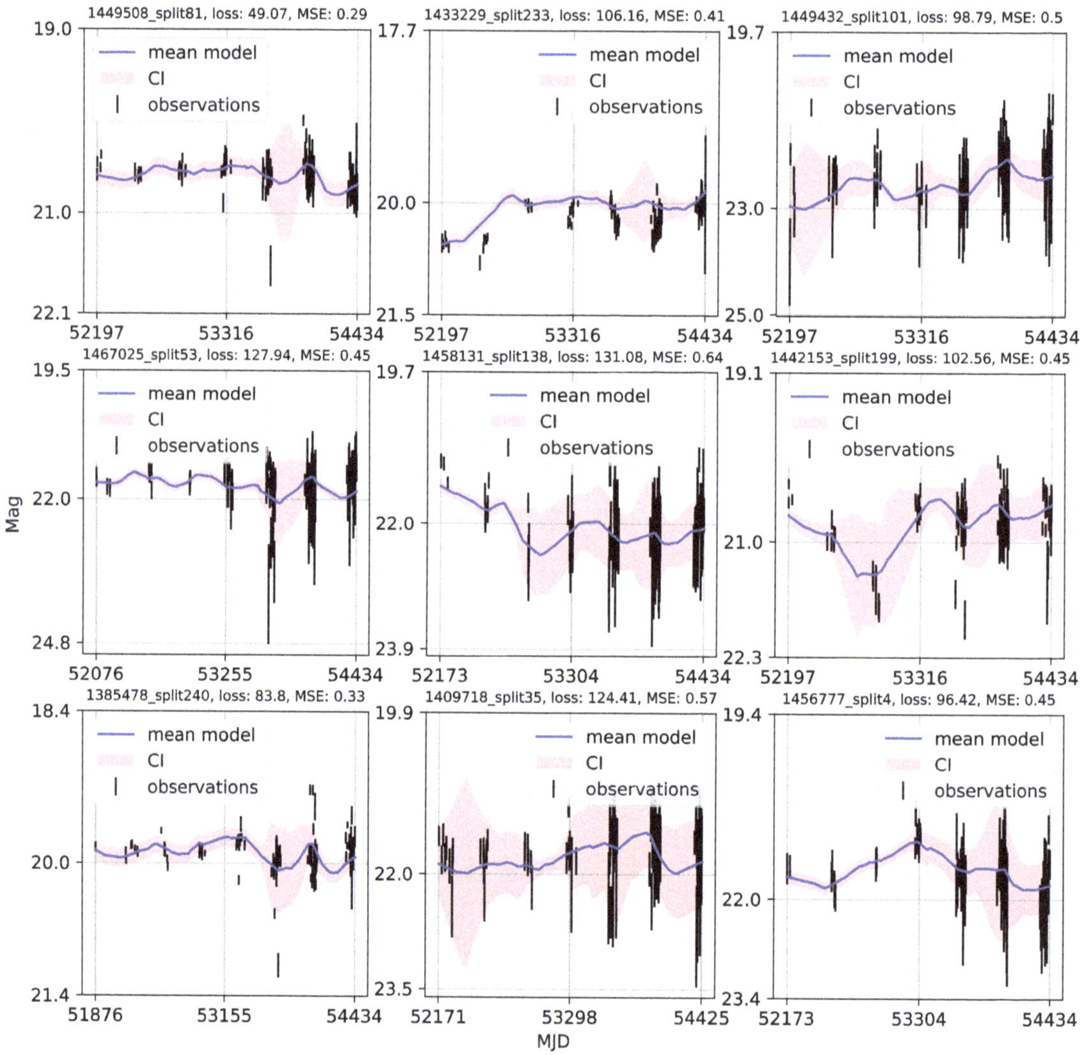

Figure A30. CNP models of light curves. The subtitle of each plot shows the object ID from the LSST AGN DC database with a tag indicating the iteration when it is selected for training, testing, or validation during the training process, the logarithmic value of probability loss, and the MSE value. Black error bars indicate observations with measurement uncertainty. Solid blue lines are model fits to the data. The red band represents the 1 σ confidence interval.

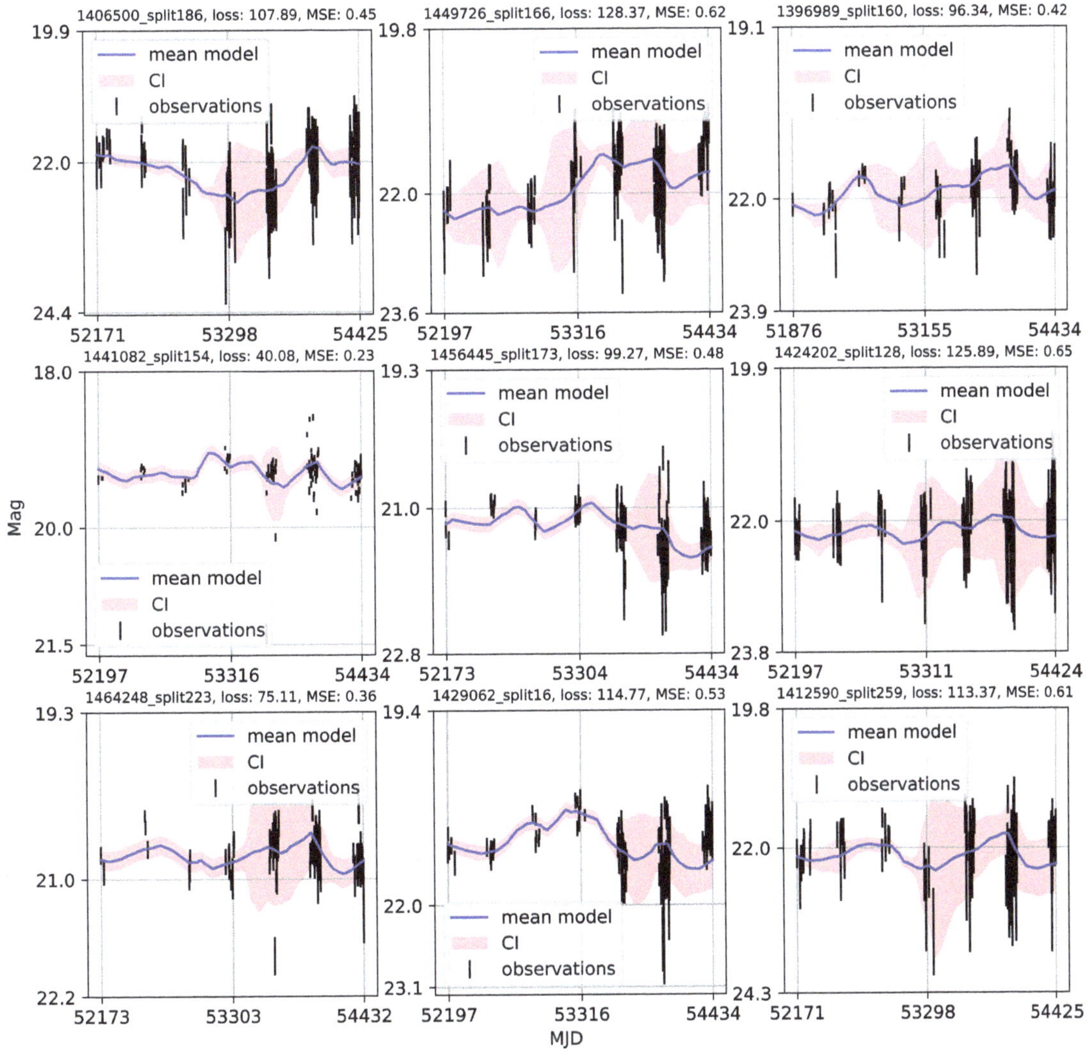

Figure A31. CNP models of light curves. The subtitle of each plot shows the object ID from the LSST AGN DC database with a tag indicating the iteration when it is selected for training, testing, or validation during the training process, the logarithmic value of probability loss, and the MSE value. Black error bars indicate observations with measurement uncertainty. Solid blue lines are model fits to the data. The red band represents the 1 σ confidence interval.

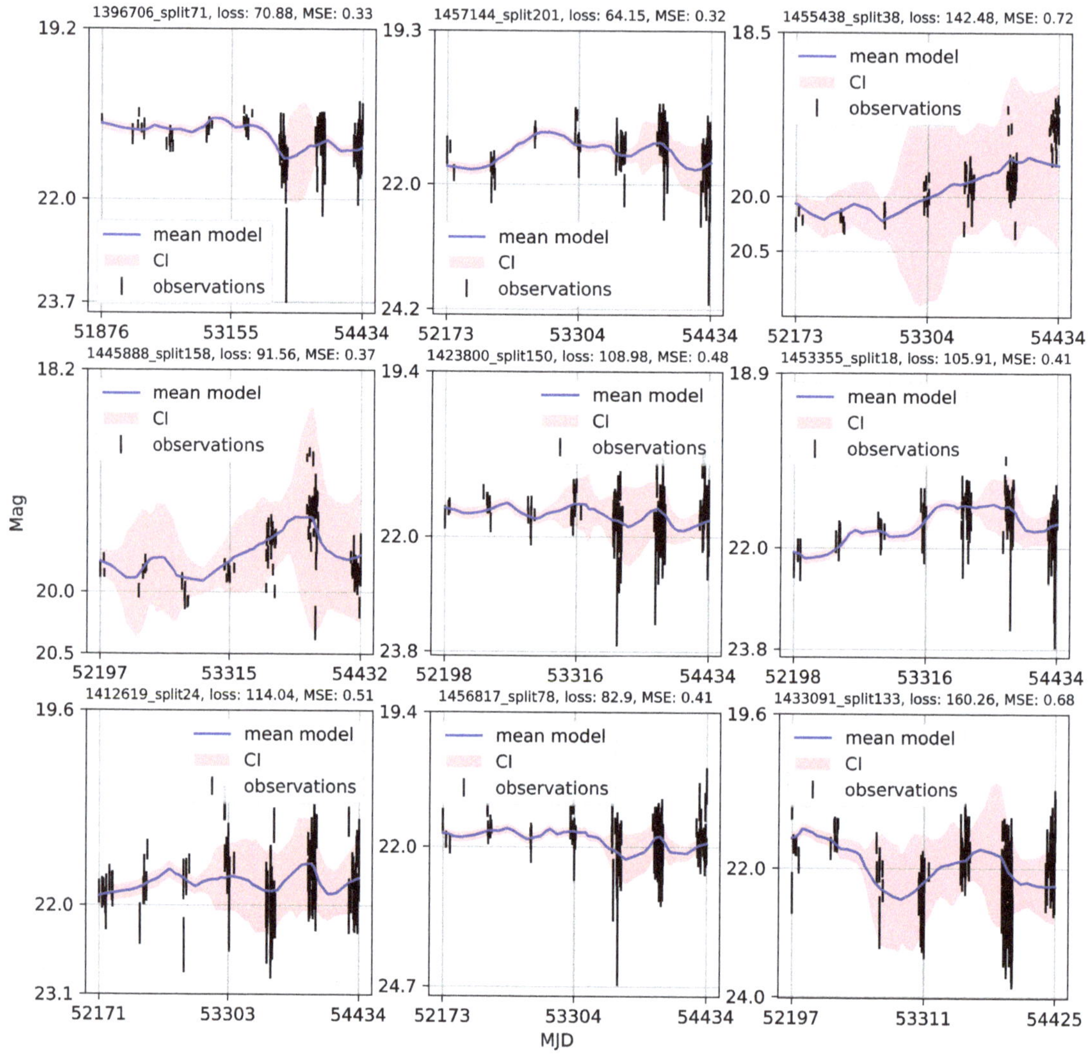

Figure A32. CNP models of light curves. The subtitle of each plot shows the object ID from the LSST AGN DC database with a tag indicating the iteration when it is selected for training, testing, or validation during the training process, the logarithmic value of probability loss, and the MSE value. Black error bars indicate observations with measurement uncertainty. Solid blue lines are model fits to the data. The red band represents the $1\,\sigma$ confidence interval.

Notes

1 Gaining knowledge from large astronomical databases is a complex procedure, including various deep learning algorithms and procedures so we will use 'deep learning' in that wide context.

2 SER-SAG is Serbian team that contributes to AGN investigation and participates in the LSST AGN Scientific Collaboration

3 F_{var} behaves as normalized variance so it is more robust to outliers, flares

4 We underline the distinction between NPs and both classic neural networks and GPs previously applied to quasars' light curves. Classical neural network fit a single model across points based on learning from a large data collection, whereas GP fits a

distribution of curves to a single set of observations (i.e., one light curve). NP combines both approaches, taking use of neural network ability to train on a large collection and GP's ability to fit the distribution of curves because it is a metalearner.

5 In neural processes, the predicted distribution over functions is typically a Gaussian distribution, parameterized by a mean and variance.

6 We will use a subscript θ to denote all the parameters of the neural network such as number of layers, learning rate, size of batches, etc.

7 An assumption on $\mathcal{P}$ is that all finite sets of function evaluations of f are jointly Gaussian distributed. This class of random functions are known as Gaussian Processes (GPs).

8 We note that the our loss function works quite similarly to the Cross-Entropy. In the `PyTorch` ecosystem, Cross-Entropy Loss is obtained by combining a log-softmax layer and loss.

9 Data were transformed using min-max scaler adapted to the range $[a, b]$

$$\text{scaled\_value} = \frac{(\text{original\_value} - \text{original\_min\_value}) \times (b - a)}{\text{original\_max\_value} - \text{original\_min\_value}} + a$$

where $a = -2$, $b = 2$ and original_value stands for the input data (time instances, magnitudes, magnitudes errors), original_max_value is the maximum of the original_value, and original_min_value is the minimum of the original_value. This linear transformation (or more precisely affine) preserves the original distribution of data, does not reduce the importance of outliers, and preserves the covariance structure of the data. We used the range of $[-2, 2]$ for enabling direct comparisons with [33] original testing data, ensuring consistency in our analysis.

10 The N is the maximum number of points in the light curves in the given batch; missing values are zeropadded for shorter light curves. We emphasize that our sample of light curves is well balanced as a result of SOM clustering (see for an counter example [58]), so that the number of points per light curve covers a fairly limited range [103, 127] points, requiring negligible padding.

References

1. Ivezić, Ž.; Kahn, S.M.; Tyson, J.A.; Abel, B.; Acosta, E.; Allsman, R.; Alonso, D.; AlSayyad, Y.; Anderson, S.F.; Andrew, J.; et al. LSST: From Science Drivers to Reference Design and Anticipated Data Products. *Astrophys. J.* **2019**, *873*, 111. [CrossRef]
2. Anderson, S.F.; Haggard, D.; Homer, L.; Joshi, N.R.; Margon, B.; Silvestri, N.M.; Szkody, P.; Wolfe, M.A.; Agol, E.; Becker, A.C.; et al. Ultracompact AM Canum Venaticorum Binaries from the Sloan Digital Sky Survey: Three Candidates Plus the First Confirmed Eclipsing System. *Astron. J.* **2005**, *130*, 2230–2236. [CrossRef]
3. Bloom, J.S.; Starr, D.L.; Butler, N.R.; Nugent, P.; Rischard, M.; Eads, D.; Poznanski, D. Towards a real-time transient classification engine. *Astron. Nachr.* **2008**, *329*, 284. [CrossRef]
4. Scolnic, D.; Kessler, R.; Brout, D.; Cowperthwaite, P.S.; Soares-Santos, M.; Annis, J.; Herner, K.; Chen, H.Y.; Sako, M.; Doctor, Z.; et al. How Many Kilonovae Can Be Found in Past, Present, and Future Survey Data Sets? *Astrophys. J. Lett.* **2018**, *852*, L3. [CrossRef]
5. Nuttall, L.K.; Berry, C.P.L. Electromagnetic counterparts of gravitational-wave signals. *Astron. Geophys.* **2021**, *62*, 4.15–4.21. [CrossRef]
6. Kaspi, S.; Brandt, W.N.; Maoz, D.; Netzer, H.; Schneider, D.P.; Shemmer, O. Reverberation Mapping of High-Luminosity Quasars: First Results. *Astrophys. J.* **2007**, *659*, 997–1007. [CrossRef]
7. MacLeod, C.L.; Ivezić, Ž.; Kochanek, C.S.; Kozłowski, S.; Kelly, B.; Bullock, E.; Kimball, A.; Sesar, B.; Westman, D.; Brooks, K.; et al. Modeling the Time Variability of SDSS Stripe 82 Quasars as a Damped Random Walk. *Astrophys. J.* **2010**, *721*, 1014–1033. [CrossRef]
8. Graham, M.J.; Djorgovski, S.G.; Drake, A.J.; Mahabal, A.A.; Chang, M.; Stern, D.; Donalek, C.; Glikman, E. A novel variability-based method for quasar selection: Evidence for a rest-frame ∼54 d characteristic time-scale. *Mon. Not. R. Astron. Soc.* **2014**, *439*, 703–718. [CrossRef]
9. Chapline, G.F.; Frampton, P.H. A new direction for dark matter research: Intermediate-mass compact halo objects. *J. Cosmol. Astropart. Phys.* **2016**, *2016*, 042. [CrossRef]
10. Burke, C.J.; Shen, Y.; Blaes, O.; Gammie, C.F.; Horne, K.; Jiang, Y.F.; Liu, X.; McHardy, I.M.; Morgan, C.W.; Scaringi, S.; et al. A characteristic optical variability time scale in astrophysical accretion disks. *Science* **2021**, *373*, 789–792. [CrossRef]
11. Risaliti, G.; Lusso, E. A Hubble Diagram for Quasars. *Astrophys. J.* **2015**, *815*, 33. [CrossRef]
12. Marziani, P.; Dultzin, D.; del Olmo, A.; D'Onofrio, M.; de Diego, J.A.; Stirpe, G.M.; Bon, E.; Bon, N.; Czerny, B.; Perea, J.; et al. The quasar main sequence and its potential for cosmology. *Proc. Nucl. Act. Galaxies Cosm. Time* **2021**, *356*, 66–71. [CrossRef]
13. Tachibana, Y.; Graham, M.J.; Kawai, N.; Djorgovski, S.G.; Drake, A.J.; Mahabal, A.A.; Stern, D. Deep Modeling of Quasar Variability. *Astrophys. J.* **2020**, *903*, 54. [CrossRef]
14. Kawaguchi, T.; Mineshige, S.; Umemura, M.; Turner, E.L. Optical Variability in Active Galactic Nuclei: Starbursts or Disk Instabilities? *Astrophys. J.* **1998**, *504*, 671–679. [CrossRef]
15. Hawkins, M.R.S. Timescale of variation and the size of the accretion disc in active galactic nuclei. *Astron. Astrophys.* **2007**, *462*, 581–589. [CrossRef]

16. Zakharov, F.; Popović, L.Č.; Jovanović, P. On the contribution of microlensing to X-ray variability of high-redshifted QSOs. *Astron. Astrophys.* **2004**, *420*, 881–888. [CrossRef]
17. Kelly, B.C.; Bechtold, J.; Siemiginowska, A. Are the variations in quasar optical flux driven by thermal fluctuations? *Astrophys. J.* **2009**, *698*, 895. [CrossRef]
18. Sesar, B.; Ivezić, Ž.; Lupton, R.H.; Jurić, M.; Gunn, J.E.; Knapp, G.R.; DeLee, N.; Smith, J.A.; Miknaitis, G.; Lin, H.; et al. Exploring the Variable Sky with the Sloan Digital Sky Survey. *Astron. J.* **2007**, *134*, 2236–2251. [CrossRef]
19. MacLeod, C.L.; Ivezić, Ž.; Sesar, B.; de Vries, W.; Kochanek, C.S.; Kelly, B.C.; Becker, A.C.; Lupton, R.H.; Hall, P.B.; Richards, G.T.; et al. A description of quasar variability measured using repeated sdss and poss imaging. *Astrophys. J.* **2012**, *753*, 106. [CrossRef]
20. Kozłowski, S. Limitations on the recovery of the true AGN variability parameters using damped random walk modeling. *Astron. Astrophys.* **2017**, *597*, A128. [CrossRef]
21. Kelly, B.C.; Becker, A.C.; Sobolewska, M.; Siemiginowska, A.; Uttley, P. Flexible and scalable methods for quantifying stochastic variability in the era of massive time-domain astronomical data sets. *Astrophys. J.* **2014**, *788*, 33. [CrossRef]
22. Graham, M.J.; Djorgovski, S.G.; Drake, A.J.; Stern, D.; Mahabal, A.A.; Glikman, E.; Larson, S.; Christensen, E. Understanding extreme quasar optical variability with CRTS—I. Major AGN flares. *Mon. Not. R. Astron. Soc.* **2017**, *470*, 4112–4132. [CrossRef]
23. Xin, C.; Haiman, Z. Ultra-short-period massive black hole binary candidates in LSST as LISA 'verification binaries'. *MNRAS* **2021**, *506*, 2408–2417. [CrossRef]
24. Amaro-Seoane, P.; Audley, H.; Babak, S.; Baker, J.; Barausse, E.; Bender, P.; Berti, E.; Binetruy, P.; Born, M.; Bortoluzzi, D.; et al. Laser Interferometer Space Antenna. *arXiv* **2017**, arXiv:1702.00786.
25. Haiman, Z.; Kocsis, B.; Menou, K. The Population of Viscosity- and Gravitational Wave-Driven Supermassive Black Hole Binaries among Luminous Active Galactic Nuclei. *Astrophys. J.* **2009**, *700*, 1952. [CrossRef]
26. Emmanoulopoulos, D.; McHardy, I.M.; Papadakis, I.E. Generating artificial light curves: Revisited and updated. *Mon. Not. R. Astron. Soc.* **2013**, *433*, 907–927. [CrossRef]
27. Kelly, B.C.; Treu, T.; Malkan, M.; Pancoast, A.; Woo, J.H. Active Galactic Nucleus Black Hole Mass Estimates in the Era of Time Domain Astronomy. *Astrophys. J.* **2013**, *779*, 187. [CrossRef]
28. Mushotzky, R.F.; Edelson, R.; Baumgartner, W.; Gandhi, P. Kepler Observations of Rapid Optical Variability in Active Galactic Nuclei. *Astrophys. J. Lett.* **2011**, *743*, L12. [CrossRef]
29. Smith, K.L.; Mushotzky, R.F.; Boyd, P.T.; Malkan, M.; Howell, S.B.; Gelino, D.M. The Kepler Light Curves of AGN: A Detailed Analysis. *Astrophys. J.* **2018**, *857*, 141. [CrossRef]
30. Yu, W.; Richards, G.T.; Vogeley, M.S.; Moreno, J.; Graham, M.J. Examining AGN UV/Optical Variability beyond the Simple Damped Random Walk. *Astrophys. J.* **2022**, *936*, 132. [CrossRef]
31. Zhang, S.Q.; Wang, F.; Fan, F.L. Neural Network Gaussian Processes by Increasing Depth. *IEEE Trans. Neural Netw. Learn. Syst.* **2022**, 1–6. [CrossRef] [PubMed]
32. Danilov, E.; Ćiprijanović, A.; Nord, B. Neural Inference of Gaussian Processes for Time Series Data of Quasars. *arXiv* **2022**, arXiv:2211.10305.
33. Garnelo, M.; Schwarz, J.; Rosenbaum, D.; Viola, F.; Rezende, D.; Eslami, S.; Teh, Y. Neural Processes. In Proceedings of the Theoretical Foundations and Applications of Deep Generative Models Workshop, International Conference on Machine Learning (ICML), Stockholm, Sweden, 14–15 July 2018; pp. 1704, 1713.
34. Yu, W.; Richards, G.; Buat, V.; Brandt, W.N.; Banerji, M.; Ni, Q.; Shirley, R.; Temple, M.; Wang, F.; Yang, J. *LSSTC AGN Data Challenge 2021*; Zenodo: Geneve, Switzerland, 2022. [CrossRef]
35. Čvorović-Hajdinjak, I.; Kovačević, A.B.; Ilić, D.; Popović, L.Č.; Dai, X.; Jankov, I.; Radović, V.; Sánchez-Sáez, P.; Nikutta, R. Conditional Neural Process for nonparametric modeling of active galactic nuclei light curves. *Astron. Nachr.* **2022**, *343*, e210103. [CrossRef]
36. Oguri, M.; Marshall, P.J. Gravitationally lensed quasars and supernovae in future wide-field optical imaging surveys. *Mon. Not. R. Astron. Soc.* **2010**, *405*, 2579–2593. [CrossRef]
37. Neira, F.; Anguita, T.; Vernardos, G. A quasar microlensing light-curve generator for LSST. *Mon. Not. R. Astron. Soc.* **2020**, *495*, 544–553. [CrossRef]
38. Savić, D.V.; Jankov, I.; Yu, W.; Petrecca, V.; Temple, M.; Ni, Q.; Shirley, R.; Kovačević, A.; Nikolić, M.; Ilić, D.; Popović, L. Č.; et al. The LSST AGN Data Challenge: Selection methods. *Astrophys. J.* **2022**, *submitted*.
39. Richards, J.W.; Starr, D.L.; Butler, N.R.; Bloom, J.S.; Brewer, J.M.; Crellin-Quick, A.; Higgins, J.; Kennedy, R.; Rischard, M. On Machine-learned Classification of Variable Stars with Sparse and Noisy Time-series Data. *Astrophys. J.* **2011**, *733*, 10. [CrossRef]
40. Kovacevic, A.; Ilic, D.; Jankov, I.; Popovic, L.C.; Yoon, I.; Radovic, V.; Caplar, N.; Cvorovic-Hajdinjak, I. LSST AGN SC Cadence Note: Two metrics on AGN variability observable. *arXiv* **2021**, arXiv:2105.12420.
41. Kovačević, A.B.; Radović, V.; Ilić, D.; Popović, L.Č.; Assef, R.J.; Sánchez-Sáez, P.; Nikutta, R.; Raiteri, C.M.; Yoon, I.; Homayouni, Y.; et al. The LSST Era of Supermassive Black Hole Accretion Disk Reverberation Mapping. *Astrophys. J. Suppl.* **2022**, *262*, 49. [CrossRef]
42. Kasliwal, V.P.; Vogeley, M.S.; Richards, G.T. Are the variability properties of the Kepler AGN light curves consistent with a damped random walk? *Mon. Not. R. Astron. Soc.* **2015**, *451*, 4328–4345. [CrossRef]

43. Vettigli, G. MiniSom: Minimalistic and NumPy-Based Implementation of the Self Organizing Map. 2018. Available online: https://github.com/JustGlowing/minisom (accessed on 8 June 2023).
44. Edelson, R.A.; Krolik, J.H.; Pike, G.F. Broad-Band Properties of the CfA Seyfert Galaxies. III. Ultraviolet Variability. *Astrophys. J.* **1990**, *359*, 86. [CrossRef]
45. Vaughan, S.; Edelson, R.; Warwick, R.S.; Uttley, P. On characterizing the variability properties of X-ray light curves from active galaxies. *Mon. Not. R. Astron. Soc.* **2003**, *345*, 1271–1284. [CrossRef]
46. Solomon, R.; Stojkovic, D. Variability in quasar light curves: Using quasars as standard candles. *J. Cosmol. Astropart. Phys.* **2022**, *2022*, 060. [CrossRef]
47. Condon, J.J.; Matthews, A.M. ΛCDM Cosmology for Astronomers. *Publ. Astron. Soc. Pac.* **2018**, *130*, 073001. [CrossRef]
48. Planck Collaboration; Aghanim, N.; Akrami, Y.; Ashdown, M.; Aumont, J.; Baccigalupi, C.; Ballardini, M.; Banday, A. J.; Barreiro, R. B.; Bartolo, N.; et al. Planck 2018 results—VI. Cosmological parameters. *Astron. Astrophys.* **2020**, *641*, A6. [CrossRef]
49. Rodrigo, C.; Solano, E.; Bayo, A. *SVO Filter Profile Service*, Version 1.0; IVOA Working Draft: València, Spain, 15 October 2012. [CrossRef]
50. Rodrigo, C.; Solano, E. The SVO Filter Profile Service. In Proceedings of the XIV. 0 Scientific Meeting (Virtual) of the Spanish Astronomical Society, Virtual, 5–9 September 2020.
51. Dexter, J.; Agol, E. Quasar Accretion Disks are Strongly Inhomogeneous. *Astrophys. J. Lett.* **2011**, *727*, L24. [CrossRef]
52. Zu, Y.; Kochanek, C.S.; Kozłowski, S.; Udalski, A. Is Quasar Optical Variability a Damped Random Walk? *Astrophys. J.* **2013**, *765*, 106. [CrossRef]
53. Caplar, N.; Lilly, S.J.; Trakhtenbrot, B. Optical Variability of AGNs in the PTF/iPTF Survey. *Astrophys. J.* **2017**, *834*, 111. [CrossRef]
54. Ruan, J.J.; Anderson, S.F.; MacLeod, C.L.; Becker, A.C.; Burnett, T.H.; Davenport, J.R.A.; Ivezić, Ž.; Kochanek, C.S.; Plotkin, R.M.; Sesar, B.; et al. Characterizing the Optical Variability of Bright Blazars: Variability-based Selection of Fermi Active Galactic Nuclei. *Astrophys. J.* **2012**, *760*, 51. [CrossRef]
55. Foong, A.; Bruinsma, W.; Gordon, J.; Dubois, Y.; Requeima, J.; Turner, R. Meta-learning stationary stochastic process prediction with convolutional neural processes. *Adv. Neural Inf. Process. Syst.* **2020**, *33*, 8284–8295.
56. Tak, H.; Mandel, K.; Van Dyk, D.A.; Kashyap, V.L.; Meng, X.L.; Siemiginowska, A. Bayesian estimates of astronomical time delays between gravitationally lensed stochastic light curves. *Ann. Appl. Stat.* **2017**, *13*, 1309–1348. [CrossRef]
57. Breivik, K.; Connolly, A.J.; Ford, K.E.S.; Jurić, M.; Mandelbaum, R.; Miller, A.A.; Norman, D.; Olsen, K.; O'Mullane, W.; Price-Whelan, A.; et al. From Data to Software to Science with the Rubin Observatory LSST. *arXiv* **2022**, arXiv:2208.02781.
58. Sánchez-Sáez, P.; Lira, H.; Martí, L.; Sánchez-Pi, N.; Arredondo, J.; Bauer, F.E.; Bayo, A.; Cabrera-Vives, G.; Donoso-Oliva, C.; Estévez, P.A.; et al. Searching for Changing-state AGNs in Massive Data Sets. I. Applying Deep Learning and Anomaly-detection Techniques to Find AGNs with Anomalous Variability Behaviors. *Astron. J.* **2021**, *162*, 206. [CrossRef]
59. Holmstrom, L.; Koistinen, P. Using additive noise in back-propagation training. *IEEE Trans. Neural Netw.* **1992**, *3*, 24–38. [CrossRef] [PubMed]
60. Matsuoka, K. Noise injection into inputs in back-propagation learning. *IEEE Trans. Syst. Man Cybern.* **1992**, *22*, 436–440. [CrossRef]
61. Bishop, C. Training with noise is equivalent to Tikhonov regularization. *Neural Comput.* **1995**, *7*, 108–116. [CrossRef]
62. Wang, C.; Principe, J. Training neural networks with additive noise in the desired signal. *IEEE Trans. Neural Netw.* **1999**, *10*, 1511–1517. [CrossRef]
63. Zur, R.M.; Jiang, Y.; Pesce, L.L.; Drukker, K. Noise injection for training artificial neural networks: A comparison with weight decay and early stopping. *Med. Phys.* **2009**, *36*, 4810–4818. [CrossRef]
64. Goodfellow, I.; Bengio, Y.; Courville, A. Adaptive computation and machine learning. In *Deep Learning*; MIT Press: Cambridge, MA, USA, 2016.
65. Zhang, X.; Yang, F.; Guo, Y.; Yu, H.; Wang, Z.; Zhang, Q. Adaptive Differential Privacy Mechanism Based on Entropy Theory for Preserving Deep Neural Networks. *Mathematics* **2023**, *11*, 330. [CrossRef]
66. Reed, R.D.; Marks, R.J. *Neural Smithing: Supervised Learning in Feedforward Artificial Neural Networks*; MIT Press: Cambridge, MA, USA, 1998.
67. Naul, B.; Bloom, J.S.; Pérez, F.; van der Walt, S. A recurrent neural network for classification of unevenly sampled variable stars. *Nat. Astron.* **2018**, *2*, 151–155. [CrossRef]
68. Kasliwal, V.P.; Vogeley, M.S.; Richards, G.T.; Williams, J.; Carini, M.T. Do the Kepler AGN light curves need reprocessing? *Mon. Not. R. Astron. Soc.* **2015**, *453*, 2075–2081. [CrossRef]
69. De Cicco, D.; Bauer, F.E.; Paolillo, M.; Sánchez-Sáez, P.; Brandt, W.N.; Vagnetti, F.; Pignata, G.; Radovich, M.; Vaccari, M. A structure function analysis of VST-COSMOS AGN. *Astron. Astrophys.* **2022**, *664*, A117. [CrossRef]
70. Hawkins, M. Variability in active galactic nuclei: Confrontation of models with observations. *Mon. Not. R. Astron. Soc.* **2002**, *329*, 76–86. [CrossRef]
71. Lewis, G.F.; Miralda-Escude, J.; Richardson, D.C.; Wambsganss, J. Microlensing light curves: A new and efficient numerical method. *Mon. Not. R. Astron. Soc.* **1993**, *261*, 647–656. [CrossRef]
72. Hopkins, P.F.; Richards, G.T.; Hernquist, L. An Observational Determination of the Bolometric Quasar Luminosity Function. *Astrophys. J.* **2007**, *654*, 731. [CrossRef]

73. Shen, X.; Hopkins, P.F.; Faucher-Giguère, C.A.; Alexander, D.M.; Richards, G.T.; Ross, N.P.; Hickox, R.C. The bolometric quasar luminosity function at z = 0–7. *Mon. Not. R. Astron. Soc.* **2020**, *495*, 3252–3275. [CrossRef]
74. De Paolis, F.; Nucita, A.A.; Strafella, F.; Licchelli, D.; Ingrosso, G. A quasar microlensing event towards J1249+3449? *Mon. Not. R. Astron. Soc.* **2020**, *499*, L87–L90. [CrossRef]
75. Graham, M.J.; McKernan, B.; Ford, K.E.S.; Stern, D.; Djorgovski, S.G.; Coughlin, M.; Burdge, K.B.; Bellm, E.C.; Helou, G.; Mahabal, A.A.; et al. A Light in the Dark: Searching for Electromagnetic Counterparts to Black Hole-Black Hole Mergers in LIGO/Virgo O3 with the Zwicky Transient Facility. *Astrophys. J.* **2023**, *942*, 99. [CrossRef]
76. Jovanović, P.; Zakharov, A.F.; Popović, L.Č.; Petrović, T. Microlensing of the X-ray, UV and optical emission regions of quasars: Simulations of the time-scales and amplitude variations of microlensing events. *Mon. Not. R. Astron. Soc.* **2008**, *386*, 397–406. [CrossRef]
77. Wang, J.; Smith, M.C. Using microlensed quasars to probe the structure of the Milky Way. *Mon. Not. R. Astron. Soc.* **2010**, *410*, 1135–1144. [CrossRef]
78. Wambsganss, J. Microlensing of Quasars. *Publ. Astron. Soc. Aust.* **2001**, *18*, 207–210. [CrossRef]

universe

MDPI

Article

Gas and Stars in the Teacup Quasar Looking with the 6-m Telescope

Alexei V. Moiseev [1,*] and Alina I. Ikhsanova [2]

[1] Special Astrophysical Observatory, Russian Academy of Sciences, 369167 Nizhny Arkhyz, Russia
[2] Dipartimento di Fisica e Astronomia "G. Galilei", Università di Padova, Vicolo dell'Osservatorio 3, 35122 Padova, Italy
* Correspondence: moisav@gmail.com

Abstract: New results on the radio-quiet type 2 quasar, known as the Teacup galaxy (SDSSJ1430+1339), based on the long-slit and 3D spectroscopic data obtained at the Russian 6-m telescope, are presented. The ionized gas giant nebula, which extends up to $r = 56$ kpc in the [O III] emission line, was mapped with the scanning Fabry–Perot interferometer. The direct estimation of the emission line ratios confirmed that the giant nebula is ionized by the AGN. Stars in the inner $r < 5$ kpc are significantly younger than the outer host galaxy and have a solar metallicity. The central starburst age ($\sim$1 Gyr) agrees with possible ages for the galactic merger events and the previous episode of the quasar outflow produced two symmetric arcs visible in the [O III] emission at the distances $r = 50$–55 kpc. The ionized gas velocity field can be fitted by the model of a circular rotating disk significantly inclined or even polar to the stellar host galaxy.

Keywords: interstellar medium (ISM); nebulae; galaxies; individual; teacup

1. Introduction

Extended Emission-Line Regions (EELRs) detected in the outskirts of some active galaxies are considered as a result of Active Galactic Nuclei (AGN) feedback in the radiative and kinetic (jet and wind outflows) forms. The Seyfert galaxy Mrk 6 gives us a nearby, recent, and a very typical example of EELR related with AGN ionization cone observed outside of stellar host up to a projected distance 40 kpc [1]. A study of such structures allows us to better understand both the history of nuclear radiative output on the scales 0.01–0.1 Myr and the distribution of intergalactic medium (and references therein [2–4]).

One of the largest EELR among low-redshift radio-quiet AGN was recently found around the Teacup galaxy (SDSS J143029.88+133912.0). The galaxy was discovered by volunteers of the Galaxy Zoo project, and its EELR was confirmed in follow-up spectroscopic observations [5]. This type-2 quasar was nicknamed due to the morphology of the ionized gas bubbles that extend 10 kpc from the galactic center in the shape of a 'handle'. Keel et al. [6] considered Teacup as a fading AGN, whereas X-ray data have indicated that it is possible that no fading is required [7].

The Hubble Space Telescope (HST) imaging shows that the Teacup is a bulge-dominated galaxy, with a shell-like structure and a tidal tail, which have been interpreted as an indicative of merger 1–2 Gyr ago with a cold disk system and 1/10 mass ratio [8]. Another interesting feature of this system is the presence of a giant outflow generated by small-scale radio jets and/or quasar winds (Keel et al. [6], Harrison et al. [9]). This outflow appears to be responsible for the bubble-like morphology of the galaxy. The ionized gas kinematics of the outflow in the inner 15 kpc were studied several times using long-slit and 3D-spectroscopic technique [6,10–12]. However, the stellar kinematics of the host galaxy is still unknown.

The long-slit Gran Telescopio Canarias (GTC) optical spectroscopy by Villar-Martín et al. [13] reveals that the ionizied gas around Teacup extends up to 50 kpc in the Hα

emission line. This giant nebula has been considered by the authors as reservoir of the circumgalactic medium populated by tidal debris produced by galactic merger events. The external gas is most likely photoionized by the nuclear radiation and is significantly dynamical cold comparing with the inner 10–20 kpc region. The recent GTC deep image in the Hα emission line clearly demonstrates that the giant Teacup EELR (155 $\times$ 87 kpc in a total size) elongates in the same direction with the main axis of the inner bubble and radio jet [14]. Several arcs and emission knots are visible up to $\sim$56 kpc from the AGN. However, details of gas rotation pattern are still unknown (it has been proposed that the gas maybe settled in a giant rotating disk [13]).

The giant nebula is most likely photoionized by the nuclear radiation, however to construct their diagnostic diagrams, Villar-Martín et al. [13] accepted the I([O III] λ 5007)/I(Hβ) flux ratio similar with the value obtained early for the internal regions, due to the lack of a green lines in the GTC spectrum.

In this work we present the results of new long-slit and 3D spectroscopic observations at the 6-m telescope of the Special Astrophysical Observatory of the Russian Academy of Sciences (SAO RAS) performed to solve the above-mentioned puzzles of the Teacup galaxy: the properties of the host stellar population and the structure and the velocity field of the extended nebula, including direct estimation of its ionization state. Following Ref. [13], we accepted Teacup redshift $z = 0.085$, which corresponds to the distance 360 Mpc (H_0 = 71 kms^{-1} Mpc^{-1}) and a spatial scale 1.58 kpc arcsec^{-1}.

2. Observations and Data Analysis

The observations were carried out at the prime focus of the SAO RAS 6 m telescope with the SCORPIO-2 multi-mode focal reducer [15]. The detector was CCD $2K \times 4.5K$ E2V 42-90 . The observations in the long-slit mode were performed with the $6' \times 1''$ slit and the scale $0.36''$/px, other parameters are given in the Table 1 (T_{exp}—total exposure, θ—mean seeing value, $\Delta\lambda$ and $\delta\lambda$ are the spectral range and resolution). The position angle $PA = 60^\circ$ corresponds to the major axis of the nebula (Figure 1). The initial data reduction was performed in a standard way, as described in our previous papers, e.g., Ref. [16].

The 3D-spectroscopy mapping in the [O III]λ5007 emission line was carried out in the scanning Fabry–Perot interferometer (FPI) mode of SCORPIO-2 with the same low-resolution FPI that is usually used for the tunable filter imaging in the MaNGaL device [17]. During the observations we subsequently obtained a narrow band (the bandwidth $\sim$13 Å) images with different central wavelength: five frames spanned the spectral range around the redshifted [O III]λ5007 line with the step 5.9 Å and one frame in continuum at the central wavelength shifted on 24 Å from this emission line. The field of view at the new detector CCD $2K \times 4K$ E2V 261-80 was $6.8'$ sampled with a pixel scale $0.78''$ and $0.39''$ at the nights 19 April 2020 and 24 April, respectively; other parameters are given in Table 1.

The preliminary data reduction (bias, flat-field correction, combining individual 90 s exposures with cosmic-ray hits cleaning) was performed with IDL-based software as described in Moiseev et al. [17] for the MaNGaL scanning mode. The air-glow lines subtraction, photometric correction and phase-shift wavelength calibration were done with our software for SCORPIO-2/FPI data reduction (see description and references in Ref. [18]). The data obtained during two nights were merged in the single data cube containing low-resolution [O III] spectra in each pixel with the size $0.78''$. The total exposure was about 2.4 h. The surface brightness of the most faint detected emission filaments is about 2×10^{-18} erg s^{-1} cm^{-2} arcsec^{-2}.

Table 1. Log of SCORPIO-2/6 m telescope observations.

Mode	Date of Obs.	T_{exp}, s	θ, $''$	$\Delta\lambda$, Å	$\delta\lambda$, Å
Long-slit	11 February 2018	4 $\times$ 1200	2.2	3500–7220	5
FPI	19 April 2020	41 $\times$ 90	3.2	5410–5470	13
FPI	24 April 2020	50 $\times$ 90	1.6	5410–5470	13

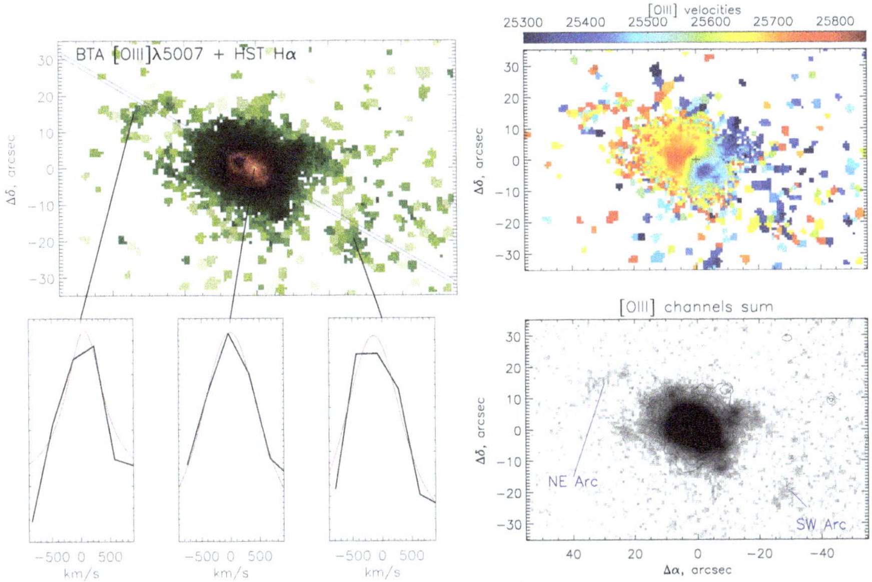

Figure 1. The left panel: the [O III] emission map taken with FPI after Voronoi tessellation binning (the green scale) combined with ACS HST Hα image [8] of the central bubble region (in the red scale). The purple lines mark the position of the spectrograph slit (1$''$ in the width). The bottom plots show the emission-line profiles in the selected regions, the Voigt fitting is marked by the red. The right panels display the ionized gas velocity field (right top) and the sum of three channels centered at the emission line in the original 0.78$''$ pixel scale with the isophotes of the image in the channel correspond to the stellar continuum. The most distance emission filaments are marked: the NE Arc (firstly mapped in Ref. [14]) and SW Arc.

In order to improve signal-to-noise (S/N) ratio in the spectra of the faint outer region the Voronoi tessellation was used [19]. The emission line in the binned data cube was fitted with Voigt function, which provides a good approximation of the observed FPI spectra [20]. The example of emission line profiles together with the [O III] flux maps (in the binned and original resolution) and line-of-sight velocity field is shown in Figure 1.

3. The Stellar Population Properties and Kinematics

The observed long-slit spectra contain a combination of the ionized gas emission and stellar absorption lines. We used the penalized pixel-fitting (pPXF) method to fit a stellar population spectrum [21], using MILES stellar spectral library by Vazdekis et al. [22] covering the range 3525–7500 Å with a twice higher spectral resolution than in the SCORPIO-2 data. In order to improve the signal-to-noise ratio, the spectra used for the pPXF analysis were binned along the slit using the step exponentially increasing with radius: from 2 px (0.7$''$) bins in the nucleus up to 10 px (3.6$''$) at the distances $r \sim 10''$ from the center. An example of the observed Teacup spectrum fitted by pPXF model is shown in Figure 2.

Figure 3 shows the distribution along the spectrograph's slit of the stellar line-of-sight velocities, age, and metallicity derived from the description of the SCORPIO-2 spectra by the single stellar population (SSP) pPXF model. This plot demonstrates that the inner $r < 3$–$4''$ (5–6.5 kpc) region is decoupled both in kinematic and stellar population properties from the outer host galaxy. The luminosity-averaged age of stars in the central region is significantly younger compared with the outer part of the galaxy ($\sim$1 vs. 7–9 Gyr) and has a near solar metallicity typical for the recent burst of star formation, whereas [M/H] ≈ -1 in the outer region, which is typical for the "red-sequence" early-type galaxy.

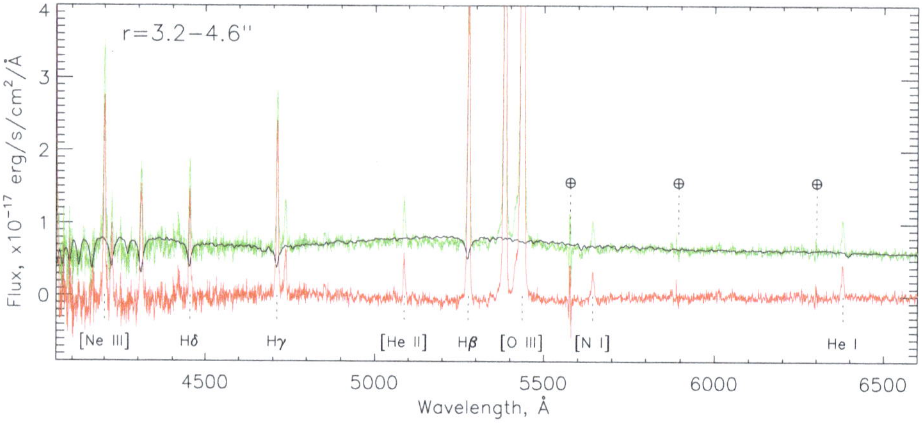

Figure 2. pPXF analysis of the observed spectrum (the green line) in the radial bin corresponds to the distance from the Teacup nucleus r = 3.2–4.6″. The stellar population model is shown in black, the red spectrum is a residual after model subtraction. The main ionized gas emission lines and position of the airglow lines are labeled.

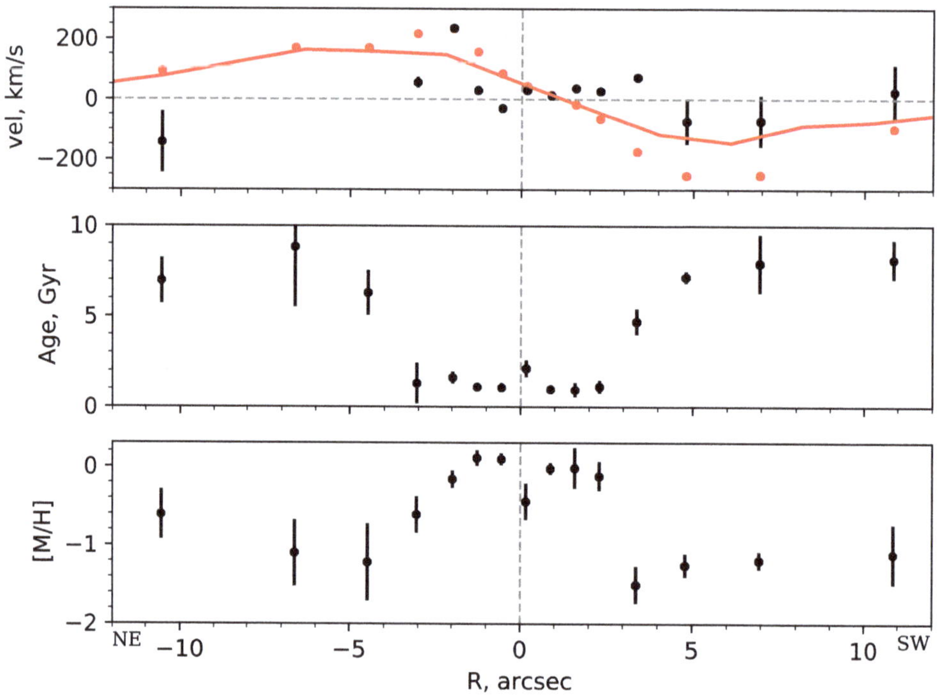

Figure 3. Gas/stars kinematics and stellar population properties derived from SCORPIO-2 observations at PA = 60°. From top to bottom: line-of-sight velocity distribution for stars (black dots) and ionized gas (red), the red line shows the velocities in this direction according to FPI data, the systemic velocity 25,482 km s^{-1} was subtracted; radial distribution of the SSP age; the same for the stellar metallicity.

The flux and velocity of the ionized gas emission lines were derived via Gaussian fitting of galactic spectra after subtraction of the pPXF model in each radial bin. During this procedure we accepted the same line-of-sight velocity separately for the system of the forbidden ([O III], [N II], [S II]) and the Balmer hydrogen lines. The top panel in the Figure 3 demonstrates an agreement between velocities of the forbidden lines in the SCORPIO-2 long-slit spectrum (red points) and a pseudo-slit cut through the FPI velocity field (red line) if the differences in the spatial and spectral resolutions will be taken into account.

According to our Gaussian fitting results, the mean flux ratio of the brightest Balmer lines at $r < 12''$ $I(\mathrm{H}\alpha)/I(\mathrm{H}\beta)= 2.9 \pm 0.1$ is in a good agreement with the 'standard' value of an intrinsic Balmer decrement both for AGN and H II-regions ($\sim$3.1 and 2.86, see Groves et al. [23] for a review). For this reason, we did not correct the observed spectrum for interstellar dust extinction. The reddening map presented in Gagne et al. [11] reveals that the significant extinction $E(B-V) > 0.2$ (possibly related to a circumnuclear dust lane) is detected in the inner $r < 2''$. This region is poorly resolved in our long-slit data, with a seeing value of $2.2''$.

The kinematics of the stars significantly differs from the gaseous component: in the inner $r < 5''$ the radial gradient of stellar line-of-sight velocities is near zero with a hint of slow counter-rotation in the SW part of the velocity distribution. Such a kinematic feature is indicative of a multi-spin galaxy where the gas and stars rotate in different planes and/or in different directions. Indeed, an absence of a velocity gradient should be observed if we put a slit along the rotating disk minor axis, or if an unresolved counter-rotating stellar component presents in the kinematically distinct core [24,25]. Both features are unsurprising after galactic merging or accretion of the external gas by early type host galaxy [26].

4. The Gas Ionization

To discriminate the gas excitation mechanism the BPT (after [27]) diagnostic diagrams of the emission line flux ratios were widely used. In a previous work, Ref. [13] argued that even the very external regions of the Teacup giant nebula is illuminated by the AGN ionized radiation. However, the GTC spectra did not cover the green region. This is the reason why in Ref. [13] a very important $I([\mathrm{O\,III}]\,\lambda\,5007)/I(\mathrm{H}\beta)$ line ratio for the giant EELR was accepted as the same nuclear value.

In our 6-m telescope spectrum we were able to detect the [O III] line in the emission arcs up to $r \approx 55$ kpc (Figure 4, left). The flux in the Hβ was also estimated in the integrated spectrum on distances r = 12–26$''$ (19–41 kpc). This gives a direct estimation of the $I([\mathrm{O\,III}]\,\lambda\,5007)/I(\mathrm{H}\beta)= 0.80 \pm 0.08$, which agrees in the errors with the prediction in Ref. [13]. The diagram $I([\mathrm{O\,III}]\,\lambda\,5007)/I(\mathrm{H}\beta)$ vs $I([\mathrm{N\,II}])/I(\mathrm{H}\alpha)$ (Figure 4, right) clearly demonstrates that the observed line ratios both for the central and external parts of the giant nebula correspond to AGN-type ionizaion. Here we used $I([\mathrm{O\,III}]\,\lambda\,5007)/I(\mathrm{H}\beta)$ and $I([\mathrm{N\,II}])/I(\mathrm{H}\alpha)$ ratios derived from the 6 m and GTC observations, respectively, and accepted that the $I([\mathrm{N\,II}])/I(\mathrm{H}\alpha)$ ratio in the NE region $PA = 60^\circ$ at $r = 12-26''$ is the same as in the integrated GTC spectral measurements in two apertures at $r = 11.8 \pm 2.4''$ and $16.6 \pm 1.8''$, as according to Villar-Martín et al. [13].

A similar situation in other BPT-diagrams is presented by Villar-Martín et al. [13]: the line ratios of the emission lines in the Teacup EELR clearly corresponds to the ionization by UV-radiation of the central QSO.

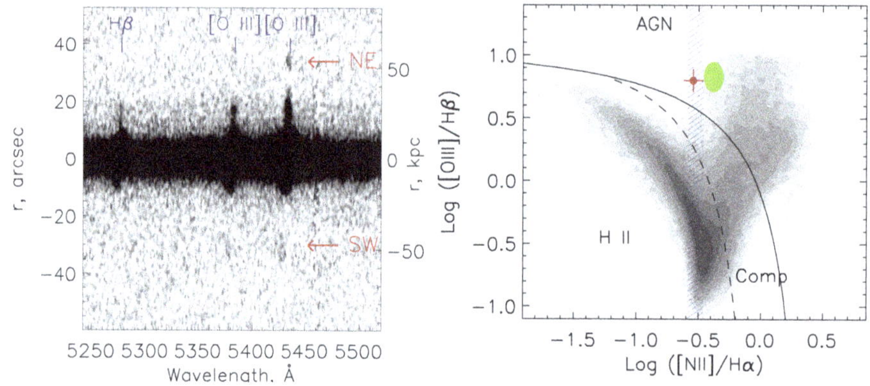

Figure 4. Ionized gas in the Teacup giant EELR. **Left**: a fragment of the 2D SCORPIO-2 spectrum along $PA = 60°$ around the Hβ+[O III]λλ4959, 5007 region. The red arrows mark the NE and WS Arcs. **Right**: a BPT diagram similar to one the presented in Figure 3 [13]. The gray contours represent the density distribution of the SDSS galaxies line ratios based on the data from Kewley et al. [28]. The division lines between the star-forming galaxies, AGN, and composite nucleus are taken from Kewley et al. [29], Kauffmann et al. [30]. The vertical blue shadowed area represents the range of values measured for the giant nebula at GTC [13]. The green filled ellipse shows the line ratios measured by Gagne et al. [11] for the areanearest the AGN regions ($r < 15$ kpc). The red point corresponds to both our SCOPRIO-2 (I([O III] λ 5007)/I(Hβ)) and GTC (I([N II])/I(Hα)) measurements for the distances r = 19–41 kpc.

5. The Extended Nebula: Morphology and Kinematics

At the first glance, the distribution of the ionized gas derived in SCORPIO-2 observations looks similar to the deep Hα emission line images recently published in Villar Martín et al. [14]: the whole emission structure up to $r \approx 18''$ (∼28 kpc) possesses the inner bubble (∼10 kpc) and distant emission patches and filaments at $r \approx 30$–$35''$ (∼47–55 kpc). The giant emission nebula is elongated in the same position angle with the radio jet direction ($PA \approx 60°$ [9]). However, the image in the hydrogen recombination line Hα reveals the characteristic Arc+Cavity structure in the NE side of the nebula, and a very faint straight filament in the SW direction. In contrast with this picture, on the image in the high excited forbidden [O III] line (Figure 1), we see the arc-like structure in the SE side as well. It seems to be symmetric with the NE arc being in agreement with Villar Martín et al. [14], who suggested that the large scale morphology of the nebula is influenced by AGN.

We tried to describe the observed line-of-sight ionized gas velocity distribution by the model of a regular circular rotation using our adaptation of a classical 'tilted-ring' technique (see Refs. [18,26] and the references therein). The mean parameters of the gaseous disk orientation were determined from the central part of the velocity field ($r < 25''$): inclination $i_0 = 43 \pm 7°$ and the position angle $PA_0 = 62 \pm 4°$. Figure 5 shows the radial variations of the main model parameters: the circular rotation velocity V_{ROT} and kinematic position angle PA_{kin} for the fixed values of the inclination ($i_{kin} = i_0$) and systemic velocity. The PA_{kin} was also fixed for large radii ($r > 21''$) in order to avoid unstable approximation in the sparse velocity field regions.

It is possible that the rotation curve presented in Figure 5 is affected by non-circular motions, first of all in the central $r < 15$ kpc (location of the giant bubble—'teacup handle'). This fact is indicated by both the PA_{kin} radial changes and maps of the residual velocities (Figure 5, right) with well-ordered patches of negative and positive values. The maximal value of the deviations from the mean rotation (±70 km s^{-1}) could be considered as the lower limit on the speed of a large-scale outflow; the projection effect can increase this value. Note that the amplitude of these non-circular motions along line-of-sight is significantly smaller than V_{ROT} at the corresponding distances.

The spread of velocity measurements in the faint distant regions prevents us from interpretation of its velocity residuals. Nevertheless, we can conclude that the gas in both the NE and SW arcs lies on the common flat rotation curve, with an amplitude of 100–140 $\mathrm{km\,s^{-1}}$.

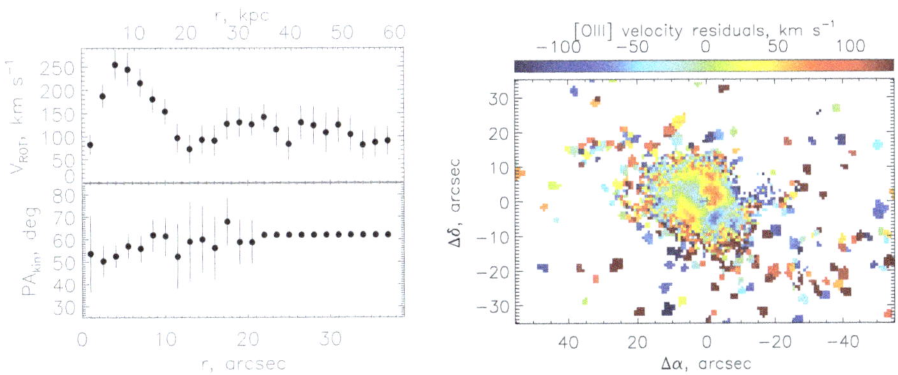

Figure 5. Radial variations of the tilted-ring model parameters (**left panel**): the circular rotation velocity (**top**) and the kinematic position angle (**bottom**). The **right panel** shows residual velocity field after subtraction of the disk circular rotation model.

6. Discussion

There is no doubt that the stellar morphology of the Teacup galaxy reveals the footprint of a merger event (Section 1). The observed shell-like narrow tidal features could be either a result of a minor merger (mass ratio $\sim$1/10) of a cold disk system with a massive elliptical galaxy [31,32], or a moderate merger of two disk galaxies (mass ratio $\sim$1/2, [33]). In both cases, the lifetime of the observed stellar arcs could be about 1–2 Gyr.

In Section 3 we find out that the stellar population in the center of the galaxy ($r < 5$ kpc) is significantly younger (SSP age $\sim$ 1 Gyr) and richer in metal ([M/H] ≈ 0) compared to the outer host (SSP age $\sim$ 7 Gyr, [M/H] ≈ -1). The age of the central starburst puts it in the above range of the merger age. This indicates that the star formation is possible triggered by interaction, as has been predicted in numerous simulations (Ref. [33] and references therein). It appears that the same interaction event fed the AGN. In this case, we do not exclude that the central star formations were partly a result of AGN outflow positive feedback. For instance, Ref. [8] found several candidates for young star clusters in the inner 8 kpc region, outside of the ionization cone boundaries.

The ionized gas distribution agrees with the picture suggested by Ref. [14] for this and other optically selected Type-2 quasars: in the systems undergoing interaction or merging, the ionized gas was spread over large spatial scales (10–90 kpc) and was illuminated by the AGN. Using the long-slit spectroscopy we directly measured the indicative line ratio $I([\mathrm{O\,III}]\,\lambda\,5007)/I(\mathrm{H}\beta)$ in the external regions (19–41 kpc) of the Teacup nebula for the first time. Our result fully confirmed the previous suggestion about domination of the AGN ionization, even on these distances [13].

The analysis of the [O III] velocity field demonstrates that the gas kinematics can be described in the model of a global rotating disk with the line-of-nodes major axis $PA_0 = 62 \pm 4°$. This value is significantly different from the orientation of the stellar continuum isophotes (Figure 1): $PA_* = 157 \pm 3°$ according to our estimation at r = 18–20″. The same orientation of the red continuum isophotes also appears in Figure A.16 in Ref. [22]. It implies that the gaseous disk is significantly inclined or even orthogonal to the stellar one, which also agrees with kinematically decoupling stars/gas according to long-slit data (Section 3): the stellar radial velocity gradient is near zero and significantly smaller than gaseous one, because the slit crosses the stellar host near its minor axis.

The domination of a regular circular rotation in the gaseous kinematics together with the multi-spin gas/stars configuration might constrain the parameters of the interaction produced by the Teacup galaxy. It is possible that in most of the nebula we are observing not the main galaxy gas, spread by galactic interaction and outflow, but the matter accreted from a companion with a corresponding spin orientation.

The deep [O III] emission line image reveals that the arc-like faint emission structure elongated according to the radio jet direction exists not only in the NE direction, as follows from the previous Hα map by Villar Martín et al. [14], but also appears in SW at the same distance from the AGN (50–55 kpc). The shape of both the SW and NE arcs partly repeats the well-known inner emission bubbles, suggesting its possible origin: Could it be related to a previous AGN outflow?

There are different estimations of current AGN outflow velocities in the region of 10–12 kpc radio bubbles, spanning the range of $V_{out} = 50\ldots150\,\mathrm{km\,s^{-1}}$ [9,11,12]. If we accepted a conservative estimation $V_{out} > 70\,\mathrm{km\,s^{-1}}$ according to the [O III] residual velocity map (Section 5), then the dynamical age of the NE and SW arcs is <0.8 Gyr. This value seems to be in good agreement with the age of a central star formation and minor/moderate merging. Therefore, our estimation does not contradict the fact that the most distant emission arcs around the Teacup galaxy are related with the first QSO activity episode triggered by galactic interaction, which also started a central burst of star formation. Moreover, it is possible that a circumnuclear starburst also contributed to the formation of the external emission arcs via a galactic wind (see the paper by López-Cobá et al. [34] for a review and observational examples). A more deep study of the external gas ionization properties and its kinematics (including the velocity dispersion distribution) are needed to separate the possible influence of AGN and starburst-driven wind and radio jet action on a formation of the emission arcs.

7. Conclusions

3D spectroscopy with scanning FPI is an old but very powerful technique used to study different astrophysical objects. In this work we present new observational capabilities of the SAO RAS 6 m telescope with low-resolution FPI, which was early used by our team as a tunable filter at 1–2.5 m telescopes. The example of the giant nebula related with radio-quiet quasar known as the Teacup galaxy demonstrates that with this device we are able to map emission lines at the surface brightness level $(few) \times 10^{-18}\,\mathrm{erg\,s^{-1}\,cm^{-2}\,arcsec^{-2}}$ during 2 h of exposures and even study the gas kinematics if the amplitude of velocity changes exceeds 50–20 $\mathrm{km\,s^{-1}}$.

The Teacup galaxy has been well studied, including multiwavelength data from X-ray to radio and integral-field spectroscopy in optical and near infrared. Nevertheless, using SCORPIO-2 long-slit and 3D spectroscopy we obtained the following new results:

- The indicative line ratio $log(I([\mathrm{O\,III}]\,\lambda\,5007)/I(\mathrm{H}\beta))$ was directly estimated for the external regions (19–41 kpc) of the Teacup nebula. The obtained value 0.80 ± 0.08 lies in the range 0.7–0.9 presented by Gagne et al. [11] for the nearest to the AGN region. Together with the $I([\mathrm{N\,II}])/I(\mathrm{H}\alpha)$ ratio obtained early on in the GTC observations it confirms that the domination of AGN radiation in gas ionization is in good agreement with the conclusion of the paper [13];
- The stars in the inner $r < 5$ kpc are significantly younger and richer in metal than the outer host galaxy. The starburst age ($\sim$1 Gyr) agrees with the timescale of a merger event, as proposed by Keel et al. [8];
- The ionized gas velocity field can be described in the term of a circular rotating disk with a flat rotation curve up to distances 50–60 kpc. This disc appears to be significantly inclined or even polar to the stellar host galaxy;
- The deep map of the [O III] emission reveals two symmetric arcs in the external region of the EELR (r = 50–55 kpc). It might be a remnant of the previous AGN outflow (it may be in a combination with a starburst-driven galactic wind) with the age < 0.8 Gyr.

An intriguing puzzle is the alignment of the line-of-nodes of the global rotating gaseous disk and radio jet (and ionization cone) direction. Is this a coincidence or a manifestation of a more powerful AGN influence on the surrounding gas than we expected? In any case, we hope that the results presented here will be useful for further detailed simulations of the Teacup system formation, including interaction with a companion and AGN feedback.

Author Contributions: Conceptualization, A.V.M.; Observations and data reduction, A.V.M.; methodology, A.V.M. and A.I.I.; validation, A.V.M. and A.I.I.; investigation, A.V.M. and A.I.I.; writing—original draft preparation, A.V.M. and A.I.I.; writing—review and editing, A.V.M. and A.I.I.; visualization, A.V.M. and A.I.I.; supervision, A.V.M.; funding acquisition, A.V.M. All authors have read and agreed to the published version of the manuscript.

Funding: This research was funded by grant No075-15-2022-262 (13.MNPMU.21.0003) of the Ministry of Science and Higher Education of the Russian Federation.

Institutional Review Board Statement: Not applicable.

Informed Consent Statement: Not applicable.

Data Availability Statement: The data underlying this article will be shared on reasonable request to the corresponding author. The raw data are available in the General Observation Archive of the SAO RAS telescopes: https://www.sao.ru/oasis/cgi-bin/fetch.

Acknowledgments: We obtained the observed data on the unique scientific facility "Big Telescope Alt-azimuthal" of SAO RAS. The long-slit observations were peformed by Dmitry Oparin. We thank the anonymous referees and Cristina Ramos Almeida for their constructive comments and Aleksandrina Smirnova for her help in preparing the text. This work is dedicated to the memory of Victor Afanasiev, whose enthusiasm and work helped make these observations. Some of the data presented in this paper were obtained from the Mikulski Archive for Space Telescopes (MAST). This research made use of NASA's Astrophysics Data System and the the NASA/IPAC Extragalactic Database (NED), which is operated by the Jet Propulsion Laboratory, California Institute of Technology, under contract with the National Aeronautics and Space Administration.

Conflicts of Interest: The authors declare no conflict of interest.

Abbreviations

The following abbreviations are used in this manuscript:

ASC	Advanced Camera for Surveys
EELR	Extended Emission-Line Regions EELR
FPI	Fabry–Perot interferometer
GTC	Gran Telescopio Canarias
HST	Hubble Space Telescope
SAO RAS	Special Astrophysical Observatory of the Russian Academy of Sciences
SSP	Single stellar population

References

1. Smirnova, A.A.; Moiseev, A.V.; Dodonov, S.N. A close look at the well-known Seyfert galaxy: Extended emission filaments in Mrk 6. *Mon. Not. R. Astron. Soc.* **2018**, *481*, 4542–4547. [CrossRef]
2. Morganti, R. Archaeology of active galaxies across the electromagnetic spectrum. *Nat. Astron.* **2017**, *1*, 596–605. [CrossRef]
3. Knese, E.D.; Keel, W.C.; Knese, G.; Bennert, V.N.; Moiseev, A.; Grokhovskaya, A.; Dodonov, S.N. An [O III] search for extended emission around AGN with H I mapping: A distant cloud ionized by Mkn 1. *Mon. Not. R. Astron. Soc.* **2020**, *496*, 1035–1050. [CrossRef]
4. Keel, W.C.; Moiseev, A.; Kozlova, D.V.; Ikhsanova, A.I.; Oparin, D.V.; Uklein, R.I.; Smirnova, A.A.; Eselevich, M.V. The TELPERION survey for distant [O III] clouds around luminous and hibernating AGN. *Mon. Not. R. Astron. Soc.* **2022**, *510*, 4608–4625. [CrossRef]
5. Gagne, J.; Crenshaw, D.M.; Keel, W.C.; Fischer, T.C. Optical Spectra of the Teacup AGN. *Am. Astron. Soc. Meet. Abstr.* **2011**, *217*, 142.12.
6. Keel, W.C.; Lintott, C.J.; Maksym, W.P.; Bennert, V.N.; Chojnowski, S.D.; Moiseev, A.; Smirnova, A.; Schawinski, K.; Sartori, L.F.; Urry, C.M.; et al. Fading AGN Candidates: AGN Histories and Outflow Signatures. *Astrophys. J.* **2017**, *835*, 256. [CrossRef]

7. Lansbury, G.B.; Jarvis, M.E.; Harrison, C.M.; Alexander, D.M.; Del Moro, A.; Edge, A.C.; Mullaney, J.R.; Thomson, A.P. Storm in a Teacup: X-ray View of an Obscured Quasar and Superbubble. *Astrophys. J. Lett.* **2018**, *856*, L1. [CrossRef]
8. Keel, W.C.; Maksym, W.P.; Bennert, V.N.; Lintott, C.J.; Chojnowski, S.D.; Moiseev, A.; Smirnova, A.; Schawinski, K.; Urry, C.M.; Evans, D.A.; et al. HST Imaging of Fading AGN Candidates. I. Host-galaxy Properties and Origin of the Extended Gas. *Astrophys. J.* **2015**, *149*, 155. [CrossRef]
9. Harrison, C.M.; Thomson, A.P.; Alexander, D.M.; Bauer, F.E.; Edge, A.C.; Hogan, M.T.; Mullaney, J.R.; Swinbank, A.M. Storm in a "Teacup": A Radio-quiet Quasar with ≈10 kpc Radio-emitting Bubbles and Extreme Gas Kinematics. *Astrophys. J.* **2015**, *800*, 45. [CrossRef]
10. Keel, W.C.; Chojnowski, S.D.; Bennert, V.N.; Schawinski, K.; Lintott, C.J.; Lynn, S.; Pancoast, A.; Harris, C.; Nierenberg, A.M.; Sonnenfeld, A.; et al. The Galaxy Zoo survey for giant AGN-ionized clouds: Past and present black hole accretion events. *Mon. Not. R. Astron. Soc.* **2012**, *420*, 878–900. [CrossRef]
11. Gagne, J.P.; Crenshaw, D.M.; Kraemer, S.B.; Schmitt, H.R.; Keel, W.C.; Rafter, S.; Fischer, T.C.; Bennert, V.N.; Schawinski, K. Spatially Resolved Spectra of the "Teacup" Active Galactic Nucleus: Tracing the History of a Dying Quasar. *Astrophys. J.* **2014**, *792*, 72. [CrossRef]
12. Ramos Almeida, C.; Piqueras López, J.; Villar-Martín, M.; Bessiere, P.S. An infrared view of AGN feedback in a type-2 quasar: The case of the Teacup galaxy. *Mon. Not. R. Astron. Soc.* **2017**, *470*, 964–976. [CrossRef]
13. Villar-Martín, M.; Cabrera-Lavers, A.; Humphrey, A.; Silva, M.; Ramos Almeida, C.; Piqueras-López, J.; Emonts, B. A 100 kpc nebula associated with the 'Teacup' fading quasar. *Mon. Not. R. Astron. Soc.* **2018**, *474*, 2302–2312. [CrossRef]
14. Villar Martín, M.; Emonts, B.H.C.; Cabrera Lavers, A.; Bellocchi, E.; Alonso Herrero, A.; Humphrey, A.; Dall'Agnol de Oliveira, B.; Storchi-Bergmann, T. Interactions between large-scale radio structures and gas in a sample of optically selected type 2 quasars. *Astron. Astrophys.* **2021**, *650*, A84. [CrossRef]
15. Afanasiev, V.L.; Moiseev, A.V. Scorpio on the 6 m Telescope: Current State and Perspectives for Spectroscopy of Galactic and Extragalactic Objects. *Balt. Astron.* **2011**, *20*, 363–370. [CrossRef]
16. Egorov, O.V.; Lozinskaya, T.A.; Moiseev, A.V.; Smirnov-Pinchukov, G.V. Star formation complexes in the 'galaxy-sized' supergiant shell of the galaxy HolmbergI. *Mon. Not. R. Astron. Soc.* **2018**, *478*, 3386–3409. [CrossRef]
17. Moiseev, A.; Perepelitsyn, A.; Oparin, D. Mapper of Narrow Galaxy Lines (MaNGaL): New tunable filter imager for Caucasian telescopes. *Exp. Astron.* **2020**, *50*, 199–214. [CrossRef]
18. Moiseev, A.V. Scanning Fabry-Perot Interferometer of the 6-m SAO RAS Telescope. *Astrophys. Bull.* **2021**, *76*, 316–339. [CrossRef]
19. Cappellari, M.; Copin, Y. Adaptive spatial binning of integral-field spectroscopic data using Voronoi tessellations. *Mon. Not. R. Astron. Soc.* **2003**, *342*, 345–354. [CrossRef]
20. Moiseev, A.V.; Egorov, O.V. Reduction of CCD observations made with the Fabry-Perot scanning interferometer. II. Additional procedures. *Astrophys. Bull.* **2008**, *63*, 181–192. [CrossRef]
21. Cappellari, M. Improving the full spectrum fitting method: Accurate convolution with Gauss-Hermite functions. *Mon. Not. R. Astron. Soc.* **2017**, *466*, 798–811. [CrossRef]
22. Vazdekis, A.; Sánchez-Blázquez, P.; Falcón-Barroso, J.; Cenarro, A.J.; Beasley, M.A.; Cardiel, N.; Gorgas, J.; Peletier, R.F. Evolutionary stellar population synthesis with MILES—I. The base models and a new line index system. *Mon. Not. R. Astron. Soc.* **2010**, *404*, 1639–1671. [CrossRef]
23. Groves, B.; Brinchmann, J.; Walcher, C.J. The Balmer decrement of Sloan Digital Sky Survey galaxies. *Mon. Not. R. Astron. Soc.* **2012**, *419*, 1402–1412. [CrossRef]
24. Young, L.M.; Serra, P.; Krajnović, D.; Duc, P.A. The diversity of atomic hydrogen in slow rotator early-type galaxies. *Mon. Not. R. Astron. Soc.* **2018**, *477*, 2741–2759. [CrossRef]
25. Gasymov, D.; Katkov, I. Detailed study of galaxies with the stellar counter-rotation phenomenon. *arXiv* **2022**, arXiv:2209.11240.
26. Sil'chenko, O.K.; Moiseev, A.V.; Egorov, O.V. The Gas Kinematics, Excitation, and Chemistry, in Connection with Star Formation, in Lenticular Galaxies. *Astrophys. J. Suppl. Ser.* **2019**, *244*, 6. [CrossRef]
27. Baldwin, J.A.; Phillips, M.M.; Terlevich, R. Classification parameters for the emission-line spectra of extragalactic objects. *Publ. Astron. Soc. Pac.* **1981**, *93*, 5–19. [CrossRef]
28. Kewley, L.J.; Groves, B.; Kauffmann, G.; Heckman, T. The host galaxies and classification of active galactic nuclei. *Mon. Not. R. Astron. Soc.* **2006**, *372*, 961–976. [CrossRef]
29. Kewley, L.J.; Dopita, M.A.; Sutherland, R.S.; Heisler, C.A.; Trevena, J. Theoretical Modeling of Starburst Galaxies. *Astrophys. J.* **2001**, *556*, 121–140. [CrossRef]
30. Kauffmann, G.; Heckman, T.M.; Tremonti, C.; Brinchmann, J.; Charlot, S.; White, S.D.M.; Ridgway, S.E.; Brinkmann, J.; Fukugita, M.; Hall, P.B.; et al. The host galaxies of active galactic nuclei. *Mon. Not. R. Astron. Soc.* **2003**, *346*, 1055–1077. [CrossRef]
31. Thomson, R.C. Shell formation in elliptical galaxies. *Mon. Not. R. Astron. Soc.* **1991**, *253*, 256. [CrossRef]
32. Feldmann, R.; Mayer, L.; Carollo, C.M. Tidal Debris in Elliptical Galaxies as Tracers of Mergers with Disks. *Astrophys. J.* **2008**, *684*, 1062–1074. [CrossRef]

33. Petersson, J.; Renaud, F.; Agertz, O.; Dekel, A.; Duc, P.A. From starburst to quenching: Merger-driven evolution of the star formation regimes in a shell galaxy. *arXiv* **2022**, arXiv:2210.16333.
34. López-Cobá, C.; Sánchez, S.F.; Anderson, J.P.; Cruz-González, I.; Galbany, L.; Ruiz-Lara, T.; Barrera-Ballesteros, J.K.; Prieto, J.L.; Kuncarayakti, H. The AMUSING++ Nearby Galaxy Compilation. I. Full Sample Characterization and Galactic-scale Outflow Selection. *AJ* **2020**, *159*, 167. [CrossRef]

universe MDPI

Article

Polarimetric Reverberation Mapping in Medium-Band Filters

Elena Shablovinskaya [1,*], Luka Č. Popović [2,3], Roman Uklein [1], Eugene Malygin [1], Dragana Ilić [3,4], Stefano Ciroi [5], Dmitry Oparin [1], Luca Crepaldi [5], Lyuba Slavcheva-Mihova [6], Boyko Mihov [6] and Yanko Nikolov [6]

1 Special Astrophysical Observatory of RAS, 369167 Nizhny Arkhyz, Russia
2 Astronomical Observatory, Volgina 7, 11000 Belgrade, Serbia
3 Department of Astronomy, Faculty of Mathematics, University of Belgrade, Studentski trg 16, 11158 Belgrade, Serbia
4 Humboldt Research Fellow, Hamburger Sternwarte, Universitat Hamburg, Gojenbergsweg 112, D-21029 Hamburg, Germany
5 Dipartimento di Fisica e Astronomia, Università di Padova, 35122 Padova, Italy
6 Institute of Astronomy and NAO, Bulgarian Academy of Sciences, 72 Tsarigradsko Chaussee Blvd., 1784 Sofia, Bulgaria
* Correspondence: e.shablie@yandex.com

Abstract: Earlier, we suggested the "reload" concept of the polarimetric reverberation mapping of active galactic nuclei (AGN), proposed for the first time more than 10 years ago. We have successfully tested this approach of reverberation mapping of the broad emission line on the galaxy Mrk 6. It was shown that such an idea allows one to look at the AGN central parsec structure literally in a new light. However, the method originally assumed the use of spectropolarimetric observations, expensive in terms of telescope time, and implemented on rare large telescopes. Currently, we propose an adaptation of the polarimetric reverberation mapping of broad lines in medium-band filters following the idea of the photometric reverberation mapping, when filters are selected so that their bandwidth is oriented to the broad line and the surrounding continuum near. In this paper, we present the progress status of such monitoring conducted jointly at the Special astrophysical observatory and Asiago Cima Ekar observatory (OAPd/INAF) with support from Rozhen National Astronomical Observatory (NAO), some first results for the most frequently observed AGNs Mrk 335, Mrk 509, and Mrk 817, and the discussion of the future perspectives of the campaign.

Keywords: polarization; active galactic nuclei; reverberation mapping

1. Introduction

According to the unified model of active galactic nuclei (AGN) [1,2], the central parts of the central machine are surrounded by a gas–dust region, the so-called dusty torus. The presence of a dusty region is key to explaining the observed dichotomy of type 1 and type 2 AGN (see [3]). Characteristics of dust surrounding AGN, e.g. its location and chemical composition determine the accretion properties of the AGN. Thanks to high-angular-resolution observations of local active galaxies in the infrared (IR) (e.g., [4–8]) and molecular lines (e.g., [9,10]) now it becomes possible to obtain direct images of the dusty region, while in the optical range, this structure is still unresolvable. The development of observational capabilities made it possible to determine the geometry of the dusty region, which turned out to be different from the toroidal (see [11] for a review), and also to move from the simple models of a clumpy [12] and smooth-distributed [13] dusty "torus" to more complex ones ([7,14,15] etc.).

The equatorial scattering observed in the optical range in many central regions of type 1 AGN [16–18] is also associated with the presence of a dusty region. This process is responsible for the specific polarization signatures along the emission line profiles: an *S*-shaped swing in the polarization angle and a dip in the polarization degree along the emission line profile [19], which cannot be explained by any other polarization mechanisms in AGN. Scattering by particles of the medium (mainly electrons) occurs in the plane of rotation of the AGN at a distance of R_{sc}, where the optical depth becomes greater than 1 [16,18,20]. Based on physical assumptions, R_{sc} is consistent with the dust sublimation radius. In particular, refs. [21–23] use IR measurements as R_{sc} for estimations of supermassive black hole (SMBH) masses by spectropolarimetric data of angle swings in broad lines. However, there are no direct observations of the equatorial scattering region, and the regions observed in IR may be located farther from the AGN center at a greater optical depth than the optical radiation scattering region.

Since the scattering and emitting regions are spatially separated, polarimetric reverberation mapping can be used to determine their sizes. The simulation done by Goosmann et al. [24] showed that the equatorially scattered polarized emission of the AGN must lag behind the continuum emission. However, the first observational test for NGC 4151 showed $R_{sc} < R_{BLR}$ [25]. Due to the use of broad-band filters covering mostly continuum, the polarization observed in NGC 4151 was contributed not only by equatorial scattering but also by sources of the polarized continuum, such as the accretion disk or the base of the jet. Shablovinskaya et al. [26] revised the approach and proposed the idea of AGN reverberation mapping in polarized broad lines. When using a polarized flux in an emission line with a continuum flux subtracted from it, the influence of other polarization mechanisms is minimized, which allows us to measure the time delay that occurs precisely due to scattering by the equatorial region. The new approach was applied to the analysis of data from spectropolarimetric monitoring of the Seyfert galaxy Mrk 6. The detected delay between the polarized emission in the broad Hα line and the continuum at a wavelength of 5100 Å was about 100 days, which is close to the theoretical value ($\sim$115 days, [27]), but about two times less than the expected delay according to the spatial estimate of the size of the dust region, obtained by IR-interferometry ($\sim$214 days, [5]). This discrepancy requires a detailed analysis, but without static reinforcement based on monitoring other galaxies, it cannot clarify the physics of AGNs.

Spectropolarimetric monitoring, which initially underlay the broad line polarimetric reverberation method, requires not only a large amount of time on a large telescope but also the use of a device equipped with this mode, which is implemented only in a few observatories. Small telescopes are best suited for monitoring a large sample of objects. To adapt the technique for small instruments, it was necessary to switch from spectroscopy to direct images, as was done in the case of photometric reverberation mapping [28][1]. Similarly, AGN reverberation mapping in polarized broad lines at small telescopes can be implemented using image-polarimetry in mid-band filters oriented to the emission line and continuum nearby.

In this paper, we consider the adaptation of the method of polarimetric reverberation mapping of broad lines to observations with small telescopes and some preliminary results. The paper structure is as follows. Section 2 describes the observational technique used on 1- and 2-m class telescopes and the sample of the AGNs chosen for the first stage of the monitoring project. In Section 3, the first results for the three most frequently observed AGNs—Mrk 335, Mrk 509, and Mrk 817—are given, which are then discussed and compared with other estimations in Section 4. The perspectives of the observational approach are described in Section 5. The summary of the current project state is in Section 6.

2. Observational Technique and Sample

Since the beginning of 2020, we have been conducting polarimetric monitoring of a sample of type 1 AGN with equatorial scattering at the 1-m telescope Zeiss-1000 [30] of the Special Astrophysical Observatory of the Russian Academy of Sciences (SAO RAS), at the Copernico 1.82-m telescope of the Asiago-Cima Ekar Observatory and at 2-m telescope of Rozhen National Astronomical Observatory (NAO).

On the Zeiss-1000 telescope of the SAO RAS at different times, we used two instruments "StoP" and "MAGIC" using medium-band 250 Å-wide filters from the SED-set[2] and 100 Å-wide filters called Sy671 and Sy685. In 2020, observations were made on the photometer-polarimeter "StoP" [31]. Using a double Wollaston prism [32] as a polarization analyzer, the device made it possible to simultaneously registered four images in the detector plane corresponding to electric vector oscillations in the directions 0°, 45°, 90° and 135° and, consequently, three Stokes parameters I, Q, and U within one exposure. This method of observation makes it possible to minimize the effect of atmospheric depolarization and increase the accuracy of polarimetric observations (for more details, see [33]). In the polarimetry mode of the "StoP" device with a CCD system (2k × 2k px) Andor iKon-L 936 (a detailed study of the detector is described in [34]) in the 2 × 2 binning mode the scale is $0''.42$/pix with the field of view (FoV) for each direction of polarization $0'.9 \times 6'.1$.

Since the end of 2020, we have switched to a new device—the "MAGIC" multimode focal reducer [34,35]. Retaining all the advantages of the predecessor instrument in the technique of polarimetric observations (the linear polarization measurement accuracy is up to 0.1% for stellar sources up to 16 mag.), the new device has a large FoV: a Wollaston quadrupole prism [36], used as polarization analyzer, projects onto the CCD-detector 4 images of the input mask, corresponding to the directions of oscillation of the electric vector 0°, 90°, 45° and 135°, each with a size of $6'.5 \times 6'.5$. This allows using several local standard stars in the field of an object in differential polarimetry. When using the focal reducer with the same Andor iKon-L 936 CCD system in the 1 × 1 binning mode, the scale was $0''.45$/pix.

At Asiago Cima Ekar observatory (OAPd/INAF), data were obtained using Hα, 671 and 680 filters (70 Å-, 100 Å-, and 100 Å-widths, respectively, and centred at 656, 671, and 680 nm, respectively) and a double Wollaston prism as a polarization analyzer, placed inside the Asiago Faint Object Spectrographic Camera[3] (AFOSC) of the 1.82-m Copernico telescope. The same technique allows simultaneously obtaining four images of the input mask in different directions of polarization in the FoV $0'.8 \times 9'.4$ with Andor iKon-L 936 CCD system with a scale of $0''.51$/pix in 2 × 2 binning mode.

The data at Rozhen National Astronomical Observatory (NAO) were obtained at the 2-m Ritchey–Chrétien–Coudé (RCC) telescope using the two-channel Focal Reducer Rozhen (FoReRo-2) [37], equipped with a double Wollaston prism and a 2k × 2k px Andor iKon-L CCD camera. The images were taken through IF642 and IF672 narrow-band filters (with 26 Å and 33 Å FWHMs, centered at 6416 Å and 6719 Å, respectively). The four images have FoV of $50''.7 \times 50''.7$ each and a scale of $0''.994$/px in the 2 × 2 binning mode used. However, due to the infrequent observations and the small FoV not allowing one to apply the reduction technique shown below, we have not used further this data for the time series analysis.

During each observational night, we received calibration images (flat frames for each filter, bias) to correct data for additive and multiplicative errors. For each object, the series of images (at least 7 frames in each filter) were taken, the exposure times depend on the object brightness, and weather conditions and are usually ranged from 2 to 5 min. To correct statistics each frame is processed independently, and statistical evaluation is made by averaging the random value by robust methods giving its unbiased estimate. In this case, the polarimetric errors are the standard deviation of the robust distribution.

AGN observations were accompanied by observations of polarized standard stars and stars with zero polarization. Introducing the instrumental parameters K_Q and K_U, which characterize the transmission of polarization channels, determined from observations of unpolarized standard stars, as well as I_0, I_{45}, I_{90}, and I_{135} as the intensity at four polarization directions, we can measure three Stokes parameters:

$$I = I_0 + I_{90}K_Q + I_{45} + I_{135}K_U \tag{1}$$

$$Q = \frac{I_0 - I_{90}K_Q}{I_0 + I_{90}K_Q} \tag{2}$$

$$U = \frac{I_{45} - I_{135}K_U}{I_{45} + I_{135}K_U} \tag{3}$$

Here and below, we use Q and U to denote the normalized Stokes parameters.

Then, the degree of linear polarization P and the polarization angle φ as:

$$P = \sqrt{Q^2 + U^2} \tag{4}$$

$$\varphi = \frac{1}{2} \arctan\left(\frac{U}{Q}\right) \tag{5}$$

The observation technique and data reduction are described in more detail in [33]. Note here that the interstellar medium (ISM) polarization is corrected using only one local standard star in the field of the AGN, which may introduce a slight bias in measured polarization parameters. Yet, this bias is about to be small and stable within the monitoring campaign.

When the signal-to-noise ratio of the measured polarization in AGNs was small ($\sigma_P/P \gtrsim 0.7$, where σ_P is the error of the polarization degree P measurement), the polarization degree was corrected for the polarization bias [38]:

$$P_{\text{unbiased}} = P \cdot \sqrt{1 - (1.41 \cdot \sigma_P/P)^2}. \tag{6}$$

However, >95% of obtained data is of high signal-to-noise ratio ($\sigma_P/P < 0.7$).

Over the past two years, we have concentrated on observations of the 6 brightest (12–15 mag) objects in the sample (see Table 1) with equatorial scattering, confirmed by spectropolarimetric observations at the 6-m BTA SAO RAS. All type 1 AGNs are observed sequentially in the polarimetry mode in several mid-band filters, the passbands of which are spectrally oriented toward the emission of the broad Hα line and the continuum near the line. Note that in all cases, it is the broad Hα line that we observe since the equatorial scattering effect is most detectable there. The selection of filters for three objects from the sample Mrk 335, Mrk 509, and Mrk 817 is shown in Figures 1 and 2. Depending on the available filter sets one filter or the combination of two filters was used for obtaining the emission line flux. Observations were carried out approximately once a month, depending on the weather conditions and according to the allocated telescope time for the implementation of programs.

Table 1. AGN sample and calculated expected values of the polarization parameters (Q, U, P and φ) in the filters used in the observations based on the data from [22]. It is important to note that the earlier published data for the objects Mrk 335 and Mrk 79 have been corrected after more thorough processing.

Object	Filters	Q, %	U, %	P, %	φ, °
Mrk 335	SED675	0.28	−0.16	0.32	165.1
	SED650	0.55	−0.51	0.75	158.6
	680	0.41	−0.14	0.43	170.6
	671	0.12	−0.13	0.18	156.4
	Hα	0.40	−0.34	0.52	159.8
Mrk 817	SED625	−0.69	−0.43	0.81	106.0
	Sy685	−0.01	−0.62	0.62	134.5
	Sy671	−0.82	0.36	0.89	78.1

Table 1. *Cont.*

Object	Filters	Q, %	U, %	P, %	φ, °
Mrk 6	SED675	0.16	−0.66	0.68	141.8
	SED650	0.44	−0.62	0.76	152.7
	SED625	0.33	−0.67	0.75	148.1
Mrk 79	SED675	−0.42	0.02	0.42	88.6
	SED650	−0.40	0.04	0.40	87.1
NGC 4151	SED650	−0.13	0.18	0.22	62.9
	SED600	−0.18	0.24	0.30	63.4
Mrk 509	SED675	0.74	−0.63	0.97	159.8
	SED650	0.63	−0.60	0.87	158.2

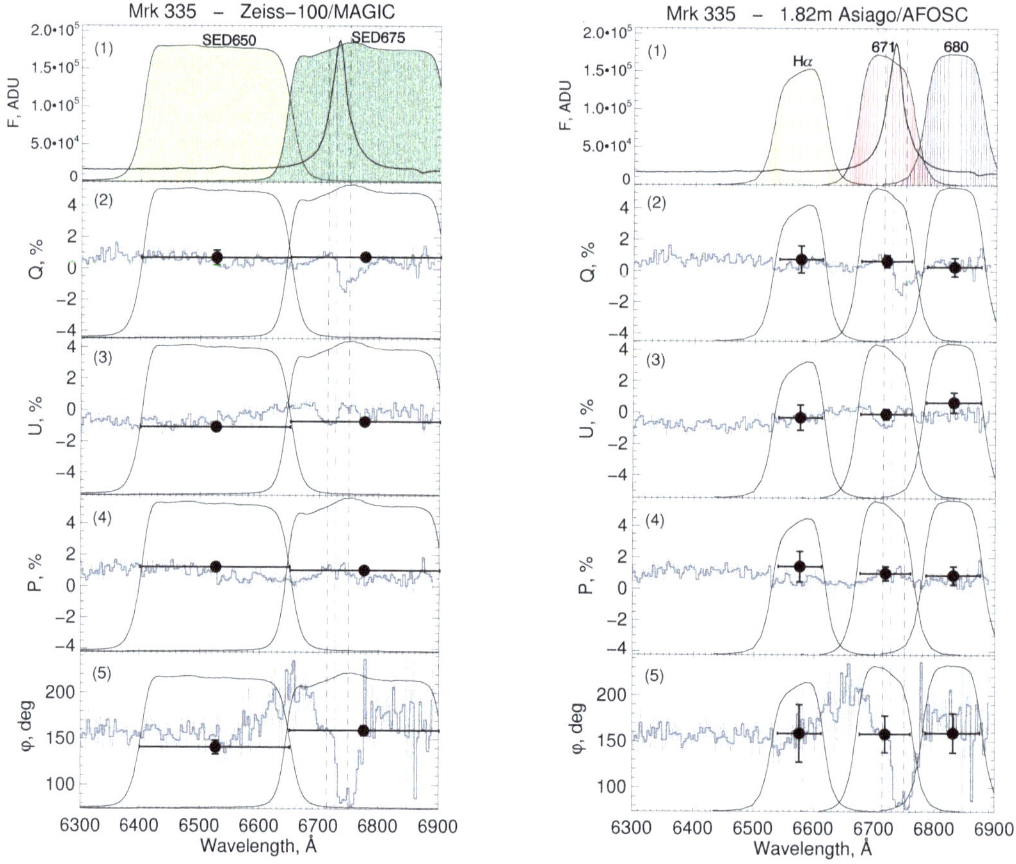

Figure 1. The spectropolarimetric data for Mrk 335 from [22] taken at 9 November 2013 with the overplotted selected filters. **Left:** SED675 (dark green) and SED650 (light green) transmission curves are given. In panels 2–5, the data of image-polarimetry obtained with Zeiss-1000/MAGIC 1 November 2022 are overplotted with black dots. **Right:** Hα (yellow), 671 (red), and 680 (purple) transmission curves are given. In panels 2–5, the data of image-polarimetry obtained with AFOSC 31 October 2022 are overplotted with black dots. In both figures: flux in ADU (1), the Stokes parameters Q (2) and U (3) in %, the polarization degree P in % (4), the polarization angle φ in degrees (5).

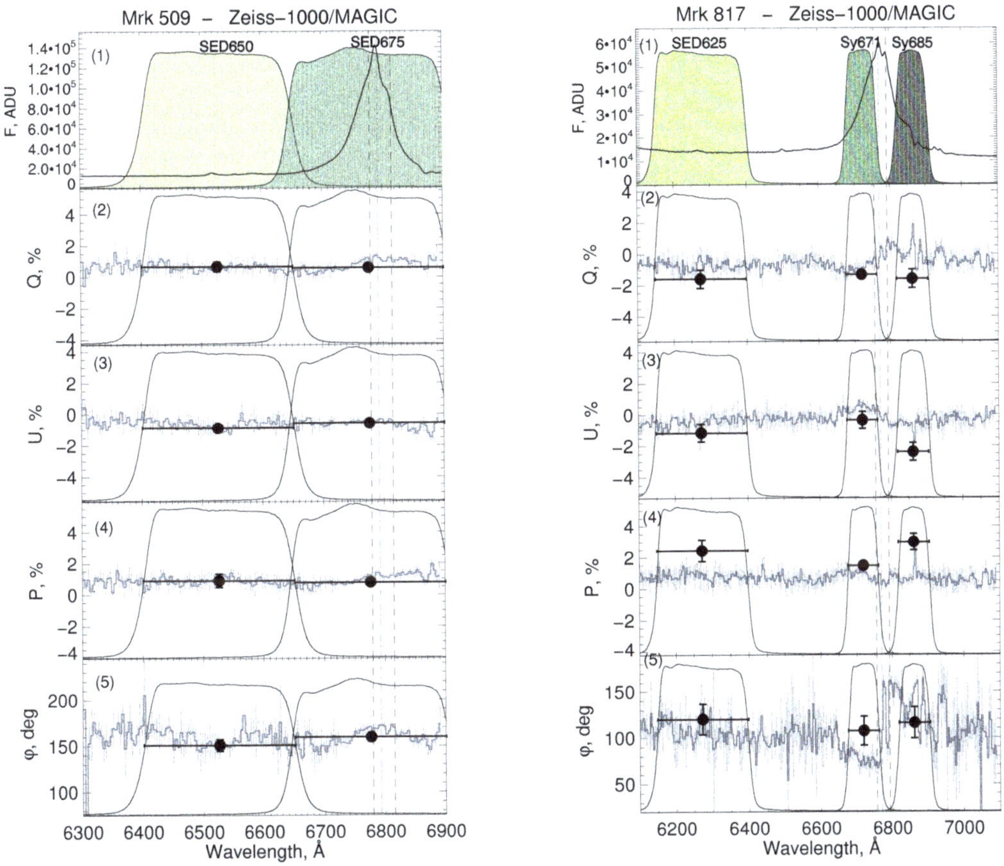

Figure 2. The spectropolarimetric data for Mrk 509 (**left**) and Mrk 817 (**right**) from [22] taken at 21 October 2014 and 29 May 2014, respectively, with the overplotted selected filters. **Left:** SED675 (dark green) and SED650 (light green) transmission curves are given for Mrk 509. In panels 2–5, the data of image-polarimetry obtained with Zeiss-1000/MAGIC 29 August 2021 are overplotted with black dots. **Right:** Sy685 (dark green), Sy671 (medium green) and SED625 (light green) transmission curves are given for Mrk 817. In panels, 2–5 the data of image-polarimetry obtained with Zeiss-1000/MAGIC 28 August 2021 are overplotted with black dots. In both figures: flux in ADU (1), the Stokes parameters Q (2) and U (3) in %, the polarization degree P in % (4), the polarization angle φ in degrees (5).

We estimated the expected values of the polarization effect due to equatorial scattering in the observations of all studied AGNs in medium-band filters for a broad line and continuum based on the previously obtained spectropolarimetric data [22]. Since the transmittance of medium-band filters is measured in the laboratory, we denote it as a filter's response function $filter(\nu)$ and multiply it by the spectral distribution of the polarization parameters ξ_ν [here we used $Q(\nu)$ and $U(\nu)$ in per cent] over the frequencies of the investigated AGNs, to determine its expected values X (in terms of Q and U) in specific filters:[4]

$$X = \frac{\int \xi_\nu \cdot filter(\nu) \cdot d\nu}{\int filter(\nu) \cdot d\nu} \quad (7)$$

The estimated values of Q and U are given in Table 1. The values of P and φ are calculated using Equations (4) and (5). It is interesting to note that in the case when the observations are carried out in two mid-band filters, one of which is oriented to the continuum, and the second is so that the flux from the broad emission Hα line falls into it, the difference between the normalized Stokes parameters between the continuum and the line is small and does not exceed ~0.3%, which is comparable to the linear polarization measurement error for AGN (0.1−0.2% in good weather conditions). The difference in the degree of polarization in the two filters is ~0.1–0.4%, and the difference in the polarization angle is no more than 10 degrees. Thus, the swing seen in the spectropolarimetric observations could not be resolved by photometric polarimetry in filters, but this could indicate a difference between the emission line and continuum polarization parameters. Here, the configuration for the Mrk 817 object deserves special attention, when a broad emission line is observed in two filters oriented to the "blue" and "red" wings of its profile. For Mrk 817 (the spectrum of the object with overplotted transmission curves of the filters used is shown in Figure 2 on the right), the difference between the normalized Stokes parameters for the continuum and the line wings reaches ~0.7%, and the deviation of the polarization angle from its value in the continuum is ±28°. It should be noted here that the Sy685 filter is also oriented to the atmospheric absorption B-band λ = 6860–6917 Å (and Table 1 shows calculations without correction for this band). Nevertheless, the Mrk 817 case most clearly shows that using medium-band filters oriented to different wings of the Hα broad line profile, we can trace the characteristic changes in the polarization angle profile, the wavelength dependence of which acquires a characteristic *S*-shaped profile during equatorial scattering on a gas–dust torus.

3. First Results

We performed polarimetric observations of the AGN sample on the 1 m and 1.82-m telescopes in 2020–2022. The weather and the time allocated within the schedules did not allow us to observe objects with a high cadence, and the total amount of data on the light curves does not exceed 25 epochs, even in the case of the most regularly observed objects. Such a meagre amount of data does not allow us to get a reliable result yet. In this section, we will consider the current status of monitoring of three objects—Mrk 335, Mrk 509, and Mrk 817. The monitoring period and the number of epochs are mean values of the nonpolarized continuum, and the broad line flux and mean polarization degree of the broad line are given in Table 2. There also, we provide the measure of variability calculated using Equation (3) from [39]. The full observational data are given in Appendix A.

Table 2. Mean values of nonpolarized continuum and broad line flux and mean polarization degree of the broad line obtained for Mrk 335, Mrk 509, and Mrk 817 during the observations: (1) object name, (2) monitoring period (dd/mm/yyyy), (3) the number of epochs, (4) mean continuum flux in mJy, (5) variability measure of the continuum flux, (6) mean broad Hα line flux with the continuum subtracted, in mJy, (7) variability measure of the broad line flux, (8) mean polarization degree of the broad line in %. For Mrk 817, the upper values of $I_{\rm line}$, $F_{\rm var}^{\rm line}$ and $P_{\rm line}$ are for the "blue" line profile wing and the bottom values are for the "red" wing.

(1)	Period (2)	N (3)	$I_{\rm cont}$, mJy (4)	$F_{\rm var}^{\rm cont}$ (5)	$I_{\rm line}$, mJy (6)	$F_{\rm var}^{\rm line}$ (7)	$P_{\rm line}$, % (8)
Mrk 335	9 September 2020–1 November 2022	23	90.2 ± 7.9	0.101	140.1 ± 9.9	0.085	0.9 ± 0.3
Mrk 509	26 May 2020–29 August 2021	11	12.2 ± 0.5	0.018	21.0 ± 2.2	0.092	1.9 ± 0.8
Mrk 817	14 December 2020–28 August 2021	8	23.7 ± 1.1	0.036	36.9 ± 1.3 23.3 ± 1.2	0.044 0.072	1.9 ± 0.4 2.1 ± 0.8

3.1. Mrk 335

Mrk 335 (z = 0.025, RA 00 06 19.5 Dec +20 12 10.6 J2000) is a well-known narrow-line Sy 1 galaxy. The signs of the equatorial scattering in broad lines were observed in Mrk 335 spectropolarimetric data for the first time in [40]. The violent polarization angle swing along the Hα line profile was confirmed in [22]; here, we present the same data obtained at 6-m BTA telescope of SAO RAS in Figure 1. As it could be seen the polarization angle variations are of about $\pm 50°$, yet the polarization degree changes relative to the continuum are minor. In Figure 1, left, the data of image-polarimetry obtained with Zeiss-1000/MAGIC 1 November 2022 are overplotted; in Figure 1, right, the data of image-polarimetry obtained with AFOSC 31 October 2022 at Asiago observatory are given. Note here that in all the cases, only slight differences of the polarization parameters between continuum and broad line bands are detected.

In total, 10 epochs of Mrk 335 polarimetric data were obtained with MAGIC and 13 epochs with AFOSC. Unfortunately, due to the different brightness of the field stars in the filters used in MAGIC and AFOSC, we were unable to use the same local standard. For the MAGIC data reduction, we used the reference star [GKG2008] 5 nearby at a distance of $\sim 1'.3$ from the source, and the star TYC 1184-771-1 at a distance of $\sim 2'.5$ for AFOSC data. The light curves polarized and broad integral line fluxes and continuum flux are shown in Figure 3. For AFOSC data, the broad line flux is the sum of fluxes measured in two filters (671 and 680). In all cases, the fluxes are given in mJy. One can see that despite the broad line curves the continuum flux light curves seem to behave in a different way in the data sets obtained with MAGIC and AFOSC. That was the reason not to merge the light curves not to introduce the systematical errors in the correlation analysis.

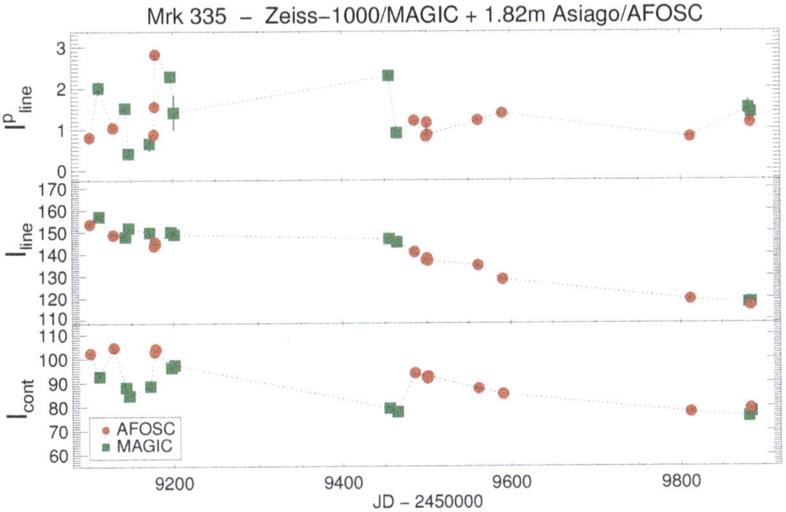

Figure 3. Mrk 335 light curves. From top to bottom: polarized broad line flux I^p_{line}, integral broad line flux I_{line} with subtracted continuum and integral continuum flux I_{cont}. Fluxes are given in mJy. Red circles are used to denote the AFOSC data, and green squares are for the MAGIC data. For AFOSC, the broad line flux is the sum of fluxes measured in two filters (671 and 680).

Additionally, we have estimated the polarization of Mrk 335 using observations at Rozhen observatory taken at 15 August 21. For IF642 oriented to the continuum $P = 1.25 \pm 0.26\%$, and $\varphi = 78°.5 \pm 6°.0$; for IF672 oriented to the Hα emission line $P = 1.13 \pm 0.13\%$, and $\varphi = 89°.0 \pm 3°.3$. Here, one can detect the slight difference of the absolute polarization level, particularly seen in the polarization angle. However, due to the lack of the local standard stars in the FoV of the source we are not able to correctly take into account the atmospheric depolarization, ISM effects, etc. Moreover, the flux calibration is also absent which makes it complicated to exclude the polarized continuum flux from the polarized line flux in a proper way. Thus, these data illustrate the possibilities of the medium-band polarimetry at 2-m telescope, yet it is not used for the further analysis.

To determine the time delay between the light curves, we performed a cross-correlation analysis using two approaches. As the main analysis tool, we used the JAVELIN code [41–43], widely used in AGN reverberation mapping campaigns. In Figure 4 the results of the JAVELIN analysis of the time delay of the polarized broad line emission $I^p_{\rm line}$ relative to the variable continuum flux $I_{\rm cont}$ for AFOSC (red histogram) and MAGIC (green histogram) data are presented. Additionally, we conducted a joint analysis of both light curves, combining them so that one of the light curves is shifted in time relative to the other by more than the duration of the entire monitoring period. The results of the analysis of this synthetic curve are shown in Figure 4 in black and provide only additional information. Similarly, in Figure 5 the time delay of nonpolarized broad line emission $I_{\rm line}$ relative to the continuum $I_{\rm cont}$ is analyzed. Also, to estimate the delay between the light curves, we used the code ZDCF [44,45]. Separately for the MAGIC and AFOSC light curves, cross-correlation analysis using ZDCF did not show results due to large uncertainties caused by a small number of points. The results of estimating the delay between the combined synthetic curves are given in Figures 4 and 5 in grey.

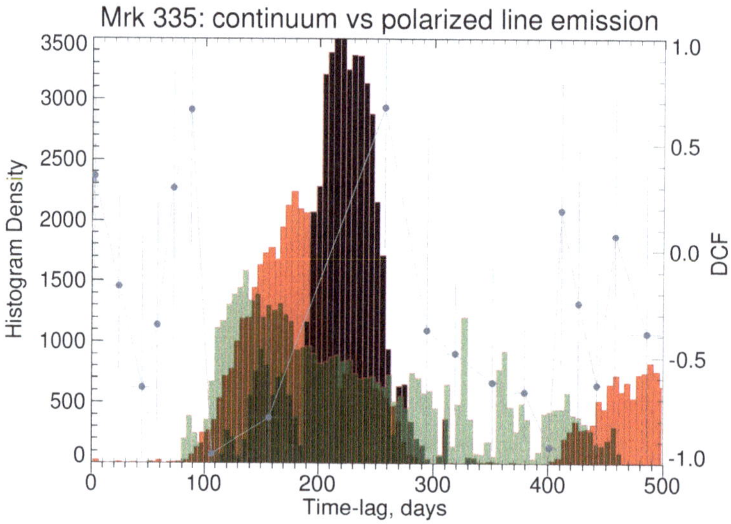

Figure 4. Time delay analysis of the polarized broad line emission $I^p_{\rm line}$ relative to the variable continuum flux $I_{\rm cont}$ for Mrk 335. Histograms show the results of JAVELIN analysis based on AFOSC data (red histogram), MAGIC data (green histogram), and combined time-shifted data (black histogram). On the left y-axis the frequency of occurrence of parameter values sets during MCMC sampling is shown. 5000 sets of parameter values were used in the simulation. The grey curve shows the results of the ZDCF analysis of the combined time-shifted data (values are given on the y-axis on the right).

Despite the number of epochs comparable to what we previously obtained for Mrk 6 in spectropolarimetric mode [26], the analysis of the delay between $I^p_{\rm line}$ and $I_{\rm cont}$ does not show an unambiguous peak for Mrk 335. In Figure 4 it can be seen that the histogram of estimates of time delays for AFOSC and MAGIC data is close, about 180 and 150 days, respectively, but the peak of the time-lag distribution has an error of 25–40%. Synthetic data show two peaks at 224 $\pm$ 24 days and 157 $\pm$ 18 days, where the given errors are formally calculated as the standard deviations of the given Gaussian-like peaks. Here, the larger peak is definitely an artifact, since it is repeated in the analysis of photometric data (Figure 5). The second peak is $\sim$4 times weaker than the first one, but its position roughly coincides with other estimates. Thus, we see the tendency of the $I^p_{\rm line}$ light curves to show a delay of about 150–180 days. However, such a time lag is close to the half of year, which characterize the typical length of the observational periods of the source, and is shorter than the gaps between these periods. This might indicate the measured value as the analysis artefact. Additionally, we performed data analysis of the time delay of $I_{\rm line}$ relatively to $I_{\rm cont}$ to estimate $R_{\rm BLR}$ if possible. JAVELIN histograms for AFOSC and MAGIC data, as well as analysis of synthetic light curves by the ZDCF method, indicate an estimated delay between 73 ± 18 and 87 ± 17 days. The AFOSC data separately demonstrate a peak at the value of 27 ± 17 days, which is close to the cadence of observations (about 1 time per month) and can be an artifact of analysis.

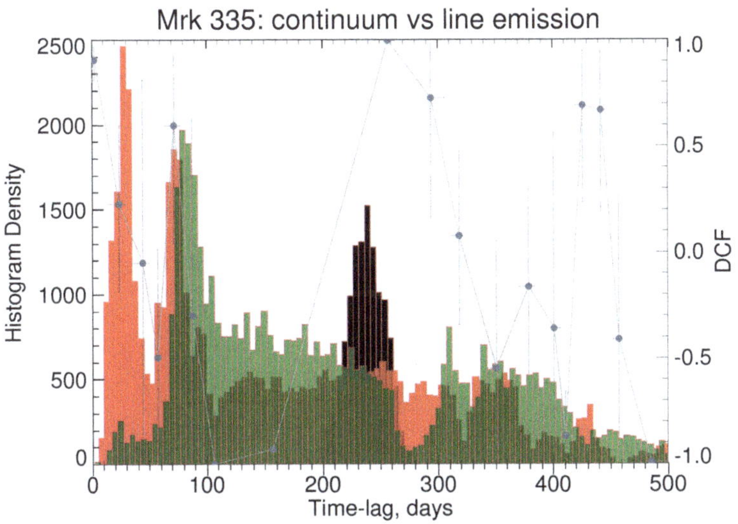

Figure 5. Time delay analysis of the broad line emission $I_{\rm line}$ relative to the variable continuum flux $I_{\rm cont}$ for Mrk 335. The coloured histograms and labels are the same as in Figure 4.

3.2. *Mrk 509*

As well as in the case of Mrk 335, Mrk 509 ($z = 0.035$, RA 20 44 09.8 Dec -10 43 24.7 J2000) was observed in spectropolarimetric mode firstly in [40] and later in [22]. The latter data were used in Figure 2 (left). As can be seen in the figure, we selected for observations two medium-band (FWHM = 250 Å) filters oriented to a broad line and a continuum near. The overplotted image-polarimetry data was obtained with Zeiss-1000/MAGIC 29 August 2021. As in the case of Mrk 335, Mrk 509 shows only a slight difference in the polarization parameters between continuum and broad line bands, more detectable in the polarization angle variations.

As far as the object can be observed for only four months a year, in 2020–2021, we gained only 11 epochs using Zeiss-1000/MAGIC. To reduce the data, we used the reference star TYC 5760-1396-1 nearby at a distance of $\sim 1'.5$ from the source. The light curves are shown in Figure 6. For all measured fluxes $I^p_{\rm line}$, $I_{\rm line}$, and $I_{\rm cont}$ the variability is observed, and the pattern of $I^p_{\rm line}$ variations differs from other light curves. The curves show a large gap between the observational epochs associated with the inability to observe the object evenly throughout the year.

To estimate the time-delay in a broad polarized line, we applied the JAVELIN code to the received data. It turned out that despite a small number of epochs, the analysis revealed an unambiguous peak at $114^{+12.7}_{-8.8}$ days (Figure 7). We have also applied the JAVELIN analysis to the data taken only in 2020 excluding the epochs from 2021, and we have obtained the same time-delay. This estimate corresponds to the size of the dusty region

~0.1 pc. Note, however, that the ZDCF analysis did not show a significant correlation. Additionally, we performed 2020 data analysis of the time delay of $I_{\rm line}$ relatively to $I_{\rm cont}$ to estimate $R_{\rm BLR}$ following the Mrk 335 case. JAVELIN histogram is shown in Figure 8 demonstrating two clear peaks at 39 ± 5 days and at 85 ± 11 days (which is approximately 39 ± 5 days × 2). Taking into account that the median cadence of observations in 2020 is ~16 days, we could not unambiguously make a conclusion about the origin of the double-peaked correlation histogram.

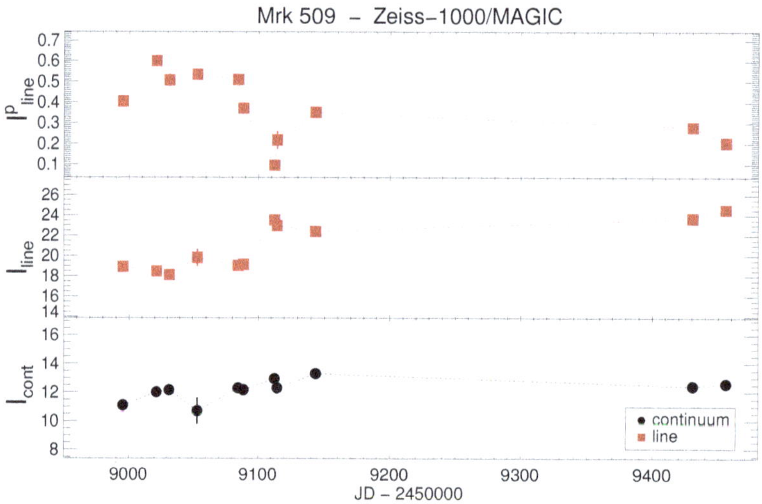

Figure 6. Mrk 509 light curves obtained with MAGIC. From top to bottom: polarized broad line flux $I^p_{\rm line}$ and integral broad line flux $I_{\rm line}$ with subtracted continuum and integral continuum flux $I_{\rm cont}$. Fluxes are given in mJy.

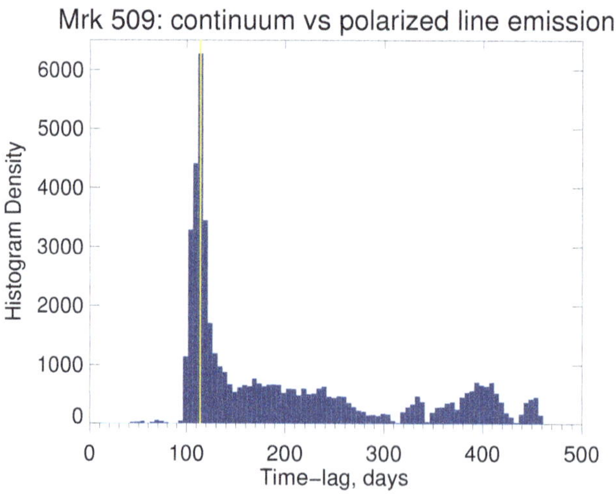

Figure 7. Time delay analysis of the polarized broad line emission $I^p_{\rm line}$ relative to the variable continuum flux $I_{\rm cont}$ for Mrk 509. Histogram show the results of JAVELIN analysis based on MAGIC data. On y-axis the frequency of occurrence of parameter values sets of during MCMC sampling is shown. 10,000 sets of parameter values were used in the simulation. The time-delay estimation equal to $114^{+12.7}_{-8.8}$ days is shown with the yellow vertical line.

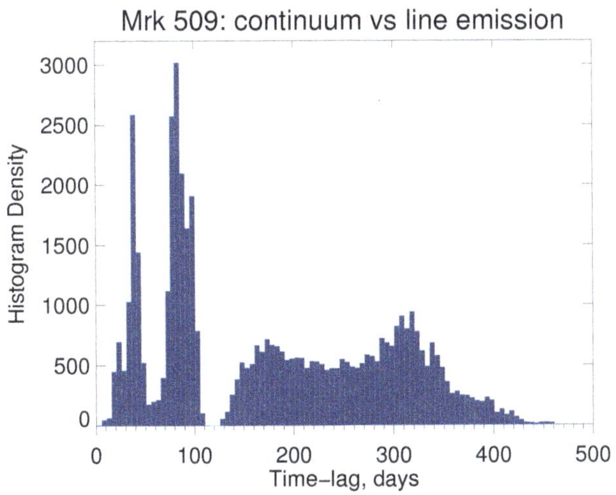

Figure 8. Time delay analysis of the broad line emission I_{line} relative to the variable continuum flux I_{cont} for Mrk 509. Histogram show the results of JAVELIN analysis based on MAGIC data. On y-axis the frequency of occurrence of parameter values sets of during MCMC sampling is shown. 10,000 sets of parameter values were used in the simulation.

3.3. Mrk 817

Mrk 817 ($z = 0.031$, RA 14 36 22.1 +58 47 39.4 J2000) is a Sy 1.2 AGN, where the equatorial scattering was discovered in [22]. The data published in that work is presented in Figure 2, right, where the transmission curves of the two filters 100 Å-width oriented to "red" and "blue" broad line wings and selected for monitoring are also shown. A broader (FWHM = 250 Å) filter was chosen for continuum polarimetry. The spectropolarimetric data demonstrate small changes in the polarization degree P and a violent switch of the polarization angle φ along the line profile. In Figure 2, right, the data of Mrk 817 image-polarimetry obtained on the MAGIC 28 August 2021 device is also plotted. Comparing the values obtained in two filters oriented to different wings of the lines, one can see a significant difference in the polarization parameters. This indicates that the use of a similar filter configuration in image-polarimetry mode may be an alternative approach for identifying signs of equatorial scattering using small-class telescopes or large instruments for observations of faint AGNs where the spectropolarimetric data show too low a signal-to-noise ratio. Note here that in Figure 2 for the Mrk 817 data, there are visible differences in the polarization parameters between the observations of 2014 and 2021, especially in Sy685 band. In this case, it is important to obtain newer spectropolarimetric data in order to confirm whether such a difference is the result of the influence of external factors (e.g., the variability of atmospheric B-band 6860–6917 Å) or internal changes in the spectrum of Mrk 817 in polarized light.

During the Mrk 817 monitoring, 8 epochs of observations were obtained using the MAGIC device in the period from December 2020 to August 2021. To reduce the data, we used a reference star of comparable brightness in the field of the object (RA 14 36 06.7 +58 50 38.4 J2000) at a distance of $\sim 3'.6$ from the source. The data were obtained relatively evenly, once or twice every two months. Unfortunately, during the monitoring period, Mrk 817 did not show significant variability either in the continuum or in the broad line. The light curves of Mrk 817 are given in Figure 9. It can be seen that I_{line} does not show differences between "red" and "blue" wings in integral light (see the middle panel in Figure 9). However, in polarized light, for several epochs of observations, both the difference in I^p_{line} between filters is visible (e.g., in the epoch of 18 December 2020), and a violent change of I^p_{line} between epochs (e.g., 2 July 2021 and 7 July 2021, see the upper panel in Figure 9). The meagre amount of data with no significant variability did not allow us to obtain any result in cross-correlation analysis.

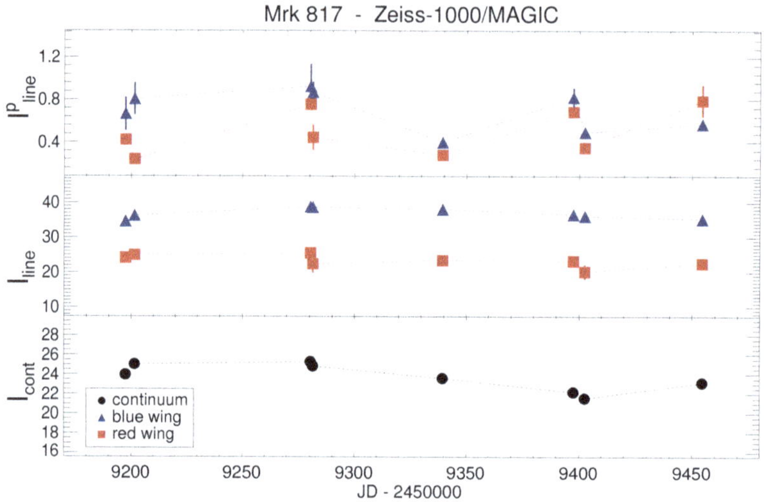

Figure 9. Mrk 817 light curves obtained with MAGIC. From top to bottom: polarized broad line flux $I^p_{\rm line}$ and integral broad line flux $I_{\rm line}$ with subtracted continuum and integral continuum flux $I_{\rm cont}$. Red and blue dots in $I^p_{\rm line}$ and $I_{\rm line}$ light curves denote the fluxes of the "red" and "blue" broad line profile wings, respectively. Fluxes are given in mJy.

4. Discussion

Since 2020 due to the lack of stable weather meeting our requirements in two observatories involved in the project (SAO RAS and Asiago) we have not reached the desired cadence when observing a sample of objects, and the total number of epochs obtained has reached 23 for only one object. Despite this, we managed to obtain some first results for three sample objects Mrk 335, Mrk 509, and Mrk 817, presented in this paper. As these AGNs are studied in deep detail in various multi-wavelength campaigns, here we discuss our results in comparison with the measurements given in the literature to investigate if the provided estimations are reliable.

Mrk 335 was intensively studied in numerous reverberation mapping campaigns. The BLR size $R_{\rm BLR}$ was measured as $16.4^{+5.2}_{-3.2}$ [46,47], $17.3^{+4.9}_{-4.3}$ [48], $15.7^{+3.4}_{-4.0}$ [49], 14.3 ± 0.7 [50,51], $10.6^{+1.7}_{-2.9}$ [52], $17.0^{+2.5}_{-3.2}$ lt days [53] in Hβ broad line and $20.5^{+2.0}_{-2.8}$ lt days [28] in Hα broad line. Given estimations of the BLR size are $\sim$10 times larger than the accretion disk size of $\sim$1 lt day [54]. According to the scale relation from [22], $R_{\rm sc} \simeq 5.1 R_{\rm BLR}$, so $R_{\rm sc}$ for Mrk 335 could be estimated as $\sim$70 lt days. IR reverberation mapping in K band provided $R_{\rm IR} \approx 166$ lt days [55] which is two times greater than a value obtained using the scale relation. Lyu et al. [56] found the size of the dusty region in *WISE W1* band $R_{W1} \approx 1300$ lt days. According to the relation of the dusty region sizes in different bands R_K:R_{W1} = 0.6:1 given in the same paper, $R_K \approx 770$ lt days which is much larger than other estimations and seems not to be reliable. Thus, the value of the polarized emission line time lag is predicted to be in $\sim$70–170 days range. Throughout our polarimetric reverberation mapping monitoring the polarized Hα emission showed a delay of about 150–180 days, which is in good agreement with the predictions of the size of the dusty region for Mrk 335. However, we prefer to refrain from the statement of such a result, primarily due to the fact that such an estimation may be caused by a correlation artefact since it is close to the value of 1/2 year. Moreover, as it was mentioned above, this estimate is longer than the length observation periods of the source, but shorter than the gaps between them. In addition, the maxima of cross-correlation functions have large uncertainties of 25–40%, which makes them unconfident. Moreover, the results of the analysis of the corresponding photometric data do not coincide with the known estimates of R_{BLR} for Mrk 335. These facts give reasons to doubt the sustainability of the results we have obtained so far.

Sy1 galaxy Mrk 509 was also studied in multiple campaigns covering the entire electromagnetic spectrum (see, e.g., [57] for a review). The BLR size R_{BLR} was measured as $76.7^{+6.3}_{-6.0}$ [46,47] and $79.6^{+6.1}_{-5.4}$ lt days [49] in Hβ broad line. From our estimations, relying on the maximum value from the double-peaked histogram in Figure 8 R_{BLR} =85 $\pm$ 11 lt days which coincides with the measurements given in the literature. However, we cannot still explain the existence of the second estimation of the time delay being two times shorter. The estimations of R_{BLR} predict a very extended BLR region, which is much larger ($\sim$40 times) than the accretion disk size of $\sim$2 lt day [58,59], on the one hand, and only $\sim$2 times less then IR reverberation mapping in K band estimations $R_{IRRM} \approx 131$ lt days [55]. GRAVITY Collaboration et al. [60] resolved the hot gas structure in Mrk 509 with VLTI/GRAVITY near-infrared interferometry and measured the size of the dusty region $R_{IRIF} \approx 296 \pm 30$ lt days. Using given R_{BLR} measurements and the scale relation from [22], $R_{sc} \approx 408$ lt days. While $R_{IRIF} > R_{IRRM}$ is usually predicted (see [26] for references), R_{sc} should be less then dusty structures in AGN. Apparently, such controversial measurements rise issues of the dusty region size. During our monitoring, we estimated $R_{sc} \approx 114^{+12.7}_{-8.8}$ lt days, or $\sim$0.1 pc. A comparison with near-IR torus interferometric data [60] shows that the equatorial scattering region is 2 times smaller than the radius of the dusty structure in the IR band, which is similar to what we previously obtained for Mrk 6 [26]. However, our estimate of R_{sc}, although consistent with the estimates of the size of the dusty region obtained by two independent methods, is only $\sim$1.3–1.6R_{BLR}. In general, all other estimates of the size of structures inside the central Mrk 509 parsec obtained independently indicate that R_{BLR} is most likely overestimated, for example, due to the presence of outflows driven by AGN observed for Mrk 509 (e.g., [61]). This reveals the necessity of more intensive homogeneous monitoring in polarized and integral light.

Mrk 817 is in the focus of several vast monitoring campaigns, e.g., AGN STORM 2 [62]. The BLR size R_{BLR} was measured as $15.0^{+4.2}_{-3.4}$ [46,47], $21.8^{+2.4}_{-3.0}$ [49], $14.0^{+3.4}_{-3.5}$ lt days[63] in Hβ broad line, and $28.3^{+2.1}_{-1.8}$, $26.8^{+2.8}_{-2.5}$, $51.7^{+14.9}_{-1.3}$ lt days simultaneously in Hβ, Hγ and FeII lines, respectively [64]. Given estimations of the BLR size are $\sim$3–6 times larger than the accretion disk size of $\sim$4.5 lt day [54]. Mrk 817 was observed within IR RM monitoring and the dusty region size was estimated as $R_{IRRM} = 89 \pm 9$ lt days [55]. This coincides with the expected estimates of R_{sc} using the scale relation $R_{sc} \approx 95 \pm 15$ lt days, which is predicted for our measurements. Due to the insignificant variability of the polarized and non-polarized fluxes, despite the monitoring period being two times longer than the expected time lag no result was obtained for Mrk 817. However, the variability of blue and red wings of polarized broad Hα line is intriguing enough to go on with the observations more intensively.

5. Future Perspectives

The new approach of polarimetric reverberation mapping in broad lines looks promising since it can provide additional information about the size of structures in AGN, and therefore, better understand the nature of processes associated with accretion onto SMBH. As we have shown in the given paper, the technique in medium-band filters together with one-shot differential polarimetry is suitable for small telescopes, yet needs a careful adaptation. It is important to note that the polarimetry of faint polarized sources, in contrast to, e.g., differential photometry, put high restrictions on permissible weather conditions. Even weak cirruses or a haze can significantly depolarize the radiation of observed objects, and the variability of atmospheric transparency between exposures significantly degrades the quality of data. Here we consider several issues related to the adaptation of observations in the framework of monitoring.

1. Filter selection. At the beginning of the monitoring observations, we selected filters from our existing sets (see Section 2), focusing on our experience of observations in the framework of photometric reverberation mapping of AGN [65,66]. Because of this, we mainly aimed to use pairs of 250 Å-width filters oriented to a broad line and a continuum near. However, as shown in Table 1 via the convolution of spectropolarimetric data with filter transmission and in Figures 1 and 2 using the example of image-polarimetric data for Mrk 335 and Mrk 509, this strategy does not always seem optimal. This is due to that since the variations of the polarization parameters along the wavelength during equatorial scattering are small, and φ has an S-shaped profile, even the 250 Å-width filter may be too broad, and the average value of the line polarization in the filter, summing up by wavelengths, will not differ markedly from the continuum. However, it is important to note that the differences are small when we consider polarization normalized by intensity. When the polarized flux in the line is considered with the subtracted polarized flux of the continuum, the behaviour of the variability in the broad line begins to differ significantly in polarized and nonpolarized light. It corresponds to the fact that we see this radiation coming from different regions of the AGN. This is what we observe in the case of objects Mrk 335 and Mrk 509. Thus, it can be argued that even if a medium-band filter covering the whole line profile is selected, the variable flux $I^p_{\rm line}$ is detectable.

In cases of bright AGNs, a more optimal strategy may be to choose narrower filters oriented on different sides from the centre of a broad emission line. In our case, we were able to implement this by using 100 Å-width filters for Mrk 817 monitoring. The light curves we obtained are not yet sufficient to reveal the approach efficiency within the monitoring framework. However, as was shown above, such a strategy is more suitable to check AGNs for signs of equatorial scattering in polarized light efficiently using telescope time.

Another problem for us was the combination of observational data obtained using a different set of filters. One can see this in the example of Mrk 335 observations, which were carried out at the 1-m telescope of the SAO RAS and 1.82-m in Asiago. Having the same trend, the variability of radiation, especially in the continuum, differs significantly between the data obtained in 250 Å-width filter SED650 and in 70 Å-width Hα. It is unlikely that the reason for this difference was the AGN emission lines being on the transmission edge of the SED650 filter, e.g., the [FeX] 6374 Å line. The more likely reason may be that different reference stars were used when processing the data sets obtained from MAGIC and AFOSC. In any case, the data obtained require additional analysis.

2. Aperture selection and host-galaxy subtraction. Two objects presented in this article, Mrk 335 and Mrk 509, are almost star-shaped sources. Their host-galaxies, which fitting can be found in [49], have a small contribution to the optical band. In our observations, taking into account the image quality, the profile of objects was indistinguishable from the profiles of stars in the field. Thus, we were able to choose the size of the aperture for photometry so that the signal-to-noise ratio was maximum. However, when the host galaxy is extended, the choice of aperture is complicated, as the larger the aperture size, the more galactic flux is recorded, lowering the contrast of the polarized radiation of the nucleus. For Mrk 817 polarimetry, a fixed aperture size of $\sim$4$''$ was chosen so that when processing data with different image quality (data with a seeing better than 3$''$ was used), the same contribution of the host-galaxy would be inside the aperture. However, greater accuracy will be achieved if, when processing AGN images with extended galaxies, a galaxy model is subtracted from the frames. If this is not so critical in the case of Mrk 817, then for, e.g., NGC 4151, where the host-galaxy has a size >3$'$ and the contrast of the nucleus is relatively small, subtraction of the galaxy fitting may be necessary to construct the light curves. This should be the subject of a separate detailed check.

3. Cadence of observations. Currently, based on simulated AGN light curves, attempts are being made to determine the optimal cadence for reverberation mapping observations. Improving cadence leads to fewer artifacts in cross-correlation analysis, but requires a large amount of telescope time. The upper limit for time resolution is the expected time delay since when observations are made with a lower frequency, the observed variability will not

be related. For example, Kovačević et al. [67] offer ∼5-days cadence for estimates of the accretion disk size of typically 1–10 lt days size using LSST. Woo et al. [68] suggested having a factor of 5 or better time resolution for a given time lag. Due to the time allocation on the telescopes, the typical cadence of our observations was planned to be about 1 month. Such a cadence is close to optimal with the expected sizes of R_{sc} for Mrk 509; for Mrk 335 and Mrk 817, a cadence of ∼20 days would be more effective. However, due to weather conditions, it turned out to be impossible to conduct observations every month, so the real-time resolution is worse. Moreover, our estimates show that it is important to simultaneously measure R_{sc} and R_{BLR} to improve the scale relation (which may, generally speaking, have a different appearance for different objects). In this case, observations should be carried out at least 1 (preferably 2) times a week. Such a cadence can be achieved using a telescope, observations on which are fully oriented only for such a task. According to the adaptation of the method of reverberation mapping of the polarized lines to observations on a 1-m-class telescope, such a project has prospects for development.

6. Conclusions

We presented the first results of reverberation mapping in polarized broad lines conducted at the 1-m telescope of SAO RAS and at 1.82-m at Astronomical observatory Asiago. Since 2020, we obtained the first results for the three most frequently observed objects from our sample of type 1 AGNs with equatorial scattering, namely, Mrk 335, Mrk 509, and Mrk 817.

- For Mrk 335, the measured dusty region size is $R_{sc} \sim 150$–180 lt days. This result coincides with the values predicted concerning the several estimations of the dusty structure in the IR band and measurements of R_{BLR} via optical reverberation mapping campaigns. However, due to the irregular observations, the monitoring is going on to check whether our result is a cross-correlation artefact.
- For Mrk 509, we obtained $R_{sc} \approx 114^{+12.7}_{-8.8}$ lt days, or ∼0.1 pc. This is 2 times smaller than the radius of the dusty structure in the IR band.
- For Mrk 817, no result is obtained due to the low variability of the object during the monitoring period. However, observations of the polarized flux in the two line profile wings demonstrate a sharp variability between epochs as well as a significant difference in the polarized flux in the two wings during one epoch. This shows the potential possibility of recording the delay of a polarized signal of a broad line in different parts of its profile.

Author Contributions: Conceptualization, E.S. and L.Č.P.; Methodology, E.S. and L.Č.P.; Software, E.S. and R.U.; Validation, D.I.; Formal analysis, L.Č.P., R.U. and E.M.; Investigation, E.S. and E.M.; Data curation, E.S., D.I., S.C., D.O., L.C., L.S.-M., B.M. and Y.N.; Writing—original draft, E.S.; Writing—review & editing, L.Č.P., E.M. and D.I.; Visualization, E.S. and E.M.; Supervision, L.Č.P.; Project administration, E.S. All authors have read and agreed to the published version of the manuscript.

Funding: E.S., E.M. and R.U. were supported by RFBR grant, project number 20-02-00048 while conducting observations on 1-m telescope of SAO RAS, reducing and analyzing the polarimetric data. L.Č.P., and D.I. acknowledge funding provided by Astronomical Observatory (the contract 451-03-68/2022-14/ 200002) and by University of Belgrade-Faculty of Mathematics (the contract 451-03-68/2022-14/200104), through the grants by the Ministry of Education, Science, and Technological Development of the Republic of Serbia. L.S.M. and B.M. acknowledge the project "Reverberation mapping of quasars in polarized light" within the agreement between Bulgarian Academy of Sciences and Serbian Academy of Sciences and Arts, 2020–2022. D.I. acknowledges the support of the Alexander von Humboldt Foundation.

Institutional Review Board Statement: Not applicable.

Informed Consent Statement: Not applicable.

Data Availability Statement: The observational data underlying this article is available on request 1 yr after the publication of this paper.

Acknowledgments: Observations with the SAO RAS telescopes are supported by the Ministry of Science and Higher Education of the Russian Federation. The renovation of telescope equipment is currently provided within the national project "Science and Universities".

Conflicts of Interest: The authors declare no conflict of interest.

Appendix A

Table A1. Observed values of nonpolarized continuum and nonpolarized and polarized broad line flux for Mrk 335, Mrk 509 and Mrk 817 : (1) date of observations (dd/mm/yyyy), (2) the same in Julian form JD-2450000, (3) continuum flux in mJy, (4) the Stokes Q parameters of the continuum flux in %, (5) the Stokes U parameters of the continuum flux in %, (6) broad Hα line flux with subtracted continuum, in mJy, (7) the Stokes Q parameters of the line flux with subtracted continuum, in %, (8) the Stokes U parameters of the line flux with subtracted continuum, in %, (9) polarized broad Hα line flux with subtracted continuum, in mJy. For Mrk 817, the upper values of I_{line}, Q_{line}, U_{line} and I^p_{line} are for the "blue" line profile wing and the bottom values are for the "red" wing. The values are calculated by robust average, and the errors are the robust standard deviation (see [69] for more details).

Date (1)	JD (2)	I_{cont} (3)	Q_{cont} (4)	U_{cont} (5)	I_{line} (6)	Q_{line} (7)	U_{line} (8)	I^p_{line} (9)
				Mrk 335 (MAGIC)				
20 September 2020	9112	92.6 ± 0.1	0.0 ± 0.1	−0.4 ± 0.4	157.1 ± 0.2	0.8 ± 0.2	−0.7 ± 0.1	2.0 ± 0.2
21 October 2020	9143	88.1 ± 0.1	−0.4 ± 0.1	0.5 ± 0.2	147.9 ± 0.1	−0.8 ± 0.1	0.1 ± 0.1	1.5 ± 0.1
25 October 2020	9147	84.6 ± 0.1	−0.5 ± 0.3	0.9 ± 0.4	151.8 ± 0.3	−0.3 ± 0.1	−0.2 ± 0.1	0.4 ± 0.1
19 November 2020	9172	88.6 ± 0.1	−0.5 ± 0.1	−0.5 ± 0.2	149.8 ± 0.1	0.1 ± 0.1	−0.1 ± 0.1	0.7 ± 0.2
14 December 2020	9197	96.2 ± 0.2	0.1 ± 0.2	−0.5 ± 0.4	150.1 ± 0.6	1.3 ± 0.2	−0.1 ± 0.4	2.3 ± 0.1
18 December 2020	9201	97.4 ± 0.1	−0.1 ± 0.1	−0.8 ± 0.6	149.0 ± 0.1	0.2 ± 0.3	0.1 ± 0.3	1.4 ± 0.4
29 August 2021	9455	79.3 ± 0.1	0.7 ± 0.1	0.6 ± 0.2	146.8 ± 0.1	0.8 ± 0.1	1.1 ± 0.1	2.3 ± 0.1
7 September 2021	9464	77.7 ± 1.1	0.2 ± 0.2	0.3 ± 2.1	145.4 ± 0.1	0.4 ± 0.5	−0.4 ± 0.1	0.9 ± 0.1
29 October 2022	9881	75.8 ± 0.1	0.8 ± 0.1	−0.2 ± 0.6	117.8 ± 0.6	0.4 ± 0.1	−0.7 ± 0.2	1.5 ± 0.2
1 November 2022	9884	77.9 ± 0.1	0.7 ± 0.1	−1.1 ± 0.4	118.0 ± 0.6	1.0 ± 0.1	−0.6 ± 0.1	1.4 ± 0.1
				Mrk 335 (AFOSC)				
9 September 2020	9101	102.3 ± 0.1	0.7 ± 0.1	−0.4 ± 0.2	153.7 ± 0.2	0.0 ± 0.1	0.5 ± 0.1	0.8 ± 0.1
7 October 2020	9129	104.6 ± 0.1	−0.2 ± 0.2	0.0 ± 0.5	148.7 ± 0.1	0.6 ± 0.3	−0.5 ± 0.4	1.0 ± 0.1
24 November 2020	9177	102.8 ± 0.2	0.2 ± 0.3	0.0 ± 0.6	143.7 ± 0.1	0.6 ± 0.2	0.2 ± 0.1	0.9 ± 0.1
25 November 2020	9178	104.1 ± 0.1	−0.2 ± 0.8	−0.3 ± 0.8	145.4 ± 0.2	0.8 ± 0.1	0.3 ± 0.4	1.6 ± 0.1
26 November 2020	9179	103.8 ± 0.1	0.1 ± 0.4	−0.2 ± 0.7	144.6 ± 0.2	0.6 ± 0.1	0.2 ± 0.2	2.8 ± 0.1
28 September 2021	9485	93.9 ± 0.1	0.6 ± 0.4	−0.1 ± 0.5	140.9 ± 0.1	0.7 ± 0.2	0.2 ± 0.3	1.2 ± 0.2
12 October 2021	9499	92.2 ± 0.1	0.3 ± 0.1	−0.3 ± 0.5	137.2 ± 0.1	−0.3 ± 0.1	0.6 ± 0.1	0.8 ± 0.1
13 October 2021	9500	91.7 ± 0.4	0.5 ± 0.1	0.2 ± 0.3	138.1 ± 0.1	0.4 ± 0.1	0.7 ± 0.1	1.2 ± 0.1
14 October 2021	9501	92.5 ± 0.1	0.4 ± 0.1	0.1 ± 0.6	137.0 ± 0.1	0.3 ± 0.1	0.3 ± 0.4	0.9 ± 0.1
13 December 2021	9561	87.5 ± 0.1	0.4 ± 0.1	0.2 ± 0.9	134.8 ± 0.3	0.4 ± 0.2	0.3 ± 0.4	1.2 ± 0.3
11 January 2022	9590	85.2 ± 0.1	0.5 ± 0.5	−0.3 ± 0.9	128.5 ± 0.1	0.0 ± 0.1	0.7 ± 0.1	1.4 ± 0.1
20 August 2022	9811	77.6 ± 0.1	0.9 ± 0.4	0.2 ± 0.6	119.3 ± 0.2	0.4 ± 0.1	0.3 ± 0.2	0.8 ± 0.1
31 October 2022	9883	79.0 ± 0.1	0.4 ± 0.2	−0.3 ± 0.8	116.3 ± 0.1	0.9 ± 0.1	0.3 ± 0.4	1.1 ± 0.1
				Mrk 509				
26 May 2020	8995	11.1 ± 0.1	0.2 ± 0.1	−0.4 ± 0.1	18.9 ± 0.1	2.3 ± 0.1	0.4 ± 0.1	0.4 ± 0.1
21 June 2020	9021	12.0 ± 0.1	0.2 ± 0.1	−0.1 ± 0.1	18.5 ± 0.1	3.1 ± 0.1	1.3 ± 0.1	0.6 ± 0.1
1 July 2020	9031	12.2 ± 0.1	0.4 ± 0.1	0.1 ± 0.3	18.1 ± 0.1	2.6 ± 0.1	1.0 ± 0.1	0.5 ± 0.1
23 July 2020	9053	10.7 ± 0.9	0.4 ± 0.1	0.6 ± 0.2	19.8 ± 0.8	2.6 ± 0.1	0.4 ± 0.1	0.5 ± 0.1
23 August 2020	9084	12.3 ± 0.1	0.9 ± 0.1	−0.6 ± 0.2	19.1 ± 0.1	2.9 ± 0.2	0.0 ± 0.1	0.5 ± 0.1
27 August 2020	9088	12.2 ± 0.1	0.5 ± 0.2	−0.3 ± 0.1	19.2 ± 0.1	1.9 ± 0.2	−0.7 ± 0.1	0.4 ± 0.1
20 September 2020	9112	13.0 ± 0.1	−0.7 ± 0.1	−0.4 ± 0.2	23.6 ± 0.1	0.1 ± 0.1	−0.5 ± 0.1	0.1 ± 0.1
22 September 2020	9114	12.3 ± 0.1	−0.3 ± 0.1	−1.4 ± 0.1	23.0 ± 0.1	−0.5 ± 0.1	−1.0 ± 0.1	0.2 ± 0.1
21 October 2020	9143	13.3 ± 0.1	1.7 ± 0.1	−0.3 ± 0.2	22.4 ± 0.1	1.8 ± 0.1	−0.9 ± 0.1	0.4 ± 0.1
4 August 2021	9430	12.4 ± 0.1	−0.1 ± 0.1	−1.0 ± 0.2	23.7 ± 0.1	0.0 ± 0.1	−1.4 ± 0.1	0.3 ± 0.1
29 August 2021	9455	12.6 ± 0.1	1.0 ± 0.1	−0.9 ± 0.2	24.6 ± 0.1	0.8 ± 0.1	−0.7 ± 0.1	0.2 ± 0.1

Table A1. *Cont.*

Date	JD	I_{cont}	Q_{cont}	U_{cont}	I_{line} (Blue) I_{line} (Red)	Q_{line} (Blue) Q_{line} (Red)	U_{line} (Blue) U_{line} (Red)	I^{p}_{line} (Blue) I_{line} (Red)
				Mrk 817				
14 December 2020	9197	24.0 ± 0.1	0.0 ± 0.1	−0.6 ± 0.2	34.7 ± 0.1 24.2 ± 0.1	−0.1 ± 0.1 −1.0 ± 0.5	−1.5 ± 0.4 −0.7 ± 0.7	0.7 ± 0.2 0.4 ± 0.1
18 December 2020	9201	25.0 ± 0.1	1.2 ± 0.1	0.2 ± 0.1	36.5 ± 0.1 24.9 ± 0.1	−1.1 ± 0.3 −2.1 ± 0.1	−0.9 ± 0.5 −0.2 ± 0.1	0.7 ± 0.1 0.6 ± 0.1
7 March 2021	9280	25.3 ± 0.1	−1.7 ± 0.1	−1.0 ± 0.1	38.9 ± 0.2 25.5 ± 0.1	−2.1 ± 0.3 −2.9 ± 0.1	−0.1 ± 0.2 −0.6 ± 0.1	0.9 ± 0.2 0.8 ± 0.1
8 March 2021	9281	24.9 ± 0.1	−1.9 ± 0.2	−1.1 ± 0.1	38.7 ± 0.1 22.4 ± 2.5	−1.9 ± 0.1 −1.7 ± 0.2	−0.5 ± 0.1 −0.6 ± 0.6	0.9 ± 0.1 0.4 ± 0.1
5 May 2021	9339	23.6 ± 0.1	−2.6 ± 0.2	−0.8 ± 0.1	38.2 ± 0.1 23.5 ± 0.1	−1.1 ± 0.1 −1.6 ± 0.6	−0.5 ± 0.3 −2.1 ± 0.1	0.4 ± 0.1 0.3 ± 0.1
2 July 2021	9397	22.2 ± 0.1	−2.4 ± 0.4	1.4 ± 0.1	36.7 ± 0.8 23.3 ± 0.2	−1.0 ± 0.3 −2.8 ± 1.6	1.7 ± 0.4 2.7 ± 0.1	0.8 ± 0.1 0.7 ± 0.1
7 July 2021	9402	21.6 ± 0.1	−1.4 ± 0.1	−0.7 ± 0.1	36.2 ± 1.3 20.3 ± 2.0	−1.0 ± 0.1 −1.6 ± 0.1	−0.3 ± 0.1 −1.1 ± 0.1	0.5 ± 0.1 0.3 ± 0.1
28 August 2021	9454	23.2 ± 0.1	−1.6 ± 0.8	−0.6 ± 0.1	35.4 ± 0.1 22.6 ± 0.1	0.1 ± 0.1 −2.0 ± 0.1	−0.2 ± 0.3 −3.9 ± 0.1	0.6 ± 0.1 0.8 ± 0.2

Notes

1 It is worth remembering that it was the approach using narrow spectral bands in the photometric observations of Seyfert galaxies that was used in the pioneering work Cherepashchuk and Lyutyi [29].

2 Details about the characteristics of medium-band filters can be found on the https://www.sao.ru/hq/lsfvo/devices/scorpio-2/filters_eng.html (accessed on 31 December 2022).

3 Details describing the instrument can be found on the https://www.oapd.inaf.it/sede-di-asiago/telescopes-and-instrumentations/copernico-182cm-telescope/afosc (accessed on 31 December 2022).

4 The multiplication of SED by the filter response function is usually called "convolution" in literature yet is not equal to the real mathematical convolution.

References

1. Antonucci, R. Unified models for active galactic nuclei and quasars. *Annu. Rev. Astron. Astrophys.* **1993**, *31*, 473–521. [CrossRef]
2. Urry, C.M.; Padovani, P. Unified Schemes for Radio-Loud Active Galactic Nuclei. *Publ. Astron. Soc. Pac.* **1995**, *107*, 803. [CrossRef]
3. Ramos Almeida, C.; Ricci, C. Nuclear obscuration in active galactic nuclei. *Nat. Astron.* **2017**, *1*, 679–689. [CrossRef]
4. Kishimoto, M.; Hönig, S.F.; Antonucci, R.; Kotani, T.; Barvainis, R.; Tristram, K.R.W.; Weigelt, G. Exploring the inner region of type 1 AGNs with the Keck interferometer. *Astron. Astrophys.* **2009**, *507*, L57–L60. [CrossRef]
5. Kishimoto, M.; Hönig, S.F.; Antonucci, R.; Barvainis, R.; Kotani, T.; Tristram, K.R.W.; Weigelt, G.; Levin, K. The innermost dusty structure in active galactic nuclei as probed by the Keck interferometer. *Astron. Astrophys.* **2011**, *527*, A121. [CrossRef]
6. Weigelt, G.; Hofmann, K.H.; Kishimoto, M.; Hönig, S.; Schertl, D.; Marconi, A.; Millour, F.; Petrov, R.; Fraix-Burnet, D.; Malbet, F.; et al. VLTI/AMBER observations of the Seyfert nucleus of NGC 3783. *Astron. Astrophys.* **2012**, *541*, L9. [CrossRef]
7. Pfuhl, O. et al. [GRAVITY Collaboration]. An image of the dust sublimation region in the nucleus of NGC 1068. *Astron. Astrophys.* **2020**, *634*, A1. [CrossRef]
8. Kishimoto, M.; Anderson, M.; ten Brummelaar, T.; Farrington, C.; Antonucci, R.; Hönig, S.; Millour, F.; Tristram, K.R.W.; Weigelt, G.; Sturmann, L.; et al. The Dust Sublimation Region of the Type 1 AGN NGC 4151 at a Hundred Microarcsecond Scale as Resolved by the CHARA Array Interferometer. *Astrophys. J.* **2022**, *940*, 28. [CrossRef]
9. Gallimore, J.F.; Elitzur, M.; Maiolino, R.; Marconi, A.; O'Dea, C.P.; Lutz, D.; Baum, S.A.; Nikutta, R.; Impellizzeri, C.M.V.; Davies, R.; et al. High-velocity Bipolar Molecular Emission from an AGN Torus. *Astrophys. J. Lett.* **2016**, *829*, L7. [CrossRef]
10. Combes, F.; García-Burillo, S.; Audibert, A.; Hunt, L.; Eckart, A.; Aalto, S.; Casasola, V.; Boone, F.; Krips, M.; Viti, S.; et al. ALMA observations of molecular tori around massive black holes. *Astron. Astrophys.* **2019**, *623*, A79. [CrossRef]
11. Hönig, S.F. Redefining the Torus: A Unifying View of AGNs in the Infrared and Submillimeter. *Astrophys. J.* **2019**, *884*, 171. [CrossRef]
12. Krolik, J.H.; Begelman, M.C. Molecular Tori in Seyfert Galaxies: Feeding the Monster and Hiding It. *Astrophys. J.* **1988**, *329*, 702. [CrossRef]
13. Pier, E.A.; Krolik, J.H. Infrared Spectra of Obscuring Dust Tori around Active Galactic Nuclei. I. Calculational Method and Basic Trends. *Astrophys. J.* **1992**, *401*, 99. [CrossRef]

14. Stalevski, M.; Fritz, J.; Baes, M.; Nakos, T.; Popović, L.Č. 3D radiative transfer modelling of the dusty tori around active galactic nuclei as a clumpy two-phase medium. *Mon. Not. R. Astron. Soc.* **2012**, *420*, 2756–2772. [CrossRef]
15. Siebenmorgen, R.; Heymann, F.; Efstathiou, A. Self-consistent two-phase AGN torus models⋆. SED library for observers. *Astron. Astrophys.* **2015**, *583*, A120. [CrossRef]
16. Smith, J.E.; Robinson, A.; Alexander, D.M.; Young, S.; Axon, D.J.; Corbett, E.A. Seyferts on the edge: Polar scattering and orientation-dependent polarization in Seyfert 1 nuclei. *Mon. Not. R. Astron. Soc.* **2004**, *350*, 140–160. [CrossRef]
17. Smith, J.E.; Robinson, A.; Young, S.; Axon, D.J.; Corbett, E.A. Equatorial scattering and the structure of the broad-line region in Seyfert nuclei: evidence for a rotating disc. *Mon. Not. R. Astron. Soc.* **2005**, *359*, 846–864. [CrossRef]
18. Goosmann, R.W.; Gaskell, C.M. Modeling optical and UV polarization of AGNs. I. Imprints of individual scattering regions. *Astron. Astrophys.* **2007**, *465*, 129–145. [CrossRef]
19. Goodrich, R.W.; Miller, J.S. Spectropolarimetry of high-polarization Seyfert 1 galaxies: Geometry and kinematics of the scattering regions. *Astrophys. J.* **1994**, *434*, 82–93. [CrossRef]
20. Marin, F.; Goosmann, R.W.; Gaskell, C.M.; Porquet, D.; Dovčiak, M. Modeling optical and UV polarization of AGNs. II. Polarization imaging and complex reprocessing. *Astron. Astrophys.* **2012**, *548*, A121. [CrossRef]
21. Afanasiev, V.L.; Popović, L.Č. Polarization in Lines—A New Method for Measuring Black Hole Masses in Active Galaxies. *Astrophys. J. Lett.* **2015**, *800*, L35. [CrossRef]
22. Afanasiev, V.L.; Popović, L.Č.; Shapovalova, A.I. Spectropolarimetry of Seyfert 1 galaxies with equatorial scattering: black hole masses and broad-line region characteristics. *Mon. Not. R. Astron. Soc.* **2019**, *482*, 4985–4999. . [CrossRef]
23. Savić, Đ.V.; Popović, L.Č.; Shablovinskaya, E. The First Supermassive Black Hole Mass Measurement in Active Galactic Nuclei Using the Polarization of Broad Emission Line Mg II. *Astrophys. J. Lett.* **2021**, *921*, L21. [CrossRef]
24. Goosmann, R.W.; Gaskell, C.M.; Shoji, M. How Polarization and Scattering can reveal Geometries, Dynamics, and Feeding of Active Galactic Nuclei. In Proceedings of the SF2A-2008, Paris, France, 30 June–4 July 2008; p. 231.
25. Gaskell, C.M.; Goosmann, R.W.; Merkulova, N.I.; Shakhovskoy, N.M.; Shoji, M. Discovery of Polarization Reverberation in NGC 4151. *Astrophys. J.* **2012**, *749*, 148. [CrossRef]
26. Shablovinskaya, E.S.; Afanasiev, V.L.; Popović, L.č. Measuring the AGN Sublimation Radius with a New Approach: Reverberation Mapping of Broad Line Polarization. *Astrophys. J.* **2020**, *892*, 118. [CrossRef]
27. Suganuma, M.; Yoshii, Y.; Kobayashi, Y.; Minezaki, T.; Enya, K.; Tomita, H.; Aoki, T.; Koshida, S.; Peterson, B.A. Reverberation Measurements of the Inner Radius of the Dust Torus in Nearby Seyfert 1 Galaxies. *Astrophys. J.* **2006**, *639*, 46–63. [CrossRef]
28. Haas, M.; Chini, R.; Ramolla, M.; Pozo Nuñez, F.; Westhues, C.; Watermann, R.; Hoffmeister, V.; Murphy, M. Photometric AGN reverberation mapping - an efficient tool for BLR sizes, black hole masses, and host-subtracted AGN luminosities. *Astron. Astrophys.* **2011**, *535*, A73. [CrossRef]
29. Cherepashchuk, A.M.; Lyutyi, V.M. Rapid Variations of Hα Intensity in the Nuclei of Seyfert Galaxies NGC 4151, 3516, 1068. *Astrophys. Lett.* **1973**, *13*, 165.
30. Komarov, V.V.; Moskvitin, A.S.; Bychkov, V.D.; Burenkov, A.N.; Drabek, S.V.; Shergin, V.S.; Emelyanov, E.V.; Komarova, V.N.; Romanenko, V.P.; Aitov, V.N. Zeiss-1000 SAO RAS: Instruments and Methods of Observation. *Astrophys. Bull.* **2020**, *75*, 486–500. [CrossRef]
31. Afanasiev, V.L.; Shablovinskaya, E.S.; Uklein, R.I.; Malygin, E.A. Stokes-Polarimeter for 1-m Telescope. *Astrophys. Bull.* **2021**, *76*, 102–108. [CrossRef]
32. Oliva, E. Wedged double Wollaston, a device for single shot polarimetric measurements. *Astron. Astrophys. Suppl. Ser.* **1997**, *123*, 589–592. .:1997175. [CrossRef]
33. Afanasiev, V.L.; Amirkhanyan, V.R. Technique of polarimetric observations of faint objects at the 6-m BTA telescope. *Astrophys. Bull.* **2012**, *67*, 438–452. [CrossRef]
34. Afanasiev, V.L.; Malygin, E.A.; Shablovinskaya, E.S.; Uklein, R.I.; Amirkhanyan, V.R.; Perepelitsyn, A.E.; Afanasieva, I.V. Small telescopes being effective: MAGIC or not? *Exp. Astron.* 2022, *in print*.
35. Afanasiev, V.L.; Amirkhanyan, V.R.; Uklein, R.I.; Perepelitsyn, A.E.; Malygin, E.A.; Shablovinskaya, E.S.; Afanasieva, I.V. Universal focal reducer for small telescopes. *Astron. Nachrichten* **2022**, *343*, e210104. [CrossRef]
36. Geyer, E.H.; Kiselev, N.N.; Chernova, G.P.; Jockers, K. Surface Polarimetry of Comet Tanaka-Machholz 1992d Using a Novel Double Wollaston Prism. In Proceedings of the Asteroids, Comets, Meteors 1993, Belgirate, Italy, 14–18 June 1993; Volume 810, p. 116.
37. Jockers, K.; Credner, T.; Bonev, T.; Kisele, V.N.; Korsun, P.; Kulyk, I.; Rosenbush, V.; Andrienko, A.; Karpov, N.; Sergeev, A.; et al. Exploration of the solar system with the Two-Channel Focal Reducer at the 2m-RCC telescope of Pik Terskol Observatory. *Kinemat. Fiz. Nebesnykh Tel Suppl.* **2000**, *3*, 13–18.
38. Simmons, J.F.L.; Stewart, B.G. Point and interval estimation of the true unbiased degree of linear polarization in the presence of low signal-to-noise ratios. *Astron. Astrophys.* **1985**, *142*, 100–106.
39. Rodríguez-Pascual, P.M.; Alloin, D.; Clavel, J.; Crenshaw, D.M.; Horne, K.; Kriss, G.A.; Krolik, J.H.; Malkan, M.A.; Netzer, H.; O'Brien, P.T.; et al. Steps toward Determination of the Size and Structure of the Broad-Line Region in Active Galactic Nuclei. IX. Ultraviolet Observations of Fairall 9. *Astrophys. J. Suppl. Ser.* **1997**, *110*, 9–20. [CrossRef]
40. Smith, J.E.; Young, S.; Robinson, A.; Corbett, E.A.; Giannuzzo, M.E.; Axon, D.J.; Hough, J.H. A spectropolarimetric atlas of Seyfert 1 galaxies. *Mon. Not. R. Astron. Soc.* **2002**, *335*, 773–798. [CrossRef]

41. Mudd, D.; Martini, P.; Zu, Y.; Kochanek, C.; Peterson, B.M.; Kessler, R.; Davis, T.M.; Hoormann, J.K.; King, A.; Lidman, C.; et al. Quasar Accretion Disk Sizes from Continuum Reverberation Mapping from the Dark Energy Survey. *Astrophys. J.* **2018**, *862*, 123. [CrossRef]
42. Zu, Y.; Kochanek, C.S.; Kozłowski, S.; Peterson, B.M. Application of Stochastic Modeling to Analysis of Photometric Reverberation Mapping Data. *Astrophys. J.* **2016**, *819*, 122. [CrossRef]
43. Yu, Z.; Kochanek, C.S.; Peterson, B.M.; Zu, Y.; Brandt, W.N.; Cackett, E.M.; Fausnaugh, M.M.; McHardy, I.M. On reverberation mapping lag uncertainties. *Mon. Not. R. Astron. Soc.* **2019**, *491*, 6045–6064. [CrossRef]
44. Alexander, T. Is AGN Variability Correlated with Other AGN Properties? ZDCF Analysis of Small Samples of Sparse Light Curves. In *Proceedings of the Astronomical Time Series*; Maoz, D., Sternberg, A., Leibowitz, E.M., Eds.; Astrophysics and Space Science Library; Springer: Dordrecht, The Netherlands, 1997; Volume 218, p. 163. [CrossRef]
45. Alexander, T. Improved AGN light curve analysis with the z-transformed discrete correlation function. *arXiv* **2013**, arXiv:1302.1508.
46. Wandel, A.; Peterson, B.M.; Malkan, M.A. Central Masses and Broad-Line Region Sizes of Active Galactic Nuclei. I. Comparing the Photoionization and Reverberation Techniques. *Astrophys. J.* **1999**, *526*, 579–591. [CrossRef]
47. Kaspi, S.; Smith, P.S.; Netzer, H.; Maoz, D.; Jannuzi, B.T.; Giveon, U. Reverberation Measurements for 17 Quasars and the Size-Mass-Luminosity Relations in Active Galactic Nuclei. *Astrophys. J.* **2000**, *533*, 631–649. [CrossRef]
48. Peterson, B.M.; Ferrarese, L.; Gilbert, K.M.; Kaspi, S.; Malkan, M.A.; Maoz, D.; Merritt, D.; Netzer, H.; Onken, C.A.; Pogge, R.W.; et al. Central Masses and Broad-Line Region Sizes of Active Galactic Nuclei. II. A Homogeneous Analysis of a Large Reverberation-Mapping Database. *Astrophys. J.* **2004**, *613*, 682–699. [CrossRef]
49. Bentz, M.C.; Peterson, B.M.; Netzer, H.; Pogge, R.W.; Vestergaard, M. The Radius-Luminosity Relationship for Active Galactic Nuclei: The Effect of Host-Galaxy Starlight on Luminosity Measurements. II. The Full Sample of Reverberation-Mapped AGNs. *Astrophys. J.* **2009**, *697*, 160–181. [CrossRef]
50. Grier, C.J.; Peterson, B.M.; Pogge, R.W.; Denney, K.D.; Bentz, M.C.; Martini, P.; Sergeev, S.G.; Kaspi, S.; Minezaki, T.; Zu, Y.; et al. Reverberation Mapping Results for Five Seyfert 1 Galaxies. *Astrophys. J.* **2012**, *755*, 60. [CrossRef]
51. Grier, C.J.; Peterson, B.M.; Pogge, R.W.; Denney, K.D.; Bentz, M.C.; Martini, P.; Sergeev, S.G.; Kaspi, S.; Zu, Y.; Kochanek, C.S.; et al. A Reverberation Lag for the High-ionization Component of the Broad-line Region in the Narrow-line Seyfert 1 Mrk 335. *Astrophys. J. Lett.* **2012**, *744*, L4. [CrossRef]
52. Du, P.; Hu, C.; Lu, K.X.; Wang, F.; Qiu, J.; Li, Y.R.; Bai, J.M.; Kaspi, S.; Netzer, H.; Wang, J.M.; et al. Supermassive Black Holes with High Accretion Rates in Active Galactic Nuclei. I. First Results from a New Reverberation Mapping Campaign. *Astrophys. J.* **2014**, *782*, 45. [CrossRef]
53. Hu, C.; Li, S.S.; Yang, S.; Yang, Z.X.; Guo, W.J.; Bao, D.W.; Jiang, B.W.; Du, P.; Li, Y.R.; Xiao, M.; et al. Supermassive Black Holes with High Accretion Rates in Active Galactic Nuclei. XII. Reverberation Mapping Results for 15 PG Quasars from a Long-duration High-cadence Campaign. *Astrophys. J. Suppl. Ser.* **2021**, *253*, 20. [CrossRef]
54. Jha, V.K.; Joshi, R.; Chand, H.; Wu, X.B.; Ho, L.C.; Rastogi, S.; Ma, Q. Accretion disc sizes from continuum reverberation mapping of AGN selected from the ZTF survey. *Mon. Not. R. Astron. Soc.* **2022**, *511*, 3005–3016. [CrossRef]
55. Koshida, S.; Minezaki, T.; Yoshii, Y.; Kobayashi, Y.; Sakata, Y.; Sugawara, S.; Enya, K.; Suganuma, M.; Tomita, H.; Aoki, T.; et al. Reverberation Measurements of the Inner Radius of the Dust Torus in 17 Seyfert Galaxies. *Astrophys. J.* **2014**, *788*, 159. [CrossRef]
56. Lyu, J.; Rieke, G.H.; Smith, P.S. Mid-IR Variability and Dust Reverberation Mapping of Low-z Quasars. I. Data, Methods, and Basic Results. *Astrophys. J.* **2019**, *886*, 33. [CrossRef]
57. Kaastra, J.S.; Petrucci, P.O.; Cappi, M.; Arav, N.; Behar, E.; Bianchi, S.; Bloom, J.; Blustin, A.J.; Branduardi-Raymont, G.; Costantini, E.; et al. Multiwavelength campaign on Mrk 509. I. Variability and spectral energy distribution. *Astron. Astrophys.* **2011**, *534*, A36. [CrossRef]
58. Pozo Nuñez, F.; Gianniotis, N.; Blex, J.; Lisow, T.; Chini, R.; Polsterer, K.L.; Pott, J.U.; Esser, J.; Pietrzyński, G. Optical continuum photometric reverberation mapping of the Seyfert-1 galaxy Mrk509. *Mon. Not. R. Astron. Soc.* **2019**, *490*, 3936–3951. [CrossRef]
59. Edelson, R.; Gelbord, J.; Cackett, E.; Peterson, B.M.; Horne, K.; Barth, A.J.; Starkey, D.A.; Bentz, M.; Brandt, W.N.; Goad, M.; et al. The First Swift Intensive AGN Accretion Disk Reverberation Mapping Survey. *Astrophys. J.* **2019**, *870*, 123. [CrossRef]
60. Dexter, J. et al. [GRAVITY Collaboration]. The resolved size and structure of hot dust in the immediate vicinity of AGN. *Astron. Astrophys.* **2020**, *635*, A92. [CrossRef]
61. Zanchettin, M.V.; Feruglio, C.; Bischetti, M.; Malizia, A.; Molina, M.; Bongiorno, A.; Dadina, M.; Gruppioni, C.; Piconcelli, E.; Tombesi, F.; et al. The IBISCO survey. I. Multiphase discs and winds in the Seyfert galaxy Markarian 509. *Astron. Astrophys.* **2021**, *655*, A25. [CrossRef]
62. Kara, E.; Mehdipour, M.; Kriss, G.A.; Cackett, E.M.; Arav, N.; Barth, A.J.; Byun, D.; Brotherton, M.S.; De Rosa, G.; Gelbord, J.; et al. AGN STORM 2. I. First results: A Change in the Weather of Mrk 817. *Astrophys. J.* **2021**, *922*, 151. [CrossRef]
63. Denney, K.D.; Peterson, B.M.; Pogge, R.W.; Adair, A.; Atlee, D.W.; Au-Yong, K.; Bentz, M.C.; Bird, J.C.; Brokofsky, D.J.; Chisholm, E.; et al. Reverberation Mapping Measurements of Black Hole Masses in Six Local Seyfert Galaxies. *Astrophys. J.* **2010**, *721*, 715–737. [CrossRef]
64. Lu, K.X.; Wang, J.G.; Zhang, Z.X.; Huang, Y.K.; Xu, L.; Xin, Y.X.; Yu, X.G.; Ding, X.; Wang, D.Q.; Feng, H.C. Reverberation Mapping Measurements of Black Hole Masses and Broad-line Region Kinematics in Mrk 817 and NGC 7469. *Astrophys. J.* **2021**, *918*, 50. [CrossRef]

65. Uklein, R.I.; Malygin, E.A.; Shablovinskaya, E.S.; Perepelitsyn, A.E.; Grokhovskaya, A.A. Photometric Reverberation Mapping of AGNs at $0.1 < z < 0.8$. I. Observational Technique. *Astrophys. Bull.* **2019**, *74*, 388–395. [CrossRef]
66. Malygin, E.A.; Shablovinskaya, E.S.; Uklein, R.I.; Grokhovskaya, A.A. Measurement of the Supermassive Black Hole Masses in Two Active Galactic Nuclei by the Photometric Reverberation Mapping Method. *Astron. Lett.* **2020**, *46*, 726–733. [CrossRef]
67. Kovačević, A.B.; Radović, V.; Ilić, D.; Popović, L.Č.; Assef, R.J.; Sánchez-Sáez, P.; Nikutta, R.; Raiteri, C.M.; Yoon, I.; Homayouni, Y.; et al. The LSST Era of Supermassive Black Hole Accretion Disk Reverberation Mapping. *Astrophys. J. Suppl. Ser.* **2022**, *262*, 49. [CrossRef]
68. Woo, J.H.; Son, D.; Gallo, E.; Hodges-Kluck, E.; Jeon, Y.; Shin, J.; Bae, H.J.; Cho, H.; Cho, W.; Kang, D.; et al. Seoul National University AGN Monitoring Project. I. Strategy And Sample. *J. Korean Astron. Soc.* **2019**, *52*, 109–119. [CrossRef]
69. Jiang, B.W.; Marziani, P.; Savić, Đ.; Shablovinskaya, E.; Popović, L.Č.; Afanasiev, V.L.; Czerny, B.; Wang, J.M.; del Olmo, A.; D'Onofrio, M.; et al. Linear spectropolarimetric analysis of fairall 9 with VLT/FORS2. *Mon. Not. R. Astron. Soc.* **2021**, *508*, 79–99. [CrossRef]

MDPI AG
Grosspeteranlage 5
4052 Basel
Switzerland
Tel.: +41 61 683 77 34

Universe Editorial Office
E-mail: universe@mdpi.com
www.mdpi.com/journal/universe

www.ingramcontent.com/pod-product-compliance
Lightning Source LLC
Chambersburg PA
CBHW040751250926
51850CB00007BB/1469

* 9 7 8 3 7 2 5 8 7 2 3 0 5 *